Introduction to Statistics for

Geographers and Earth Scientists

Other Macmillan titles of related interest

R. Till, *Statistical Methods for Earth Scientists*
R. B. G. Williams, *Intermediate Statistics for Geographers and Earth Scientists* (companion volume)

Introduction to Statistics for Geographers and Earth Scientists

R. B. G. Williams
University of Sussex
Brighton

MACMILLAN

First published 1984 by
Higher and Further Education Division
MACMILLAN PUBLISHERS LTD
London and Basingstoke
Companies and representatives
throughout the world

Printed in Hong Kong

British Library Cataloguing in Publication Data

Williams, R.B.G.
Introduction to statistics for geographers
and earth scientists.
1. Geography—Statistical methods
I. Title
519.5′02491 G70.3

ISBN 0–333–35275–0

Contents

Preface

This book aims to provide a comprehensive introduction to the subject of Statistics. It describes a wide range of elementary techniques, including those most commonly selected by research workers or taught in undergraduate courses at Universities and Polytechnics. The value of each technique is discussed, and attention is drawn to the pitfalls that must be avoided if the technique is to be correctly applied.

All the techniques are illustrated by carefully worked examples utilising real data. The examples are drawn from geography, planning and the earth sciences, but readers with backgrounds in other disciplines should not find them difficult to understand. All the calculations can be carried out using pencil and paper or using a simple pocket calculator, and only an elementary knowledge of mathematics is assumed.

The manuscript for this book has been used as a photostated text by geography students at the University of Sussex taking undergraduate statistics courses or engaged in postgraduate research. The manuscript has been revised and extended as a result of many useful suggestions from colleagues and students. I am particularly indebted to Dr. D.C. Funnell, Dr. T. Browne and Professor K.J. Tinkler for helping with sections of the text where their knowledge is more extensive than mine. I should also like to express my gratitude to Susan Rowland for undertaking the drawing of the diagrams. Needless to say I am responsible for any errors that remain either in the text or the diagrams, and I will be grateful to readers if they will draw them to my attention.

Sections of the text marked with an * may be skimmed over on first reading or omitted altogether if time is short. Although the theoretical background of each technique is explained before a worked example is given, some readers may prefer to study the example first and examine the theoretical background only when they have understood how the technique is applied.

More advanced topics in statistics, especially in regression and correlation, are examined in "Intermediate Statistics for Geographers and Earth Scientists", the companion volume to the present book. This volume contains a detailed discussion of the logic of significance testing, and examines the various ways in which statistical techniques have been used and misused in scientific research.

Computer program listings have not been included in the text because of space limitations, but it is hoped shortly to issue software cassettes with programs in BASIC that will run on the BBC Model B microcomputer.

University of Sussex
December 1983

Acknowledgements

The author and publishers wish to thank the following who have kindly given permission for the use of copyright material:

American Journal of Science for an adapted diagram from article by Arthur N. Strahler in Volume 248, October 1950.

Cambridge University Press for two adapted diagrams from *The Lognormal Distribution* (1969), by J. Aitchison and J.A.C. Brown.

Holden-Day Inc. for tables adapted from *Nonparametrics* (1975) by E.H. Lehmann

The Institute of British Geographers for adapted diagrams from article 'The geomorphological importance of jointing in the Dartmoor granite' by A.J. W. Gerrard, No. 7, June 1974; from article 'The redistribution of parliamentary seats in the United Kingdom: themes and opinions' by Gwyn Rowley in *Area*, Volume 7, No. 1, 1975; from article 'Suspended sediment delivery rates and the solute concentration of stream discharge in two Welsh catchments' by N.C. Oxley; and from article 'The geometry of shore platforms in England and Wales' by A.S. Trenhaile, No. 62, July 1974.

The *Journal of Geology* for a diagram adapted from article 'Statistical analysis in geomorphic research' by A.N. Strahler in Volume 62, 1-25, 1954.

Longman Group Limited for a table adapted from *Statistical Tables for Biological, Agricultural and Medical Research*, 1974, 6th Edition by Fisher and Yates.

Isamu Matsui for map and grid adapted from article in *Japanese Journal of Geology and Geography*, Volume 9, No. 3-4.

McGraw-Hill Book Company for tables adapted from *Schaum's Outline of Theory and Problems of Statistics* by Murray R. Spiegel, and *Introduction to Statistical Analysis* by Wilfred J. Dixon and Frank J. Massey, Jr.

The Royal Geographical Society for figures from 'Aspects of the morphometry of a "poly-cyclic" drainage basin' by Professor R.J. Chorley, in *The Geographical Journal*, Volume 124, 1958.

Universe Books for an adapted graph from *The Limits of Growth: A report for The Club of Rome's Project on the Predicament of Mankind* (Potomac Associates, 1972) by Donella H. Meadows, Dennis L. Meadows, Jørgen Randers, William W. Behrens III.

John Wiley & Sons Inc. for a table adapted from *Practical Nonparametric Statistics* (1971) by W.J. Conover.

Zambia Geographical Association for two figures adapted from *The Climate of Zambia* (Occasional Study No. 7) by Peter Hutchinson.

We also thank F.E. Croxton for permission to produce Table B - an adapted version of tables that were originally included in *Applied General Statistics* by F.E. Croxton, D.J. Cowden and S. Klein (Pitman, 3rd edn).

Every effort has been made to trace all the copyright holders but if any have been inadvertently overlooked the publishers will be pleased to make the necessary arrangement at the first opportunity.

1 The Aims of Statistics

CURRENT AIMS

The term *statistics* used with a plural verb means numerical data, as in Government statistics, vital statistics and so on.[1] Used with a singular verb statistics means the study of numerical data, and not the data themselves. It is a branch of mathematics with its own concepts, methods and values, which find many applications in geography and other disciplines.

The basic aim of statistics in this sense of a subject of study is to provide methods of organising and simplifying data so that their significance is comprehensible. A large array of figures makes tedious reading, and is difficult to interpret; but if the same figures are presented in the form of a graph or histogram, their essentials may be grasped virtually at a glance. Alternatively, the figures can be summarised by using percentages, averages and other numerical measures. These methods of simplifying data form the subject matter of what is commonly called *descriptive statistics*.

A further aim of statistics is to provide methods of drawing valid inferences from samples. No sample, however selected, can be altogether trusted. There must inevitably be some uncertainty about any inference based on a limited number of observations. A random sample of holidaymakers on a beach might by chance include no families with children, despite the fact that children are much in evidence on the beach. A random sample of pebbles on the beach might miss all the really large pebbles and be biased towards the smaller pebbles. A major preoccupation of modern statistics is deciding what degree of reliance can reasonably be placed on particular samples. This branch of the subject is known as *inferential statistics*.

HISTORICAL BACKGROUND

Few subjects have undergone such a major change of outlook as statistics. In the mid-eighteenth century statistics was the study of States - their populations, industries, natural advantages, and so forth. As such, it was considered to be a part of politics, although it could almost equally well have been included with geography. There was little or no connection with mathematics: the emphasis was almost entirely on verbal as opposed to numerical description.

Statistics received perhaps most attention in Germany, which was then divided into a great number of semi-independent States of vary-

[1] As a plural statistics has another more technical meaning discussed in Chapter 3.

ing size. These States were anxious to realise their economic potentialities and thus to ensure their long term survival. Statistics grew in importance because it attempted to provide information useful to politicians, merchants and others concerned with national aggrandisement. The University of Göttingen became a leading centre for statistical studies. Gottfried Achenwall (1719-72), Professor of Politics, gave lectures in statistics in which he discussed the physical and human resources of nations, their political alignments, and future prospects. In 1749 he published a book, *Abriss der Statwissenschaft der heutigen vornehmsten europäischen Reiche und Republiken* which contained statistical descriptions of the States of Middle Europe.

Geography in the early eighteenth century had been largely occupied with the task of compiling and revising maps. Johann Franz (1700 - 61), Professor of Geography at Göttingen, felt strongly that geography ought to be more ambitious. Like statistics, it should set out to provide information of practical advantage to the State. It should be interested in flood control, soil fertility, food supplies, manufacturing, exports and other matters. Its aim should be to teach people about their land. Although Franz published little, he exerted an important influence on the development of geography. His main work, *Der Deutsche Staatsgeographus* (Leipzig, 1753), appeared the year before he received his professorship. Franz was supported in his views by A.J. Büsching (1724-93), a lecturer at Göttingen in theology but very much a geographer in interests. Unlike Franz, Büsching was a prolific writer, publishing 13 geographical books while at Göttingen. The most famous, the *Neue Erdbeschreibung* of 1754, became the standard book on the economic geography of Europe, remaining in print until 1809. It set out the noteworthy characteristics of the different States and greatly influenced the writings of later geographers, such as Freidrich Ratzel.

During the course of the nineteenth century, statisticians grew increasingly impatient with verbal description and began to devote more and more time to the organised collection and presentation of numerical data. This change reflected a growing demand for more precise information about national resources. Nations previously little interested in record-keeping were seized with a new spirit of enquiry. Censuses were undertaken on a scale not before known, together with detailed surveys of agriculture and industry. Statistical offices were established in many capitals to collect and disseminate numerical information. Parallel with this official stocktaking, scientific research at universities and elsewhere greatly expanded. Statistics became the tool of the scientist, who used it in the design and analysis of experiments, and in setting up sample surveys, often for purely academic ends. No longer was statistics merely the servant of the State. All kinds of numerical data were now within its scope, as well as techniques for their analysis. The same techniques could be applied irrespective of the branch of knowledge the data represented.

Geography, in sharp contrast with statistics, continued to rely heavily on verbal description throughout the nineteenth century. Although increasingly refined techniques for handling data were devised by statisticians, contemporary geographers took almost no notice of them. By 1900 geography and statistics had drifted far apart. It was not until about 1950 that geographers suddenly awoke to the advantages of statistical techniques. The ensuing "Quantitative Revolution" initiated virtually a new way of thinking. Suddenly every geographical problem seemed to be statistical in nature. Researchers vied with one another to devise imaginative applications of statistical techniques and to construct an acceptable framework of theory within which to place the new quantitative data.

Now a reaction is setting in. The enthusiasm for statistics is waning somewhat as geographers become more fully aware of the assumptions and limitations of statistical techniques. This reaction is healthy. It is absurd to imagine that the answer to every geographical problem lies in statistics, and that if something cannot be measured satisfactorily it is not worth consideration. Undoubtedly, in their eagerness to employ statistical techniques geographers have committed many mistakes. But the Quantitative Revolution, while it may have over-reached itself, has been on balance a great blessing. It has revitalised geography, giving it new directions and a new sense of purpose.

Statisticians can take justifiable pride in the extent to which their subject has successfully influenced others. At the same time they may be excused for feeling a pang of regret that, while subjects like geography have been enriched, their own has been left the poorer. Many of the ideas and techniques that were once the exclusive preserve of statistics are now common property. Geographers, for example, regularly conduct elementary statistical analyses without seeking the advice of professional statisticians. The latter now find themselves called in to help with other people's problems only when the problems are complex and difficult. Statistical journals are being filled with papers that are increasingly esoteric and theoretical. Papers making routine use of statistics that might in the past have appeared in statistical journals now find a natural place in scientific journals of every description. It is these sciences that take the credit for whatever merit the papers possess, and not the discipline of statistics.

2 Types of Numerical Data

ATTRIBUTES

An *attribute* in this book is a quality, as opposed to a quantity, whereby items, individuals, objects, locations, events etc. differ one from another. Rock-type, land-use, sex, occupation and nationality are just a few of the attributes commonly encountered in geographical studies. Attributes always occur as a number of varieties. For example, a person may be male or female; an area may be industrial, residential or agricultural; a day may be cloudy or sunny. Varieties of attributes are entirely without magnitude or amount. A male is neither more nor less than a female: he is merely different.[1]

If desired, numbers can be used instead of words to refer to varieties of attributes. Thus males and females may be denoted as 1 and 0 respectively. Or residential areas may be designated 0, industrial areas 1, and agricultural areas 2. Such numbers are purely arbitrary, with little or no mathematical value. They are labels, like the numbers on hotel bedroom doors or on rugby players' shirts. They serve solely to state whether items (individuals, objects, locations, etc.) are similar or different.[2] Only in certain limited circumstances, for example in correlation analysis, are they susceptible to mathematical manipulation.

Besides being used as labels, numbers can be used to express the frequency with which varieties of attributes occur. Thus a count could be made of the numbers of French and German tourists passing through Dover, or the numbers of wheatfields, barleyfields, etc. in Avon. Such counts express the *absolute frequencies* of the items having the attributes in question. Turned into proportions, ratios or percentages they constitute *relative frequencies*.

Data in the form of frequencies are familiar in virtually all fields of geographical investigation. A great many statistical methods are available for dealing with them, as discussed later in this book.

---

[1]Statisticians, it should be noted, are not entirely agreed on how to define an attribute. Some employ the word to refer not to generalised characteristics such as sex or nationality but to the varieties into which the characteristics are divisible, e.g. male and female, French and Italian. Other statisticians use phrases such as "nominal variable" or "qualitative variable" in place of attribute.

[2]Some statisticians describe this classificatory use of numbers as a "nominal scale of measurement". It seems better, however, to avoid using the term "measurement" for such an arbitrary process of assigning numbers. "Nominal system of numbering" might be a more appropriate description.

MEASURED VARIABLES

In statistics anything that varies in magnitude from one item to another is called a *variable*. A variable can take on a number of different values, referred to as *variates* or *variate values*. Examples of variables are distance, area, slope angle, temperature and income. The velocity of a river is a variable, but the velocity of light is not, because, although it has magnitude, it is unvarying. It is an example of a *constant*.

Most variables are *measured*, in other words, they are assigned numerical values relative to some standard - the *unit of measurement*. The values are meaningless unless accompanied by the unit of measurement. To refer to a distance, for example, as 5.23 without making clear the units (miles, kilometres, or whatever) is to invite confusion. This dependence upon stated units is a distinguishing characteristic of measurements.

Measured variables are not all alike in their mathematical properties. Ideal or perfect variables have an *unlimited range:* they are capable of assuming any value from minus infinity through zero to plus infinity. Real variables are seldom like this, tending to have only a *restricted range*. For instance, distance can assume any positive value, but it cannot be negative, except perhaps in a looking-glass world. Temperature cannot fall below -273.2 degrees on the Centigrade scale (-458.0 degrees Fahrenheit), though it can take on any value greater than this. Slope angle can vary only between 0 degrees (flat) and 90 degrees (vertical).

Many of the techniques employed in inferential statistics assume for simplicity that the variables they treat possess an unlimited range. Although in any particular series of observations only a limited set of values will appear, it is assumed that theoretically any value could appear, even one infinitely large or infinitely small. It might seem that these techniques ought not to be employed on a variable with a restricted range, but nevertheless they very often are. The results are affected to an appreciable extent only if the variable under consideration yields values that tend to concentrate around either its upper or its lower limit. A sample of slope angles in a moderately hilly area would therefore be easy to treat, but samples from a very flat or very precipitous area would cause problems. Direction (compass bearing) is an example of a measured variable that is *circular*. Its lower limit of 0° is identical with its upper limit of 360°. Circular variables are difficult to deal with in inferential statistics and require special techniques.

A distinction can be made between measured variables with an *arbitrary zero* and those with an *absolute zero*. Most variables encountered in geographical studies belong to the second variety. Altitude and temperature (measured in degrees Centigrade or degrees Fahrenheit, but not degrees Kelvin) are obvious exceptions. Heights of the ground surface, for example, are customarily determined with respect to average sea-level, a convenient though entirely arbitrary zero datum.

Variables possessing an arbitrary zero do not allow the calculation of absolute ratios. Thus a hill rising to 200 metres above sea-level is twice as high with respect to sea-level as a hill rising to

100 metres, but with respect to, say, a car-park at 50 metres above sea-level it is three times as high. Variables with an absolute zero do not suffer from the disadvantage of arbitrary ratios. Thus a farm of 500 hectares is truly twice as large as one of 250 hectares.

A measured variable is said to be *continuous* if it can assume any value within its range, and *discontinuous* or *discrete* if there are gaps between successive values. A continuous variable can assume fractional values; most discontinuous variables are discontinuous through being limited to whole numbers (integers). Area, distance, rainfall, soil pH and yields per acre of crops are all continuous variables. On the other hand, the population size of cities is a discontinuous variable since it can exist only as whole numbers, fractions of a person (the unit of measurement) being impossible. Income is another discontinuous variable, there being gaps of 0.5 pence between consecutive values. Discontinuous variables are measured by *counting*.[1]

Most techniques of inferential statistics assume a variable to be continuous. However, it generally makes little difference to the result if the variable is, in fact, discontinuous. The assumption takes on practical significance only if the variable is highly discontinuous, or if the existence of discontinuities is combined with a restricted range. Suppose, for instance, that samples of families living on two estates are to be visited to determine whether the average number of children per family is the same on the two estates. The number of children cannot be less than 0, and is not very likely to exceed 10 (the record for a mother in Great Britain in recent times is 24 children). The variable "number of children" is thus not only restricted to whole numbers but also has a very restricted range. Only a limited number of statistical techniques would be precisely applicable to such a maverick variable.

RANKED VARIABLES

A ranked variable is assigned numerical values in a different way from a measured variable. Items are first arranged *(ranked)* in order of increasing or decreasing magnitude. Each occurrence is then given a number (a rank) according to its position in the sequence (first = 1, second = 2, etc.). For example, a survey of journeys-to-work made by different persons might include ranking the journeys in order of increasing distance. The longest journey might be numbered 1, the next longest 2, and so on, with the shortest journey receiving the largest number. Alternatively, the numbering could be reversed, the longest journey receiving the largest number, the next longest the next largest, down to the shortest journey which would be number one.[2]

---

[1]Measurement generally requires instruments (rulers, thermometers etc.) whereas counting does not. But counting is not fundamentally distinct from measurement. When a person measures his temperature, he is really counting degrees.

[2]Ties are a complication. The ordinary method of treating two or more occurrences that are equal in magnitude is to give each of them a rank that is one more than the rank of the previous occurrence in the sequence. Occurrences following the tie are ranked by counting each

Ranks have no units of measurement attached to them. They are "foot-loose" numbers that reduce an unruly set of measurements to order by throwing away information. Unlike measurements, ranks do not give a precise idea of magnitude. When only the rank order of occurrences is known, there is no way of determining how much one occurrence differs from another. An occurrence ranked 5, for instance, may differ vastly from one ranked 6, or it may differ so slightly as to be hardly worth distinguishing.[3]

Because measurements convey more precise information they are superior to ranks for most purposes. They are also in general less trouble to obtain. Ranks cannot usually be determined until after measurements have been made. The extra work needed to convert measurements to ranks is justified, however, if the ranking reveals interesting and unexpected relationships. For instance, an approximate relationship has been found in many countries between the size of cities and their rank. The second largest city possesses roughly half the population of the largest (or "primate") city, the third largest city possesses roughly one third, the fourth largest city one quarter, and so on. This relationship is known as the "rank-size rule".

Ranking is often undertaken to gloss over errors of measurement. The errors may be so large that the measurements are not worth keeping, but not so large that the rank ordering is affected. Substituting ranks for the measurements makes the errors unimportant.

Sometimes it is possible to carry out ranking without having first to make measurements, and valuable time may be saved. Areas, for example, tend to be time-consuming and tedious to measure, whether they are on the ground or on a map. The more irregular an area, the more difficult it is to measure satisfactorily. However, if a number of areas differ substantially in size they can be ranked immediately by eye alone, and need not be measured first. In this and similar situations, ranking is preferable to measurement unless great precision is required.

Ranking can often be accomplished when measurement is impossible, giving numerical expression to phenomena that are wholly or partly

---

of the occurrences forming the tie as a separate occurrence. For example, if six distances are 11,7,7,7,4 and 2 km, they are assigned the ranks 1,2,2,2,5 and 6 (or, alternatively, 6,5,5,5,2 and 1). Although this method of treating ties is not wrong it is rarely used in statistics. Instead the tied observations are assigned the average of the ranks that they would have had if they had been fractionally different in magnitude i.e. if no tie had occurred. Hence, if the three distances of 7 km had been 7.1, 7.0 and 6.9 km, they would have ranked 2,3 and 4 (or, alternatively, 5,4 and 3); the average of 2,3 and 4 is 3; so the six distances would be ranked 1,3,3,3,5 and 6 (or, alternatively, 6,4,4,4,2 and 1). Most people find this method of treating ties less straightforward than the ordinary method, but it has considerable advantages for statistical inference.

[3]Rank ordering creates what some statisticians refer to as an "ordinal scale of measurement". "Ordinal system of numbering" would seem a better description, since no units of measurement are involved.

subjective, such as natural beauty, social class, or stage of economic development. For example, most people would rank Switzerland as more beautiful than Belgium, the Mediterranean as more beautiful than the North Sea; but it is impossible to devise a precise scale of measurement to express the variations that exist in natural beauty. Even ranking is somewhat difficult since people tend to have different ideas about what constitutes beauty, and their ideas often change. In the eighteenth century few travellers chose to visit Switzerland. The snowy and precipitous Alps were thought to be unattractive, even terrifying, whereas the tame and orderly landscape of Flanders was universally admired.

Ranked and measured variables lend themselves to different mathematical treatment. On the whole ranked variables are easier to work with than measured variables unless lots of ties are present. Measured variables take longer to deal with, but one can do more with them, and in the end they yield more information. If precision is not important, however, and if time is short, it may be advantageous to convert measurements to ranks.

PSEUDO-ATTRIBUTES

Ranks and measured variables can always be simplified, if necessary, by introducing a classification based on magnitude. For example, areas up to a certain size (say, 10 hectares) may be arbitrarily classed as "small sized", areas between this size and another (say, 100 hectares) as "medium sized", and areas of still greater extent as "large sized". Or regions may be classed as "less than beautiful", "beautiful", and "outstandingly beautiful". In substituting descriptive terms for measurements or ranks, precision is necessarily lost. On the definition just given, both a postage stamp and Trafalgar Square would be described as small sized.

A variable that is merely classified, being neither measured nor ranked, is almost, but not quite, the same as an attribute. The varieties or classes differ from those of an attribute because they are based on magnitude and so possess a natural order. Thus to write medium/small/large would be clumsy and ill-considered. "Medium" cries out to be inserted between the two extremes of "small" and "large". Varieties of attributes do not possess magnitude and therefore do not fall into a definite sequence. For example, nationalities such as French/German/Italian can be written in any order, as can occupations such as banker, bricklayer and bus-driver.

It is difficult to know what to call a variable that has been classified. Several names have been proposed, but none has won general acceptance. *Pseudo-attribute* will be used here as it is reasonably simple and self-explanatory.

3 Populations and Samples

INTRODUCTION

Few research studies need to be exhaustive. One need seldom collect data on all items of interest in order to arrive at valid conclusions. Usually data need be collected only for a selection of items, provided these adequately reflect the characteristics of all the items to which the study refers.

The term *sample* is commonly applied to a selection of items (or a single item) drawn from a larger group all sharing the same basic characteristics. The larger group is referred to in statistics as the *population*. A sample is thus part of a whole, and the whole is the population. The population is the totality of items seen as relevant to the study and to which the conclusions supposedly refer. The sample is the fraction of the population actually examined and for which data are collected.

A population in statistics can be any collection of items, animate or inanimate. Thus it is possible to speak of the population of adult males in Great Britain, the population of households in London, the population of traffic flows in New York on Sundays, or the population of slopes in the Appalachians. The fact that "population" in statistics can mean a group of inanimate objects or events as well as living organisms offends those who like to restrict words to their original meanings and stop them becoming flabby. The statistical usage is now well established, however, and it can justly claim to fill a need, if only in statistics.

REASONS FOR SAMPLING

There are a number of reasons for choosing to study a sample rather than an entire population.

1. The population may be so large that to examine it completely might cost too much or take too much time. In order to discover the opinion of the majority of citizens of Great Britain on some issue it would not usually be necessary to organise a complete referendum. A 1% sample, if carefully drawn, would be almost as definitive, and would be much easier to arrange. An even smaller sample would suffice for many purposes, unless opinion was very finely balanced. National Opinion Polls Ltd have correctly forecast the result of six of the last seven elections in Great Britain, using on each occasion a sample of between one and two thousand voters, which is only a very small proportion of the total electorate of about 40 million voters.

2. The population may be infinitely large, making sampling unavoidable. Any period of time can be endlessly subdivided. Thus, in

theory, an infinite number of measurements of air temperature may be made in a single day, or even a single hour or minute. Likewise, every area, large or small, contains an infinite number of points (locations) within its boundaries. Sampling would therefore have to be employed to obtain data on slope angles or depths of soil at different locations in a given area. All locations could not be studied, however much time and effort were to be expended.

3. Collecting the data may adversely affect the population in some way. Sampling may be necessary in order to minimise the disturbance. Imagine, for instance, what would happen if the highway authorities stopped every car entering a big city in order to conduct a traffic survey. To avoid the traffic jams that would be created, many motorists would take alternative routes. Others, perhaps hearing warnings of the survey from radio broadcasts, would choose to stay at home, or would abandon their cars and resort to travelling by train. The traffic flow would be so disrupted that there would no longer be any point in measuring it. Obviously, the sensible procedure would be to stop only a small proportion of the cars, leaving the rest to proceed normally.

The example is not entirely imaginary. Wallis and Roberts in their book *Statistics, a new approach* (1956, pp. 475-6) describe a traffic survey conducted in New York State in 1950. Every fourth vehicle using a certain highway was stopped and its driver questioned about his starting point and destination. The 25% sample size was much larger than was necessary, and resulted in a traffic jam 16 km long, involving as many as 40,000 vehicles. The object of the traffic survey was largely defeated by the gross over-sampling.

4. Not all members of the population may be available for examination. Historical records, for instance, are frequently incomplete. Because the past cannot be resurrected, the historian must make do with the records that exist even though they may not be all he would like to have available. In the same way, the geologist is usually restricted to pre-existing exposures of rock such as quarries and river cliffs. He is scarcely in a position to blast quarries of his own.

If all available members of a population are examined, do they constitute a sample or a population? The question is not easily answered. The available members of the population are but part of a whole, and according to the definition given in the introduction may therefore be regarded as a sample. But they can also be viewed as a population - the *available population*, as distinct from the *target* or *total population*, which cannot be fully studied since not all members are available for examination. Although the target population may be the ultimate or conceptual population, it is not the one actually studied.

It is clear that the available population can be properly described as a sample if it adequately reflects the characteristics of the target population. But what if the available population is unlike the target population? In this case the available population is probably best thought of as a population in its own right rather than a sample.

STATISTICS AND PARAMETERS

The term *statistic* is commonly used to refer to a numerical measure such as an average, or a proportion, that is calculated from sample data. A numerical measure relating to a population is referred to as a *parameter*.

Statistics provide estimates of parameters. The average of a sample, for instance, is an estimate of the average of the population from which the sample is drawn (the *parent population*). Likewise, a proportion or percentage calculated from a sample is an estimate of the proportion or percentage in the parent population.

One must not assume that a statistic is necessarily an accurate estimate of the corresponding parameter of the parent population. The items that appear in a sample may not adequately reflect the characteristics of the items in the population as a whole. Thus the average of a sample may be very different from the average of the parent population.

ACCESSIBILITY SAMPLES

It is always important to try to secure a sample closely resembling the population to be studied. A sample that fails to resemble the population from which it is drawn is potentially misleading. To be worth anything, a sample must be *representative*.

A sample that includes only the most accessible or obvious members of a population is not likely to be representative. In investigating field sizes, for example, it would be foolish to restrict observations to fields next to roads, since these might well be either larger or smaller than average. Nor would it be sensible, in an attempt to determine the amount of sediment carried in suspension in a river, to sample only the surface water. More sediment is carried near the bottom than at the surface. Equally, an opinion poll that relied on volunteers coming forward to answer questions would be likely to be biased since only the most motivated would respond and their opinions might well be unrepresentative of the population as a whole. *Accessibility samples*, as they are called, are therefore best avoided, despite their convenience. There is always the possibility of serious bias, even if it is not immediately apparent.

Unfortunately, it is often difficult to select a sample that will adequately reflect the population to be studied. Part of the population may be particularly elusive or awkward to examine. A geomorphologist, for instance, may decide on a set of sample locations in order to measure slope angles, then find that some of the locations are on private property to which all access is denied. He may have to resort to trespassing in order to make the measurements.

With postal questionnaires it often happens that a large proportion of the people approached do not reply. The reasons for *non-response* are various: the persons involved may not be at home, or may simply be forgetful, or unwilling to answer questions. With some types of postal survey less than 20% of the questionnaires mailed out are returned completed.

Experience has shown that with personal interviewing many more people can be persuaded to answer questions, but almost always there are some who remain uncontactable, or who obstinately refuse to co-operate. It is difficult even with repeated visits, and the exercise of great tact, to obtain a response rate greater than about 85%. The response rate depends very much on what questions are asked. Questions concerning age and income are often not answered. A professional interviewer can achieve much better results than an amateur. Sometimes an interviewer's sex or colour is important.

Response rates generally vary from one area to another. In Great Britain persons living in urban areas tend to be less willing to answer questions than persons living in rural areas. The General Household Survey, for instance, reported a response rate in 1971 of 78.5% in conurbations, 83.9% in urban areas not in conurbations, 85.8% in semi-rural areas, and 93.4% in rural areas.

JUDGEMENTAL SAMPLES

Some investigators rely on personal judgement in deciding which members of a population to include in a sample. They may seek to strike a balance by including "a little bit of everything". Such *purposive* or *judgemental samples*, as they are called, are seldom entirely satisfactory. Their validity depends on the skill of the investigator and his knowledge of the population. In an effort to be "fair" the investigator often includes in the sample too many members of the population that are unusual or bizarre and too few that are average. The sample ends up far more diverse in relation to its size than the population from which it is drawn.

An experiment reported by Yates (1937) provides a striking illustration of the dangers inherent in judgemental sampling. A collection of 1200 stones of varying size was spread out on a table, and 12 persons were each asked to choose three samples of 20 stones representing as nearly as possible the size distribution of the whole collection. For some reason the persons in the experiment tended to choose the larger stones. The average size of the stones in the 36 samples was 22.5% greater than the true average - a considerable degree of bias considering that as many as 720 stones may have been selected from the collection.

An extreme form of judgemental sampling is the *case-study* approach, in which a detailed examination is made of a single member of the population that is believed to be "typical" or "average". The rest of the population are ignored. Case studies are widely employed in geography, despite their obvious dangers. A typical hill slope, a typical thunderstorm, a typical farm, a typical city... these are the basis of much research and teaching. Whether the examples chosen are actually typical is often doubtful. And by concentrating on what is supposedly typical, the case-study approach neglects variation, which may be of great importance.

Accessibility and judgemental samples are subject to the same limitation: there is no way of determining objectively the probable accuracy of the sample results. It is not possible to decide from the sample itself the likelihood that the sample adequately reflects the characteristics of the population from which it is drawn.

Whether inferences drawn from the sample are likely to be reliable is a matter of opinion.

Probability samples are more useful than accessibility or judgemental samples because the probable accuracy of the sample results can be evaluated mathematically. The composition of a probability sample is decided not by availability or judgement, but by chance. Each member of the population has a probability fixed in advance of being included in the sample. By referring to the mathematical "laws" of probability it is possible to calculate the probable accuracy of inferences drawn from the sample.

Various kinds of probability sample have been devised.

SIMPLE RANDOM SAMPLES

The least complicated kind of probability sample is the *simple random sample*. Each member of the population is given the same probability of being sampled.[1] A hand of playing cards dealt from a perfectly shuffled pack is a random sample because each card in the pack has exactly the same chance of being dealt as any other card.

A random sample can be drawn from a finite population by giving each member of the population a separate number. An obvious method is to write the numbers on separate cards or slips of paper which are then shuffled until they are in completely random order. The cards are dealt, one at a time, until as many members of the population have been identified as are required to make up the sample. In order to ensure equal probabilities of selection, the cards must be shuffled extremely thoroughly. As every card player knows, getting cards completely shuffled is a tedious and uncertain business. It is difficult to decide how long to continue shuffling. Since the cards tend to stick together there is a danger that the shuffling will be incomplete however long it is continued.

Instead of marking cards and shuffling them, it is much easier to refer to a *table of random digits*. Table A at the back of this book is a short extract from an extensive and widely used set of tables published by the Rand Corporation (1955). An electronic device was used to give each of the digits from 0 to 9 an equal chance of appearing anywhere in the tables. Thus taking the tables as a whole the digits occur with approximately equal frequency. Each occurrence of each digit is as far as can be determined independent of every other occurrence. Before the tables were published, they were checked and rechecked for signs of non-randomness. Some seeming departures from randomness were corrected (see Chapter 4).

Table A can be read in any direction: horizontally, vertically or diagonally. The digits are printed in blocks to facilitate reading and for no other reason. Random sampling using Table A is carried out as follows. First decide on a direction, then enter the table as randomly as possible. Picking a column and a row blindly with a pin is good enough for most purposes. Using this digit as a starting

---

[1] The meanings of "probability" and "random" are examined at length in Chapter 4. Here only an elementary understanding of the terms is required.

point, read straight along the table in the chosen direction to obtain a random sequence of digits. Breaks between blocks should be disregarded. If the population contains up to 100 members the digits should be grouped into pairs and read as 2 figure numbers. Thus the digits 09 would refer to the ninth member of the population and the digits 90 to the ninetieth. The hundredth member would be 00. If the population contains up to 1000 members, the digits should be read as three figure numbers. The digits 009 would then be the ninth member, the digits 090 the ninetieth, and the digits 000 the thousandth.

Sometimes it is impractical or inconvenient to give a number to every member of a population, for example if the population is very large or infinite in size. Whether a simple random sample can be drawn or not depends very much on the circumstances of the individual problem. It is not possible to lay down any hard and fast rules. There would appear to be no practical way of obtaining a simple random sample of leaves in a forest, or fishes in the sea. On the other hand, it is easy enough to obtain a random sample of tea from a tea pot even though the number of molecules of tea in the pot is indefinitely large. All one has to do is to stir the pot and pour out a cup. Again, it is perfectly possible to sample air temperatures at random intervals during a day, despite there being an infinite number of intervals in the day.

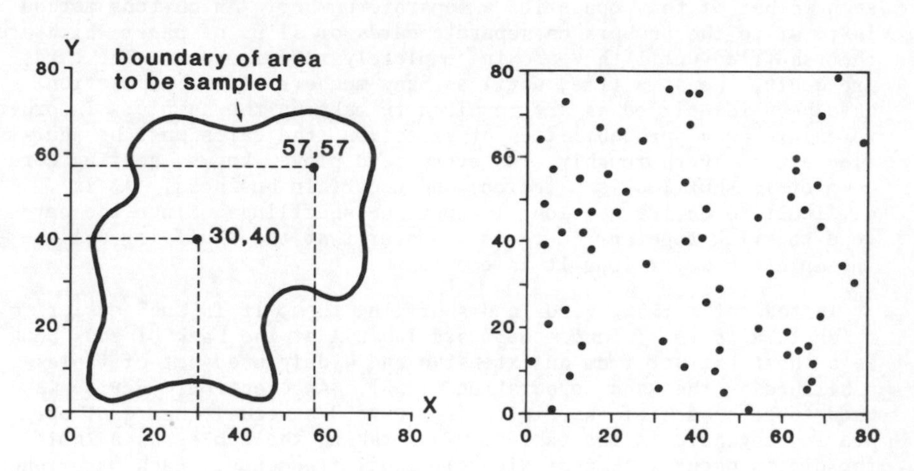

Fig. 3.1 (left) - Method of obtaining random sample points.

Fig. 3.2 (right) - Random sample of 64 points lying within a square area.

Random digits can be used to obtain a sample of randomly distributed points within a given area (Fig. 3.1). Two lines (axes) crossing at right angles are laid down so as to border the area on its southern and western sides. The axes are oriented west-east and north-south. In Fig. 3.1 they are designated X and Y respectively. The axes are subdivided into unit lengths, ruler fashion. The subdivisions are then numbered from west to east and from south to north, starting with the point where the axes intersect (the *origin*) which is numbered 0. Two random numbers are used to locate each random point. The first number is used to define the distance the point lies east of the origin (the X distance, or *X co-ordinate*), the second number the distance the point lies north of the origin (the Y distance, or *Y co-ordinate*). The numbers 30 and 40, for example, refer to a point lying 30 units east of the origin and 40 north. The numbers 57 and 57 refer to a point lying 57 units east of the origin and 57 north.

Figure 3.2 shows a square area with 64 points located using pairs of random numbers each consisting of two digits. The points are distributed irregularly, being more abundant in some parts of the area than in others. This is a typical feature of random patterns.

Because the points in Fig. 3.2 (and for that matter Fig. 3.1) are located using two digit random numbers they are restricted to positions one unit apart in an east-west and north-south direction, and cannot take up positions in between. They are, therefore, not strictly random although they are random enough for most purposes. To obtain points that are more nearly random, one would need to subdivide the axes more finely and to use random numbers of 3 or more digits.

Random samples are usually drawn *without individual replacements*. Any random number encountered once is ignored if it turns up again. Members of the population have therefore a chance of being included only once in the sample. In sampling *with individual replacements* members are included as many times as their identifying numbers are obtained from the table.

Both kinds of sample can be described as random since each member of the population has initially the same chance of appearing in the sample. Sampling with individual replacements guarantees *independence* of selection. In other words, the inclusion of a member of the population in the sample neither increases nor decreases the chance that any other member will be selected. Sampling without individual replacements secures complete independence of selection only if the population is infinitely large. With a finite population once a member of the population has been included in the sample it cannot be selected again, and therefore the remaining members have an increased chance of being selected. However, if the population is large relative to the size of the sample, any departure from independence will be negligible.

There can be no guarantee that the characteristics of a population will be reproduced accurately even by a sample drawn at random. Although 11% of the persons living in a certain village may be agricultural workers, they may not constitute 11% of a random sample, but 29%, 5% or some other percentage. They may not even be included in the sample. The mean (average) age of the persons sampled may be 32 years, but this need not be the mean of the population as a whole. Similar difficulties arise with the sampling of areas using random sample points. As Fig. 3.2 demonstrates, the points tend to be irregularly spaced and do not provide an even coverage of the study area.

A random sample is liable to be unrepresentative simply because its composition is left to chance. Although all members of the population stand an equal chance of being included, those ultimately included may well differ from the rest of the population. As a rule the larger the size of the sample the more closely it will tend to resemble the population.

If the items comprising a population vary greatly in value, a comparatively large sample may be required in order to obtain results that are reasonably accurate. If the items are much alike, a small sample may be sufficient. Unfortunately, the amount of variation in a population is seldom known in advance of sampling. It is usually a matter of guess-work to decide what size of sample is likely to yield results that can be relied upon.

The fraction of the population that is included in a sample is called the *sampling fraction*. Thus, if a population consists of 1000 items and a sample of 100 items is drawn, the sampling fraction is 1/10 or 10%. Unless the sampling fraction is close to unity (100%), its size makes no difference to the probable accuracy of the sample results. What matters is the absolute size of the sample. A sample of 100 items drawn from a population of 1000 items (sampling fraction 10%) is likely to be more representative than a sample of 50 items drawn from a population of only 200 items (sampling fraction 25%).

STRATIFIED RANDOM SAMPLES

Another kind of probability sample is the *stratified random sample*. Members of the population are grouped into classes, sometimes referred to as *strata*. A separate random sample is taken from each class as if each were a population in its own right, then the various samples are lumped together to form a single sample - the stratified random sample.

As an example, imagine that a survey is to be made to determine the proportion of households in Greater London that contain children. It seems likely that households with children tend to avoid certain classes of accommodation such as high-rise flats. Accordingly, a decision is made to set up a stratified random sample based on class of accommodation. The accommodation available in Greater London is divided into ten classes that supposedly contain different proportions of households with children. One per cent of households occupying each class of accommodation are selected at random, and inquiry is made as to whether they contain children. Then taking the total number of households with children, and dividing by the total number of households, one arrives at the proportion of households with children.

The stratified random sample is likely to give a better result than a simple random sample of the same overall size provided the supposition is correct that the proportion of households with children varies with class of accommodation. The trouble with using a simple random sample in this instance is that the selection of households is indiscriminate. Quite by chance, the sample might include an undue proportion of households living in, say, high-rise flats. If the sample fails to represent the classes of accommodation correctly, it is likely to fail to yield the correct proportion of households with children, assuming that the proportion varies according to the class of accommodation. The virtue of the stratified random sample is that its composition, unlike that of the simple random sample, is not left entirely to chance. The stratified sample is so designed that it correctly represents the proportion of households living in each class of accommodation in Greater London.

This is a very simple example of stratified random sampling, but it serves to demonstrate the main technical requirements.

1. The population must be classified *(stratified)* in advance of the sampling.

2. The classes used must be *exhaustive* and *mutually exclusive*. In other words, each member of the population must fall into one of the classes, but only one; or, to put it in still another way, the classes must not overlap, and together must include the whole of the population.

3. The classes must differ in respect of the attribute or variable under study (which is ordinarily not the same as the attribute or variable used to define the classes). If the classes do not differ, there will be no gain in precision over simple random sampling. The aim of the classification must be to secure as much uniformity as possible in the numerical values of the items within each class while maximising the differences between classes.

4. All items in the same class in the population must be given the same chance of being included in the sample. The selection of items to represent each class must be random.

Two kinds of stratified random sample may be usefully distinguished. In the case of a *proportionate* stratified random sample the same proportion of items is taken from each class in the population. The sampling fraction, in other words, is uniform. The classes are represented in the sample in the precise proportions found in the population. Each item in the population has exactly the same chance of being included in the sample regardless of the class to which it belongs. The hypothetical survey of households with children exemplifies this kind of stratified sample.

It is not essential that the same proportion of items be taken from each class in the population. Should the sampling fraction be varied from one class to another the stratified random sample is described as *disproportionate*. The chance that an item has of being selected is dependent on the class to which it belongs.

If the items in each class are about equally variable in numerical value (the amount of variation being measured by the *standard deviation*: see Chapter 7), it is best to employ a proportionate stratified sample. Also, if the amount of variation in the numerical values is unknown at the time of planning the sample, then again a proportionate sample is best. A disproportionate stratified sample is appropriate if the numerical values vary by different amounts in different classes, and these amounts are known when the sample is being planned. The reason for choosing a disproportionate sample rather than a proportionate one is that highly variable classes need to be sampled more intensively than fairly uniform classes in order to attain the same probable degree of accuracy. To take an extreme case: if all the items in a class have the same value, there is no need to select more than one item even though the class perhaps represents a large proportion of the population. If all the items are highly variable, then a large number will have to be selected in order to represent the class accurately.

The amount of variation in the numerical values in each class in a population is of course rarely known in advance of sampling. If proportionate sampling is employed and the items are not equally variable in each class, the results are unlikely to be as good as with disproportionate sampling. Nevertheless the results are likely to be better than with simple random sampling provided that the classes have been chosen intelligently.

Figure 3.3 illustrates disproportionate sampling. An estimate is required of the mean angle of slope in an area where two rock types occur: clay and sandstone. Slopes on the clay are all gentle, whereas slopes on the sandstone are more variable, some being gentle, others being steep. A stratified random sample is employed with a greater density of sampling points on the sandstone than on the clay. A quick inspection of the area indicates that slopes on the sandstone are approximately three times as variable as those on the clay; hence three times as many points per unit area are allocated to the sandstone as to the clay. The stratified random sample in effect consists of two simple random samples, the sampling fraction of one being three times the sampling fraction of the other. The sampling points are concentrated where the risk of inaccuracy is greatest.

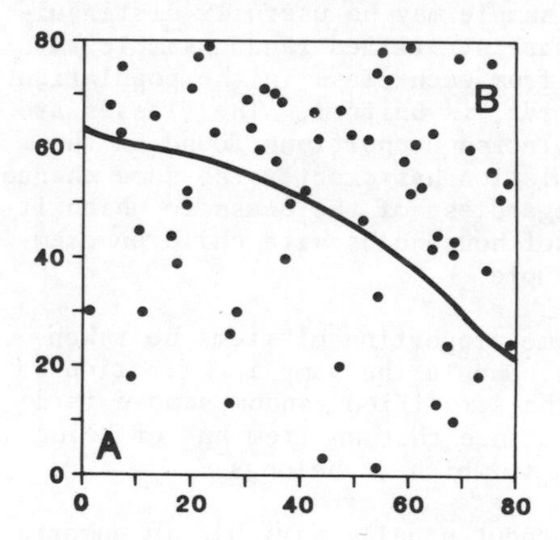

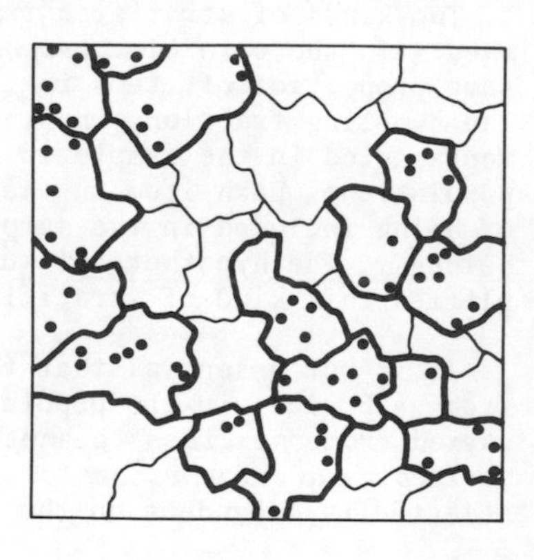

Fig. 3.3 (left) - Stratified random sample with 64 points. The density of points in area A is one third that of area B.

Fig. 3.4 (right) - Cluster sample of 64 points in 15 administrative areas.

Stratified random sampling is much used in place of ordinary random sampling for opinion polls and market research. It has been applied in many geographical contexts, from studies of tree cover and land use to household and traffic surveys.

CLUSTER SAMPLES

In cluster sampling the population is first divided into exhaustive and mutually exclusive classes, as in stratified random sampling. However, a random sample is not drawn from every class. Certain of the classes are selected at random to represent the population; then a random sample is drawn from each of the selected classes (or *clusters*). The random samples are then combined to make up the cluster sample. An alternative procedure is to include all the items in the selected classes in the cluster sample.

Cluster sampling often gives less accurate results than stratified sampling or simple random sampling. The reason is fairly obvious. If the classes selected at random are not representative of the population, then no amount of sampling within those classes will remove the bias. Cluster sampling is likely to be effective only if the classes are much alike in respect of the attribute or variable under study and if the items within each class provide a representative cross-section of the population as a whole. If the classes are very dissimilar, and if only a few are selected, the cluster sample is likely to give poor results. The sort of classification that is most appropriate for stratified random sampling is the least appropriate for cluster sampling. The greater the differences between the classes, and the less the differences within the classes, the less efficient cluster sampling becomes.

Cluster sampling is usually employed to save time or money. Figure 3.4 shows a cluster sample organised on a geographical basis. Sampling is restricted to 15 administrative areas out of a total of 35. The amount of travelling required to visit the sampling points is much less than in the case of a simple random sample or stratified sample of the same size (compare Figs. 3.2 and 3.3). Although the cluster sample is unlikely to be as reliable as the other types of sample, it has the advantage that it is much easier to carry out.

NESTED OR MULTI-STAGE SAMPLES

In nested sampling the process of dividing the population into classes is carried at least one stage further than in cluster sampling by dividing each of the classes into sub-classes. Sometimes, the sub-classes in their turn can be usefully divided into still smaller units, which it may be desirable to further divide and subdivide. Under this form of classification each member of the population belongs to a hierarchy of classes. The method of sampling is as follows: first, a number of classes are chosen at random from the complete set of classes, then a number of sub-classes are chosen at random within each of the classes that have been chosen. The process is continued until the smallest units identified in the classification are reached.

Nested sampling is readily carried out on a geographical basis. For example, several regions within a country might be selected at random, then several districts within each of the selected regions, then several towns or villages within each of the selected districts, and finally several households within each of the selected towns or villages. Figure 3.5 shows a scheme in which a square area is first divided into four large squares (stage A). Each of the large squares is divided into four medium-sized squares, two of which are selected at random (stage B). Each medium-sized square is divided into four small squares. Two of the small squares are selected at random (stage C), and each is divided into four tiny squares of which two are selected at random (stage D). The sampling points are then located in the centre of these tiny squares (stages E and F).

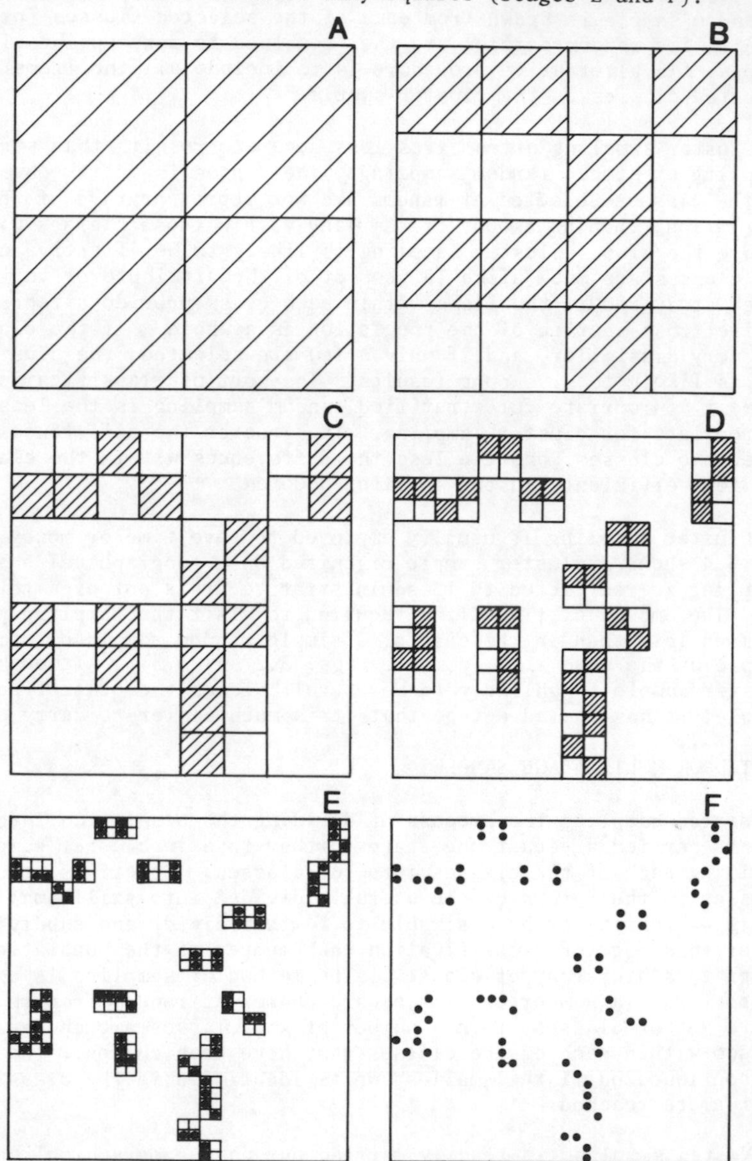

Fig. 3.5 - Stages in the development of a nested or multi-stage sample of 64 points.

Nested sampling, like cluster sampling, is often employed to keep costs of data collection to a minimum. The points shown in Fig. 3.5, for example, could be visited in far less time than the 64 randomly distributed points shown in Fig. 3.2. The main shortcoming of nested sampling is that it provides uneven coverage. Random sample points are almost always better distributed, and hence more reliably represent the population as a whole. The clustering of the nested sample points does, however, enable the population to be studied at several different scales or levels. A scheme such as that shown in Fig. 3.5 allows exploratory sampling of a large region to be combined with detailed sampling on a local scale. At the analysis stage it is possible to separate out local and regional variations.

SYSTEMATIC SAMPLES

Another commonly used method of drawing a sample is to work progressively through a population selecting members at regular rather than random intervals. A traffic survey, for example, may be conducted by stopping every tenth or hundredth car reaching a check-point. Again, in a study of the dissolved solids carried by a river, a sample of water may be taken once every 24 hours or once every seven days. Or, in the case of an area, the sampling points may be regularly spaced. Figure 3.6 shows a square area with 64 equally spaced points arranged in a grid fashion.

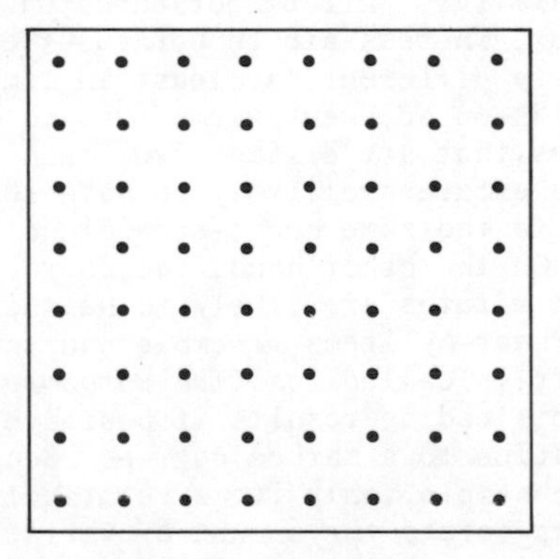

Fig. 3.6 - Systematic sample of 64 points.

Samples selected in an orderly manner are known as *systematic samples*. Their main virtue is their simplicity. Picking out every tenth car is obviously easier than picking out the 13th, 28th, 31st... car using a table of random numbers. Visiting a river at a set time each day is more convenient than visiting it all hours of the day and night. Regularly spaced sampling points are easier to locate than random points and provide a more uniform coverage of an area.

A systematic sample is likely to be as good as a random one provided the members of the population are arranged randomly. But if regularities or periodicities are present in the population, serious bias can result. Picking out every tenth bus using a particular route would be a foolish way of conducting a sample survey if the tenth bus was always bound for the same destination, a destination different from other buses. For the same reason regularly spaced points can give very misleading results if used to sample populations possessing a regular spatial pattern. Consider the problems that might arise using regularly spaced points to sample slope angles in an area of regularly spaced parallel ridges and valleys. The points might all fall on the ridge tops, or all on the valley floors. Similar problems might arise if regularly spaced points were used to sample buildings in a city laid out on a grid pattern. The buildings included in the sample might all lie on street corners.

Before drawing a systematic sample, it is always advisable to check whether a periodicity exists in the population. It is not safe to do the drawing and then check only the sample itself. The sample will give no indication that there is a periodicity if the sampling interval is equal to the period or a multiple of it. Even if preliminary checking indicates a periodicity, systematic sampling can still be used provided the sampling interval is made only a tiny fraction of the period. Thus, if every 1000th item in the population is a special case and every tenth item is selected, the sample will approximate closely to a random sample, despite the periodicity.

Another difficulty may arise if the population is not arranged randomly. It often happens that items occupying adjacent positions either in time or space are much alike, whereas items that are far apart are very dissimilar. Air temperatures on consecutive days tend to be fairly similar, whereas air temperatures on days six months apart tend to be very different, at least in temperate and arctic regions. Crop yields on adjacent farms tend to be more alike than crop yields on farms that are distant from each other. People living on the same housing estate are likely to have somewhat similar backgrounds, to belong to the same socio-economic class, and to have the same aspirations. On the other hand, the characteristics of people living on different estates are likely to be fairly dissimilar. This phenomenon whereby near-by items resemble one another more closely than items far apart is called *positive autocorrelation*. Systematic sampling can give misleading results if positive autocorrelation is present in a population to a marked degree. Because nearby items are not included in the sample, only items relatively far apart, the sample is likely to exaggerate the amount of variation in the numerical values in the population. It is not likely, however, to give a biased value for the average; it is only the spread of values around the average, or *standard deviation* (see Chapter 7), that is likely to be over-estimated.

SYSTEMATIC UNALIGNED SAMPLES

The problems caused by spatial regularities in a population can be avoided to some extent by using a *systematic unaligned sample*. The sampling points are fairly regularly spaced but not aligned (Fig. 3.7E). The study area, which for simplicity will be assumed to be square, is divided using a grid of squares equal in number to the size of sample required. A sampling point is placed randomly in the square forming the south-west corner of the grid (Fig. 3.7A). The point is placed using one random number to specify the distance separating it from the western side of the square and another random number to specify the distance separating it from the southern side. In Fig. 3.7A these distances are labelled X and Y respectively. Next, each of the squares forming the row running eastwards from the

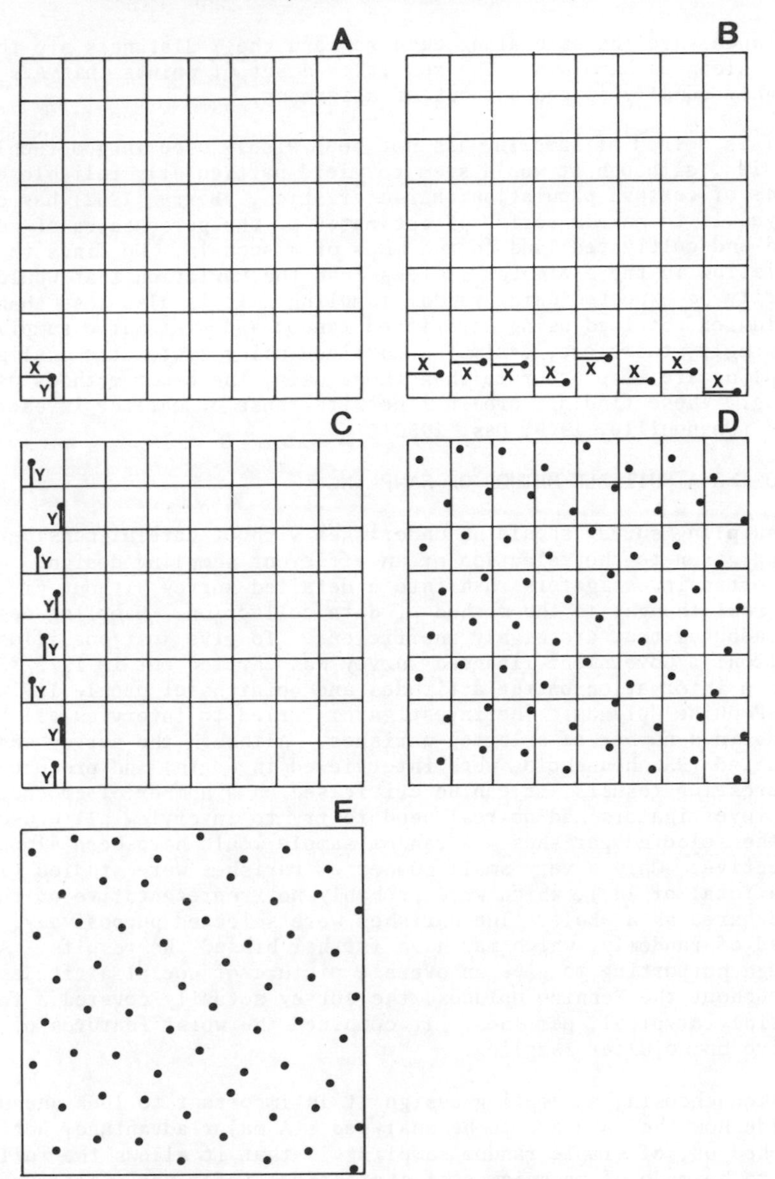

Fig. 3.7 - Stages in the development of a systematic unaligned sample of 64 points.

corner square is allocated a sampling point. The design calls for the X distance to be held constant for all the points along the row and the Y distances to be varied using additional random numbers (Fig. 3.7B). After the row has been completed, a sampling point is placed in each of the squares forming the column running northwards from the corner square. The rule here is to hold the Y distances constant while varying the X distances randomly (Fig. 3.7C). Once the column has been completed, random numbers are not used again. Sampling points are placed in the remaining squares so that the X distances are the same along each row and the Y distances are the same along each column. The result is a set of points that are roughly equally spaced but out of alignment.

This method of sampling has not been widely used in geographical studies, although it would seem to yield particularly reliable estimates of certain population characteristics. Berry (1962) has employed it to make a series of estimates of the percentages of woodland and cultivated land on two maps of Wisconsin. He finds that the variation in the estimates is less than the variation that would ordinarily be expected using random sampling. It is also less than the variation obtained using stratified random and systematic sampling. Seemingly, therefore, estimates obtained using systematic unaligned sampling are more accurate than those using the other methods of sampling. These findings are in line with those of earlier investigators (Quenouille, 1949; Das, 1950).

CHOOSING A SUITABLE METHOD OF SAMPLING

No sampling survey should be undertaken without careful consideration being given to the selection of an efficient sampling design. All too often investigators rush into a detailed survey without giving any real thought to the method of data collection. Sampling designs are adopted that are highly inefficient. To give just one illustration: a Government-financed survey was carried out in 1972/73 to obtain information on the attitudes and opinions of people living in the Pennine Uplands. The investigators tried to interview all households in a number of selected parishes. Although the survey was very detailed (956 households were interviewed in depth) and produced some interesting results, it can be criticised on a number of grounds. The investigators had no real need to try to interview all households in the selected parishes - a random sample would have been almost as effective. Only a very small number of parishes were studied (11 out of a total of 181), which were probably not representative of the study area as a whole. The parishes were selected purposively, instead of randomly, which may have further biased the results. Although purporting to give an overall picture of social attitudes throughout the Pennine Uplands, the survey actually covered a few, possibly atypical, parishes. It combined the worst features of purposive and cluster sampling.

When choosing a sampling design it is important to look ahead and decide how the data are to be analysed. A major advantage, not yet touched on, of simple random sampling is that it allows the fullest use to be made of techniques of statistical inference. Although simple random samples are often not as representative as other probability samples, they are more amenable to analysis. This advantage is of the greatest importance. The number of techniques that can be applied to other forms of probability sampling, particularly systematic unaligned sampling, is unfortunately fairly limited. Moreover, the techniques that are available tend to be much more complicated mathematically. It is wise not to resort to other forms of sampling without first considering whether an appropriate method of analysis is available.

4 Probability and Randomness

THE MEANING OF PROBABILITY

Most scientists are deterministic in their outlook. They proceed on the assumption that events have causes, and are not spontaneous. Given full knowledge of the causes, all events supposedly are entirely predictable. However, owing to ignorance of the causes, which may be very complicated, scientists often find themselves unable to predict events with certainty. In place of a statement of certainty they generally can offer only a statement of probability.

A probability, then, is based on imperfect knowledge. With complete knowledge probability ceases to exist. A person who knows only that a standard pack of playing cards lies on the table may say that the probability of the bottom card being the Queen of Hearts is 1 in 52. But another person who has peeped at the bottom card may say for sure whether it is the Queen of Hearts.

What is probability? Dictionaries commonly define it by using words, such as "chance" or "likelihood", that are little more than synonyms for probability and so fail to clarify its meaning. The awkward truth is that probability is impossible to define with any precision. Indeed, some philosophers argue that it is best accepted as indefinable, being so basic a concept as to be incapable of formulation in terms of simpler ideas. However, this would seem to be going too far. One possible definition of probability is the degree of belief or expectation that it is rational to entertain concerning an uncertain event or hypothesis.

In order to understand the way in which probabilities are assigned to events or hypotheses, consider the four questions:

1. What is the probability that a playing card drawn blindly from a thoroughly shuffled pack will be black (a spade or a club)?

2. What is the probability of snow next Christmas Day?

3. What is the probability that men will land on Mars before the end of the century?

4. What is the probability that the genus *Homo* originated in Africa?

Each of these questions has to be answered in a different way.

In the case of the card the answer can be reached deductively. Because the pack is thoroughly shuffled the cards will lie in random order. Each card is therefore equally likely to be chosen. There are 13 spades and 13 clubs in the pack of 52 cards, so the chance of getting a black card is 26 in 52 or 1 in 2. Note that this result derives from *a priori* reasoning and not from actual experiments with playing cards. However, experiments can be conducted, if desired, to test whether the result is realistic.

The answer to the second question cannot be obtained by *a priori* reasoning. The two possible outcomes of snow and no snow cannot be assumed to be equally likely. Obviously the answer must depend on observational evidence. Provided the climate is not changing (a major proviso), the probability of snow next Christmas Day is simply the frequency of occurrence of snow on Christmas Days in years past: about one year in 10 in south-east England.

No appeal to observational evidence is possible for the third question, because up to now the only landings on Mars have been by unmanned space craft. The probability depends on personal assessment of technological progress, political ambitions and other factors. The fourth question has also to be answered subjectively, but it differs from the third and preceding questions in that it refers to an event that has already passed. The reason probability is involved is that not enough facts are known to allow the question to be answered with certainty.

Probabilities calculated by regarding all possible events or outcomes as equally likely are called *theoretical* or *a priori probabilities*. Probabilities calculated by observing the relative frequency of occurrence of events or outcomes are known as *empirical* or *a posteriori probabilities*. *Subjective probabilities*, as exemplified by the last two questions, are exercises of personal judgement.

What sorts of situations permit theoretical probabilities to be calculated? Mainly it would seem games of chance, lottery-type situations, random sampling, and the like. There are clearly many situations in everyday life that do not allow the calculation of theoretical probabilities, and for which only empirical and subjective probabilities can be established.

Not everyone is agreed that theoretical probabilities exist independently of empirical probabilities. Possibly they are nothing more than rationalisations of empirical probabilities. Just why, for instance, does one accept that the probability of a coin landing heads uppermost is 1 in 2? Admittedly, the coin is symmetrical (or approximately so), the centre of gravity is half-way between the two faces, and the number of spins is unplanned and haphazard. But can one properly deduce from these facts that the probability of heads is 1 in 2? Surely one accepts the figure of 1 in 2 because one knows from experience that coins come down heads on about half the occasions that they are tossed. It is because the empirical probability is about 1 in 2 that one accepts the theoretical argument that the probability may be exactly 1 in 2. Since the theory matches reality, one is prepared to accept the theory which in itself is not entirely convincing. At least this is what some mathematicians and logicians have argued.

At this point another difficulty enters the picture. No theoretical probability can be proved or disproved beyond all possible doubt using observational evidence. A coin when tossed once may come down heads. This does not alter the probability on the next throw; yet the next throw may also yield heads, and so perhaps for ten throws in a row. This freak result would not invalidate the theoretical statement that the chance of a head is 1 in 2 each time the coin is tossed. In theory, the more times the coin is tossed, the more closely the proportion of heads can be expected to approach 1 in 2, but this does not mean that in any finite number of tosses the proportion of heads will be exactly 1 in 2. Even if the coin perversely yields 50 heads in a row, or 500, this is not conclusive proof that the chance of a head is other than 1 in 2. Such a result might be obtained even if the chance of a head is as stated 1 in 2, although admittedly the probability of the result is very low, so low in fact that one would in practice assume that the chance of a head at least in relation to the coin in question is other than 1 in 2, either because the coin is biased in favour of heads (perhaps it is not perfectly balanced) or because it is being tossed incorrectly. Reasonable though such an assumption may be, it is important to recognise that it is merely an assumption. No amount of observational evidence would ever be sufficient to disprove conclusively the theoretical statement that the probability of heads is 1 in 2.

It is also worth observing that empirical probabilities cannot be demonstrated to be true or false any more conclusively than can theoretical probabilities. The calculation of the number of past occurrences may be correct, but further experience may show that some adjustment in the statement of probability is required. This is particularly so if conditions change. History does not always repeat itself. Even if snow has fallen every Christmas Day for years past, this does not guarantee that snow will fall next Christmas Day.

Subjective probabilities are also incapable of verification. No subjective probability can be conclusively proved or disproved by anything that actually happens.

Although empirical probabilities can be determined for a much wider range of situations than can theoretical probabilities, they are nevertheless subject to one major limitation: an event can be assigned an empirical probability only if it is a member of a larger group of similar events. If an event is unique in the sense that nothing resembling it is known, then it cannot be assigned an empirical probability. The only way of arriving at the probability of unique events is by subjective assessment. However, many mathematicians and logicians refuse to accept subjective assessment as a valid means of obtaining probabilities of events whether classed as unique or otherwise. They argue that a probability is simply not worth anything, unless obtained in an objective manner. This would seem to be taking rather a narrow view of probability. People in everyday life constantly make subjective assessments of probability, contending, for example, that "there is a high probability that the Government will be forced to resign before the year is out" or "the probability that the committee will reverse its decision is less than 1 in a million". Such statements are not necessarily invalid because they are expressions of opinion. The crucial question to decide is whether they are arrived at rationally. Probability is concerned with degrees of rational belief, as distinct from degrees of actual belief of persons who have incorrectly weighed the available evidence or who have constructed invalid arguments.

A problem devised by J.L. Watling (quoted by Ayer, 1965) neatly illustrates the kinds of difficulties that can arise in the calculation or subjective assessment of probabilities. A man is walking along a path that lies in a valley. He is about to reach a place where the path divides into three, one path continuing along the valley, the other two ascending the valley side. Assuming that the man will take one of the three paths, what is the probability of his selecting the valley path?

The *a priori* approach to probability, although at first sight promising, in fact fails to provide a satisfactory answer. Assuming that it is equally possible that the man will take any one of the three paths, the chance of his selecting the valley path is evidently one in three. But this assumption of equality is precarious, and if one was betting on the outcome it would not be wise to assume equality of choice. There is no compelling reason for regarding the paths as equal in the man's mind. The man may see himself as presented with only two alternatives: to climb the valley side or to stay in the valley. If he resorts to tossing a coin, the chance of his taking the valley path is one in two. But he may not regard the two alternatives as equal, in which case the *a priori* approach breaks down altogether.

The empirical approach to probability also encounters severe difficulties. All men are not equal. Although, in the past, 3 out of every 10 men may have chosen the valley path, this does not mean that it would be wise to bet on the basis that the chance is 3 out of 10 that the man presently walking will choose the same path. Perhaps he is wearing a grey flannel suit. To compare him with every tramp who has wandered along in the past, every hiker, or every shepherd with a flock, would not be reasonable. In choosing a path he may well be guided by considerations similar to those of men in grey flannel suits before him. He may not like the look of the rough, boulder-strewn paths up the valley side. He may not feel up to climbing. The probability of his choosing the valley path ought in principle to be calculated from observations of the choices made by other men in grey flannel suits. Unfortunately, few such men may have preceded him, making any reliable assessment of the probability impossible. Moreover, if the argument is carried to its logical conclusion, every aspect of the man's appearance ought to be taken into account (his hat, lack of a raincoat, etc.). In the event there will be nobody to compare him with. As has already been explained, empirical probabilities cannot be established by treating situations as unique, but only by making generalisations. The difficulty is that the level of generalisation cannot be fixed except arbitrarily, even though it may significantly affect the probabilities.

The difficulty becomes less acute the greater the level of generalisation that is attempted. To revert to the same example, suppose one was not particularly anxious to state the probability of this particular man taking the valley path, but was instead aiming to state the probability of passers-by in general doing so. Perhaps most passers-by are mountaineers who take one of the mountain paths. If one blindly says that there are 3 chances in 10 of the man in the grey flannel suit taking the valley path, ignoring the fact that he is obviously not a mountaineer, one may "wrongly" state the probability for the man in the grey flannel suit, but this would not matter provided one's concern was only to state the probability for passers-by in general.

The subjective approach to probability suffers from the obvious disadvantage that the probability cannot be fixed with anything like mathematical precision. Different observers can be expected to assess differently the probability that the man will choose the valley path. The main advantage of the subjective approach is that it allows the man's uniqueness to be taken more fully into account. If he frequently stops to mop his brow with a handkerchief, this can be judged to reduce the probability that he will choose one of the paths that climb the valley side and to increase the probability that he will choose the less steeply inclined path following the valley. The extent of the reduction, and the increase, are matters of opinion, however, and cannot be precisely quantified, except by resorting to empirical evidence.

Enough has been said to indicate some of the difficulties surrounding the concept of probability. Mathematicians and logicians repeatedly debate the meaning of probability without arriving at any general agreement. The lack of agreement must not be allowed to obscure the fact that the concept of probability is applied freely and fruitfully in everyday affairs, little thought being given to its precise meaning. Moreover, in certain situations probability can be assigned an almost precise meaning with no difficulty. Thus, in the case of events that can be repeated endlessly and whose outcome is essentially random (the tossing of a coin, the rolling of a die) *a priori* probabilities can be shown to be the same as empirical probabilities "over the long run", and consequently virtually no problems of definition arise. For example, suppose cards are drawn one by one from a thoroughly shuffled pack. Each card is inspected, then returned to the pack which is thoroughly shuffled again before the next card is drawn. In the long run one half of the cards drawn ought to be black and the other half red, although in the short run the proportions may be very different. Figure 4.1 records the results of one experiment. At the end of every tenth draw the total number of black cards secured up to that time was recorded as a proportion of the total number of draws. At the end of the first 10 draws the number of black cards stood at 4, giving a proportion of 0.40. After 20 draws the proportion was 0.55, 11 black cards having been secured. After 30 draws the proportion was down to 0.53. As the drawing continued the proportion fluctuated less and less, approaching a value of 0.50. In the long run, given an indefinitely large number of draws, the proportion can be presumed to attain a limiting value of exactly 0.50. It is this limiting value that constitutes the probability of a black card.

The "long run" definition of probability can be formalised as follows: the probability of a given event (or outcome) is the theoretical value of the proportion or fraction:

$$\frac{\text{Number of occurrences of the event}}{\text{Number of opportunities for occurrences of the event}}$$

when the number of opportunities is infinitely large. It is this definition that is customarily employed in statistical inference. For example, when a statistician says that the probability of obtaining a certain sample from a population, or a sample less represent-

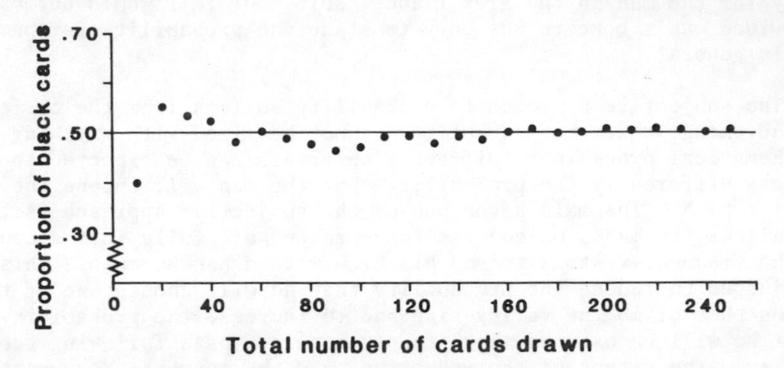

Fig. 4.1 - Changes in the proportion of black cards when cards are drawn one by one from a thoroughly shuffled pack.

ative of the population, is 5/100, he means that, if an infinite number of samples were drawn from the population, then on average 5 in every 100 samples would be identical to the sample in question, or would be even less like the population. In the first 100 samples, there might be none answering the description, in the next hundred perhaps 8, but in the long run the number would average 5 in 100.

THE ARITHMETIC OF PROBABILITY

In order to understand statistical procedures it is necessary to acquire some knowledge of the mathematical principles of probability. What follows is necessarily a very brief outline. Much fuller accounts are available in the standard mathematical texts on probability, for instance, Feller (1957) and Parzen (1960).

The probabilities discussed so far have been mainly expressed as proportions, viz. 1 in 2, 1 in a million. It is customary in mathematics, however, to express probabilities as fractions or decimals. For example, 1 in 2 becomes 1/2 or 0.5. Probability is thus assessed on a scale that starts at 0 and ends at 1. A probability of 0 means that the event or outcome in question is impossible. For example, the probability of seeing Shakespeare alive is 0. If an event or outcome is bound to happen the probability is reported as 1. The probability of not seeing Shakespeare is therefore 1. In general, if an event occurs on average once every n times that there is an opportunity for it to occur, the probability is reported as 1/n. Thus the probability of obtaining a head with a single toss of a coin that is true (unbiased) is 1/2 or 0.5. The probability of a tail is likewise 1/2 or 0.5. The probability of throwing a 6 with a die that is true is 1/6 or 0.167. The probability is the same for each of the other numbers on the die.

Events are described as *mutually exclusive* if they cannot occur together, that is to say if the occurrence of any one precludes the occurrence of the others. A coin can show a head or a tail, but not both simultaneously. Thus heads and tails are mutually exclusive.

Probabilities of mutually exclusive events may be combined by addition. For example, the probability of obtaining either a head or a tail with a single throw of a true coin is 0.5 + 0.5 = 1. This is as it should be since either a head or a tail is bound to appear with a probability of 1. The probability of throwing either a 6 or a 1 with a single throw of a true die is 1/6 + 1/6 = 1/3. The probability of obtaining any one of the four remaining numbers is 1/6 + 1/6 + 1/6 + 1/6 = 2/3. The probability of obtaining one of the six possible numbers is 1/3 + 2/3 = 1.

In general, if $e_1, e_2, e_3, \ldots, e_n$ are n mutually exclusive events having respective probabilities of occurrence $p_1, p_2, p_3, \ldots, p_n$, the probability of either e_1 or e_2 occurring at any one time is $p_1 + p_2$, the probability of either e_1, e_2 or e_3 occurring is $p_1 + p_2 + p_3$, and the probability of one of the n separate events occurring is $p_1 + p_2 + p_3 + \ldots + p_n$. Also, if p is the probability that a given event will occur and q is the probability that it will not occur, then $p + q = 1$ or $q = 1 - p$. These relationships are referred to as the *addition rule*.

Events that are able to occur together are described as *independent* if the probability of each one of them remains the same despite the occurrence or non-occurrence of any of the others. Otherwise they are described as *dependent*. Independent events may be combined using the *multiplication rule*. In the case of two such events, e_1 and e_2, having respective probabilities p_1 and p_2, the probability of obtaining both e_1 and e_2 is $p_1 \times p_2$. For example, the probability of drawing the ace of spaces from a pack of well shuffled cards is the probability of drawing an ace multiplied by the probability of drawing a spade i.e. 4/52 × 1/4 = 1/52. The separate probabilities can be multiplied together because the value of a card is independent of its suit. The probability of a card being an ace is neither increased nor decreased by its being a spade.
As an extension of this, if $e_1, e_2, e_3, \ldots, e_n$ are n independent events having respective probabilities of occurrence $p_1, p_2, p_3, \ldots, p_n$, the probability of all the events occurring is $p_1 \times p_2 \times p_3 \times \ldots \times p_n$.

The multiplication rule can be used to obtain the probability of independent events occurring in a particular order. Thus the probability of obtaining first a 6 and then a 1 with two successive throws of a die is 1/6 × 1/6 = 1/36. The multiplication rule applies because the occurrence of a 6 on the first throw does not alter the probability of a 1 on the second throw. The throws are independent. Since the probability of obtaining first a 1 and then a 6 is also 1/36, the probability of obtaining either 6 then 1 or 1 then 6 is 1/36 + 1/36 = 1/18 by the addition rule.

If an unbiased coin is tossed five times the probabilities of occurrence of 0, 1, 2, 3, 4 and 5 heads are 1/32, 5/32, 10/32, 10/32, 5/32 and 1/32 respectively. These probabilities can be obtained by identifying all the possible arrangements of heads and tails (Table 4.1).

TABLE 4.1 - POSSIBLE ARRANGEMENTS OF HEADS AND TAILS WITH FIVE TOSSES OF AN UNBIASED COIN

		Number of Heads			
0	*1*	*2*	*3*	*4*	*5*
TTTTT	HTTTT	HHTTT	TTHHH	THHHH	HHHHH
	THTTT	THHTT	HTTHH	HTHHH	
	TTHTT	TTHHT	HHTTH	HHTHH	
	TTTHT	TTTHH	HHHTT	HHHTH	
	TTTTH	HTHTT	THTHH	HHHHT	
		HTTHT	THHTH		
		HTTTH	THHHT		
		THTHT	HTHTH		
		THTTH	HTHHT		
		TTHTH	HHTHT		

H = Head T = Tail

Each arrangement has the same probability since it consists of the same number (5) of either heads or tails, the probabilities of which are the same. Since there are 32 arrangements in all, the probability of each is 1/32. (This result can also be obtained using the multiplication rule. For example, the probability of HTTHT is 1/2 × 1/2 × 1/2 × 1/2 × 1/2 = 1/32.) The probability of obtaining exactly X heads is simply the number of arrangements in which exactly X heads appear divided by the total number of arrangements, viz. 32. The probability of 3 heads, for example, is 10/32 since (as will be seen by studying Table 4.1) 10 arrangements out of the 32 contain 3 heads and under the addition rule, if each arrangement has a probability of 1/32, the total probability must be 10/32. The probability of 4 heads is 5/32, there being 5 arrangements with 4 heads in Table 4.1.

If the coin is tossed a great many times, listing all the possible arrangements of heads and tails is hardly feasible. Fortunately, the number of arrangements yielding exactly X heads can be readily calculated using the formula

$$\frac{K!}{X!(K - X)!}$$

where K is the number of tosses. The symbol K!, called *K factorial*, stands for the product K(K - 1) (K - 2) (K - 3) etc., or K multiplied by every positive integer (whole number) smaller than K. For example, if K is 5, K! is 5 (4) (3) (2) (1) = 120. In the same way X! stands for X (X - 1) (X - 2) (X - 3) etc. and (K - X)! stands for (K - X) (K - X - 1) (K - X - 2) (K - X - 3) etc. 0! is defined as 1.

With 5 tosses the number of arrangements yielding exactly 3 heads is therefore

$$\frac{5!}{3!(5-3)!} = \frac{5(4)(3)(2)(1)}{[3(2)(1)][(2)(1)]} = \frac{120}{[6][2]} = 10$$

The total number of different arrangements obtainable with 5 tosses is $2^5 = 32$. The probability of obtaining exactly 3 heads in 5 tosses is thus 10/32, the same result as is yielded by Table 4.1.

To generalise if an unbiased coin is tossed K times, the probability of obtaining exactly X heads is

$$\left[\frac{K!}{X!(K-X)!}\right] \Big/ \; 2^k = \frac{K!}{X!(K-X)!}\,(1/2)^k$$

RANDOMNESS

The term "random" was introduced briefly in the last chapter. It is a concept closely related to probability. If the various outcomes of a given process are all equally probable, then each outcome can be said to be random. A sequence of digits is therefore random if each of the digits is given the same probability of being selected for any position in the sequence. If the probability is the same, there can be no relationship between the digits otherwise than by chance. Each digit is independent of every other digit.

Because the digits are equally probable they ought in theory to appear with equal frequency in a sequence of infinite length. For a sequence of finite length, however, there may be chance differences. The following sequence was generated by rolling a die, an accepted method of producing random digits

4 1 6 3 3 4 3 4 2 4 4 1 4 6 5 6 ...

Although the die was seemingly perfectly balanced and unbiased, the sequence has an excess of certain digits and a deficiency of others. The digit 2, for example, appears only once, whereas the digit 4 appears five times. These differences are presumably only accidental. Were the sequence longer, the digits would doubtless appear in more equal proportions.

In a random sequence not only ought individual digits to appear with approximately equal frequency, but also pairs of digits such as 11, 43, 57 and 96, triplets such as 111, 274, 816 and 954, and so on.

If there is any order, pattern or relationship in a sequence of random digits, it is entirely the result of chance. Consider again the sequence of digits

4 1 6 3 3 4 3 4 2 4 4 1 4 6 5 6 ...

The digits would seem to be arranged in no particular order. If a rule exists for predicting the digits, it is by no means obvious. The sequence is seemingly incompressible - it cannot apparently be shortened or simplified. The most concise way of specifying it would seem to be the sequence itself.

The following sequence of digits, by contrast, is obviously ordered

1 2 3 4 5 6 1 2 3 4 5 6 1 2 3 4 ...

The digits 1 to 6 are arranged in order, then repeated. The sequence would seem to be entirely predictable judging from what can be seen of it. The last digit shown on the right is presumably followed by a 5, then a 6, a 1, and so on. Even if continued over many pages, the sequence would be readily compressible into a simple instruction such as "print the digits 1 to 6 and repeat n times".

If a die is rolled 16 times, it is as likely to produce the ordered sequence of digits shown immediately above as the disordered sequence of digits preceding it. Both sequences, in other words, are equally probable. Each sequence is just one outcome out of a total of 6^{16} different outcomes. It should be particularly noted, however, that the probability of obtaining a sequence as ordered as the ordered sequence is very low while the probability of obtaining a sequence as disordered as the disordered sequence is high. There are very few really orderly sequences amongst the 6^{16} possible sequences; most are largely or entirely disordered.

Many writers believe that an ordered sequence of digits ought not to be called random under any circumstances. They consider that randomness ought to be judged not by the process producing it, but by the results. What matters is not the provenance of a given sequence of digits, but whether the digits lack order, pattern or relationship. If the digits are arranged systematically, the sequence is not random. On this definition, the ordered sequence of digits shown above would not be random, even if it were the result of a process such as rolling a die 16 times.

Chaitin (1975) has attempted to improve on this definition. He considers that randomness can be equated with incompressibility. A sequence of digits is random, he argues, if it is incapable of being reduced to a more compact form. The extent to which a given sequence is compressible thus provides a measure of the extent to which it is non-random. The fact that the ordered sequence shown above is highly compressible is an indication that it is very far from random.

One objection to judging randomness on the results is that the suggested criteria (disorder, incompressibility, etc.) are somewhat vague. The sequence

3 1 4 1 5 9 2 6 5 3 5 8 9 7 9 3 ...

is seemingly disordered and patternless, using these words in a loose sense. The sequence can nevertheless be reduced to a highly compact form. It is in point of fact merely the value of π with the decimal point removed, π of course being the circumference of a circle divided by its diameter. On Chaitin's definition, the sequence is almost totally non-random, even though there is no obvious relationship between the digits, and the sequence never repeats itself, no matter how far it is continued. There are thus good grounds for regarding the sequence as fairly random despite its compressibility.

A second objection to judging randomness on the results is that no sequence of digits of any length can possibly be random on the criteria given above. Some degree of order is discernible in every long sequence of digits. With patience one can always find certain regularities or traces of a pattern. Every sequence of digits can be compressed to some extent. Randomness is thus seen to depend on there being no very obvious order, pattern, etc. The difficulty is to define what is obvious and what is not. The definition cannot avoid being arbitrary: a pattern that is obvious to an expert mathematician may not be obvious to someone whose mathematical ability is limited.

A third objection can be raised. Even if it were possible to construct a random sequence of digits, there would be no way of proving conclusively that the sequence was random in any sense above defined. One cannot prove a negative, save in special circumstances. Although the digits may seem to lack order this is not proof that no order exists. Just because a sequence appears incompressible, this does not mean that it cannot be compressed.

Defining randomness in terms of process would seem less open to objection than defining it in terms of the results. Nevertheless randomness still cannot be proven since one can never be sure that a supposedly random process is operating in an entirely random manner. For example, one can conceive of a perfectly balanced and unbiased die, but in practice no die is perfect. A real die cannot be relied upon to yield digits that are exactly random. Given a set of digits produced by a die that is seemingly perfect, all one can say is that the digits are seemingly random and then only by examining the digits. If the digits are

6 6 6 6 6 6 6 6 6 6 6 6 6 6 6 6 6 ...

this could be the result of chance, but in practice one would suspect the die of being biased whether or not it appeared to be perfectly balanced. On the other hand, if the die produced the ordered set of digits shown previously

1 2 3 4 5 6 1 2 3 4 5 6 1 2 3 4 ...

one might be very surprised, but not inclined to suspect the die, since a biased die would be no more likely to produce such an ordered set than an unbiased die. Again, if the die produced the disordered set of digits representing π

3 1 4 1 5 9 2 6 5 3 5 8 9 7 9 3 ...

one would tend to assume that the die was operating in a random or approximately random manner, but there would be no conclusive proof.

It is difficult to design a process that is seemingly random. Most mechanical processes operate in a conspicuously non-random manner. Even the accepted methods of achieving randomness, such as shuffling cards and tossing coins, are seemingly not entirely reliable. Some element of order tends to creep in despite every precaution. Cards tend to stick together when shuffled; coins are often not quite symmetrical and so when tossed tend to fall one way more than the other; dice are seldom perfectly balanced, tending to favour certain numbers. Even the elaborate electronic machines used to generate the random digits that appear in many published tables of random digits are not necessarily free from bias. The reasons are discussed briefly by Hacking (1965, pages 129-30). The randomness of the digits has to be carefully checked before they are published. The checking poses serious problems, however. The digits 0 to 9 (and for that matter the hundred pairs of digits 00 to 99, and so on) ought to appear with approximately equal frequency. But what degree of approximation is allowable? The arrangement of digits ought in theory to be completely haphazard. But what tests for order should be applied? There must not be too many since no table can be expected to pass every test that can be devised. Any machine producing random digits must periodically produce obviously ordered sets of digits simply because it is random and knows no better. Within any long series of random digits there are bound to be some apparent departures from randomness.[1] The departures cannot be taken as proof that the random digit machine has broken down, but nevertheless they suggest the possibility and are therefore disconcerting. Furthermore, if regularities are present in a table of random digits this may render the digits unusable for drawing random samples. Random sampling requires that all members of the population be given an equal probability of selection. Unfortunately, it is impossible to construct a table of random digits that will satisfy this requirement for both large and small populations. Any sequence of digits satisfactory for sampling a large population will inevitably contain stretches that are not suitable for sampling very small populations. The problem is discussed by Kendall and Stuart (1977, Vol. 1, p.232) in the following terms:

"Suppose, to take an extreme case, we constructed a table of $10^{10^{10}}$ digits. The chance of any digit being a zero is 1/10 and thus the chance that any given block of a million digits are all zeros is $10^{-10^{6}}$. Such a set should therefore arise fairly often in the set of $10^{10^{10}-6}$ blocks of a million. If it did not, the whole set would not be satisfactory for certain sampling experiments. Clearly, however, the set of a million zeros is not suitable for drawing samples in an experiment requiring less than a million digits".

If one wanted to draw a random sample with individual replacements from a population of only 1000 items, and entered the table at the beginning of the set of a million zeros, one would end up with

[1]Brown (1957, page 52) in a detailed analysis of the concept of randomness refers to this as "the monkey theorem". A monkey with a typewriter "strikes the keys at random and eventually, in the long run, writes all the Shakespeare sonnets or something equally unlikely".

only one item in the sample. Though not suitable for sampling a small population the million zeros are nevertheless best retained for sampling a large population. If one wanted to sample a population of $10^{10^{10}}$ items one would not wish to exclude the item represented by a million zeros, nor for that matter the items represented by a million ones, a million and one zeros, etc.

The compiler of a table of random digits is therefore faced with a dilemma. If his machine produces repetitions that do not look random - and in the long run it is bound to produce such repetitions - he has either to ignore them or to alter them to make the table seemingly more random. Most compilers have adopted the latter strategy. For example, the Rand Tables were found to be significantly "biased" when first completed and were adjusted before publication. The digits of most random digit tables are so carefully selected as to be strictly speaking not random at all. In a sense they are more random than random. This awkward contradiction bothers many mathematicians and must also bother users of the tables. However it is doubtful whether the rather arbitrary methods of construction of the tables seriously prejudice the results of standard statistical procedures such as the significance tests discussed later in this book. Small errors would seem likely, but not large ones. However, anyone using tables of random digits has clearly to tread warily given the peculiarities of their construction. The tables are but an attempt to attain what perhaps is unattainable.

5 Frequency Distributions

Masses of raw data are difficult to comprehend, unless carefully organised and presented. Consider, for instance, the data shown in Table 5.1. Seventy five towns and villages containing 10 or more persons were selected at random from amongst 1318 shown on a map of Iowa published by Rand McNally. The straight-line distances separating the towns and villages from their nearest neighbours were measured in the sequence listed.

TABLE 5.1 - DISTANCES IN KM BETWEEN NEAREST-NEIGHBOUR TOWNS AND VILLAGES IN IOWA

7.2, 7.3, 8.9, 6.0, 4.8, 11.6, 5.1, 4.9, 6.7, 9.3, 2.7, 6.9, 9.4, 9.7, 7.8, 6.5, 7.1, 12.7, 9.4, 5.7, 6.6, 8.8, 6.7, 9.9, 7.0, 10.6, 7.9, 3.9, 8.3, 6.2, 3.4, 8.6, 8.4, 6.6, 7.5, 7.9, 10.7, 9.5, 4.2, 7.4, 8.5, 4.4, 11.0, 8.0, 5.6, 9.0, 5.4, 6.2, 6.3, 6.7, 11.6, 5.4, 8.5, 5.4, 7.4, 5.2, 4.4, 5.2, 8.8, 6.4, 5.8, 6.9, 6.6, 9.5, 8.6, 3.6, 8.0, 7.4, 5.9, 6.4, 12.5, 10.0, 3.5, 6.0, 7.2

It is hard to get an overall picture of the variation in distances since they are not in an organised form. The smallest and longest distances can be found fairly easily by reading through the table, but it is less obvious around what value, if any, the distances tend to concentrate.

A much clearer picture of the pattern of variation is obtained by rearranging the distances in order of magnitude from smallest to largest (Table 5.2). It is at once apparent that the distances range from 2.7 to 12.7 km and tend to concentrate somewhere in the middle, around 7 km.

A rearrangement of raw data in ascending (or descending) order of magnitude is known as an *array*. An array, though easier to interpret, takes up as much space as raw data and becomes impractical if the data are extensive.

TABLE 5.2 - ARRAY OF THE NEAREST-NEIGHBOUR DISTANCES

2.7, 3.4, 3.5, 3.6, 3.9, 4.2, 4.4, 4.4, 4.8, 4.9, 5.1, 5.2, 5.2, 5.4, 5.4, 5.4, 5.6, 5.7, 5.8, 5.9, 6.0, 6.0, 6.2, 6.2, 6.3, 6.4, 6.4, 6.5, 6.6, 6.6, 6.6, 6.7, 6.7, 6.7, 6.9, 6.9, 7.0, 7.1, 7.2, 7.2, 7.3, 7.4, 7.4, 7.4, 7.5, 7.8, 7.9, 7.9, 8.0, 8.0, 8.3, 8.4, 8.5, 8.5, 8.6, 8.6, 8.8, 8.8, 8.9, 9.0, 9.3, 9.4, 9.4, 9.5, 9.5, 9.7, 9.9, 10.0, 10.6, 10.7, 11.0, 11.6, 11.6, 12.5, 12.7

FREQUENCY DISTRIBUTIONS

A *frequency distribution* is formed by grouping the data into a number of classes or categories. The classes or categories are defined numerically, as in the following table.

TABLE 5.3 - FREQUENCY DISTRIBUTION OF THE NEAREST-NEIGHBOUR DISTANCES

Distance to nearest-neighbour town or village in km	*Number of towns and villages*
2-2.9	1
3-3.9	4
4-4.9	5
5-5.9	10
6-6.9	16
7-7.9	12
8-8.9	11
9-9.9	8
10-10.9	3
11-11.9	3
12-12.9	2
	Total = 75

A frequency distribution summarises data; it does not provide the detailed information of an array, but is much easier to comprehend. In the case of Table 5.3, for example, it is immediately obvious that the class 6-6.9 km contains more towns and villages than any other class. There is a progressive decline in frequencies either side of this, the class of highest frequency, or *modal* class.

The *width* of a class, or *class interval*, is defined as the difference between the highest and lowest values that can fall in the class. The nearest neighbour distances in Table 5.1 are rounded to one significant decimal place or 0.1 km. Hence a class such as 2-2.9 km in Table 5.3 actually includes distances ranging from 1.95 to 2.95 km. The following class, 3-3.9 km, includes distances from 2.95 to 3.95 km. The width of each class is therefore 1.0 km not 0.9 km as might at first appear. The width is the real upper limit of the class minus the real lower limit. Alternatively it is the difference between the apparent lower limit of the class and the apparent lower limit of the following class.

The *mid-point* of a class is found by adding together the apparent lower limit of the class and the apparent upper limit, then dividing by two. Thus the mid-point of the class 2-2.9 km is (2 + 2.9)/2 = 2.45 km. The same result can be obtained by adding the real lower limit to the real upper limit, and dividing by two viz. (1.95 + 2.95)/2 = 2.45 km.

There are no hard and fast rules concerning the construction of frequency distributions. It is very much a matter of personal choice what classes are used. Nevertheless certain obvious points need to be borne in mind:

1. If too many classes are used, the distribution will be rendered unwieldy and will not summarise the data sufficiently. On the other hand, if the data are compressed into too few classes, important detail will be obscured. Thus to group all the nearest-neighbour distances into two or three classes would carry summarisation too far for many purposes.

The number of classes has therefore to be a compromise. It is generally desirable to distinguish between 5 and 20 classes depending on the number of observations. If there are fewer than 25 or 30 observations an array is likely to be more useful than a frequency distribution. For 25 or 30 observations a frequency distribution composed of 5 or 6 classes is usually the best choice. More observations permit the use of more classes. However, if more than about 20 classes are used, the distribution becomes difficult to grasp at a casual glance.

2. It is usually advisable to select classes of the same width, particularly if computations are to be based on the frequency distribution. Classes of unequal width e.g. 2-2.9, 3-5.9, 6-7.9 tend to be confusing and a nuisance in computations. In certain circumstances, however, unequal classes are unavoidable. For example, classes of equal width would not provide a satisfactory frequency distribution of settlement size in Western Europe. With settlements varying in size from less than 100 to over 11 million inhabitants, even a distribution of as many as 100 classes of equal width would not discriminate sufficiently between the smaller settlements. The first class, extending from less than 100 to over 11,000 inhabitants would embrace all villages, towns and smaller cities, leaving the medium and large-size cities spread thinly over the remaining 99 classes. A much better distribution could be obtained by varying the width of the classes. Villages; towns; small, medium and large cities could then be assigned to separate classes.

Open-ended classes are classes lacking either a lower or an upper limit e.g. "2.9 km and under", "12 km and over". They are often employed at the beginnings and ends of frequency distributions in order to reduce the number of classes and to save on space. Although useful in this respect, open-ended classes are best avoided because they are not precise and cause difficulties in computations.

3. Each observation in the sample or population should be assignable to one and only one class. Classes such as 2-3, 3-4, 4-5 are objectionable because the class limits overlap, and observations coinciding with the limits (e.g. 3 and 4) can be placed in two classes.

When recording a discontinuous variable, ambiguity can be readily avoided by selecting natural breaks in the values to act as class limits. For example, suitable classes for data on the number of persons per household might be 1 person, 2 persons, 3 persons, etc., or, if fewer classes were needed, 1 or 2 persons, 3 or 4 persons, etc. Because fractions of a person are impossible, the classes are unambiguous.

In the case of a continuous variable there are no natural breaks in the values to exploit, but there are breaks arising from rounding which serve almost as well. When selecting the classes, consideration must be given to the manner in which the values have been rounded. In geographical studies, rounding is usually carried out to the *nearest* whole number, tenth or hundredth rather than the *last*. A distance recorded as 6 km is greater than 5.5 km but less than 6.5 km, while one recorded as 6.3 is greater than 6.25 km but less than 6.35 km. In contrast, in everyday usage, the tendency is to round down. For example, a person giving his age as 42 usually means that it is not lower than 42 nor higher than 43.

Below are given three illustrations of how the classes might be written for a variable recorded to (a) the nearest whole number, (b) the nearest tenth, (c) the nearest hundredth.

(a) 1-2, 3-4, 5-6, 7-8, etc.
(b) 1-2.9, 3-4.9, 5-6.9, 7-8.9, etc.
(c) 1-2.99, 3-4.99, 5-6.99, 7-8.99, etc.

Writing classes in this way obscures the real class limits, however. For instance, the class 1-2 means in reality 0.5 - 2.5, the class 3-4 means 2.5 - 3.5, and so on. Moreover, there is still ambiguity unless it be made clear how items like 2.5 are recorded. Are they rounded up or down? If the digit to be discarded is a 5, followed by zeros, the usual practice is to round the measurement so that the last digit retained is even. For example, 2.5 becomes 2 when rounded to one digit, 3.5 becomes 4, and 4.5 becomes 4. In the same way 15.25 becomes 15.2 when rounded to two digits, and 15.75 becomes 15.8.

For a variable rounded down to the last whole number or fraction, the classes might be

(d) 1 to less than 3, 3 to less than 5, 5 to less than 7, etc.

or more briefly

(e) 1-, 3-, 5-, etc.

Note that it would not be correct to write the classes in this manner if the rounding was carried out to the nearest whole number or fraction rather than the last. For example, an item that measured 2.97 units would be recorded as 3.0 rather than 2.9. Under classifications (d) and (e) the item would be assigned to the second class, whereas in reality it would belong to the first.

RELATIVE FREQUENCIES

Sometimes it is useful to express the number of observations in each case as a proportion or percentage of the total number of observations. For example, the 16 distances in the class 6-6.9 km. in Table 5.3 represent a proportion of 16/75 or 0.213. This is equivalent to 21.3%. As Table 5.4 demonstrates, proportional frequencies add up to 1, percentage frequencies to 100%.

Proportional and percentage frequencies are called *relative frequencies*. Frequencies representing actual numbers of observations are called *ordinary* or *absolute frequencies* when they need to be distinguished from relative frequencies. Otherwise they are called simply frequencies.

TABLE 5.4 - RELATIVE FREQUENCY DISTRIBUTION OF THE NEAREST-NEIGHBOUR DISTANCES

Distance to nearest-neighbour town or village in km	*Proportion (p)*	*Percentage (%)*
2-2.9	0.013	1.3
3-3.9	0.053	5.3
4-4.9	0.067	6.7
5-5.9	0.133	13.3
6-6.9	0.213	21.3
7-7.9	0.160	16.0
8-8.9	0.147	14.7
9-9.9	0.107	10.7
10-10.9	0.040	4.0
11-11.9	0.040	4.0
12-12.9	0.027	2.7
	Total 1.000	100.0

Samples of difference size may be readily compared as relative frequency distributions. All that is necessary is that the class boundaries of the distribution coincide. Frequency distributions, i.e. absolute frequency distributions, are more difficult to compare since allowance has to be made for sample size.

Proportional frequencies, it should be noted, represent empirical probabilities. Thus, if a town or village in the Iowa sample is picked at random, the empirical probability that it will lie between 6 and 6.9 km from its nearest neighbour is 0.213. The probability of its being located over 10 km from its neighbour is 0.040 + 0.040 + 0.027 = 0.107.

GRAPHICAL REPRESENTATION OF FREQUENCY DISTRIBUTIONS

Frequency distributions (and also relative frequency distributions) may be illustrated graphically by means of *histograms*. The variable in question is shown on the horizontal axis of the histogram. The classes are represented by vertical rectangles or bars, the areas of which are varied in proportion to the frequencies.

An example of a histogram is provided by Fig. 5.1 which depicts the nearest-neighbour distances of Table 5.3. Because the classes in the table are of equal width, each bar of the histogram is made the same width. Consequently, not only the areas of the bars but also their heights represent frequency. A scale of absolute frequency is shown on the left, a scale of percentage frequency on the right.

The boundaries between the classes in Fig. 5.1 appear at first sight to coincide exactly with the 1 km subdivisions of the horizontal scale. In fact they are off-set by 0.05 km. To avoid giving an erroneous impression of the class limits, the mid-values of the classes can be shown on the horizontal scale, as in Fig. 5.4B. Unfortunately, this makes the scale more difficult to read because usually the class mid-values are not whole numbers: clarification in one direction makes for confusion in another.

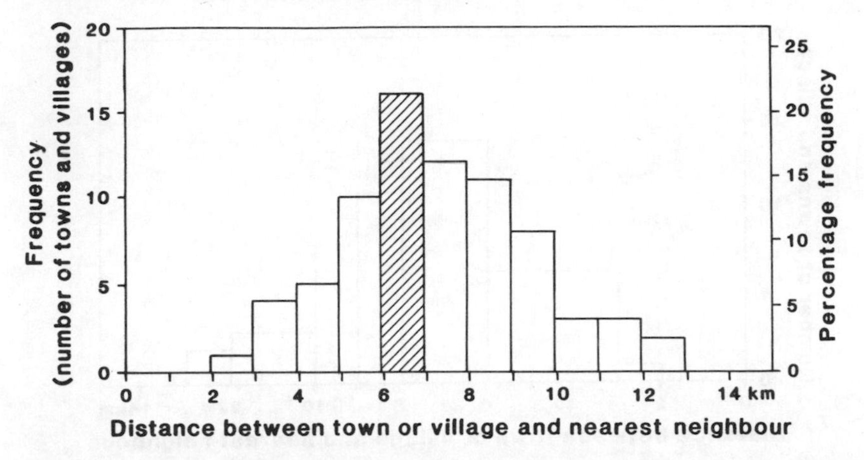

Fig. 5.1 - Histogram of the nearest-neighbour distances. (Modal class shaded).

When classes are of unequal width, as in Table 5.5, the width of the bars forming the histogram should be varied in proportion to the width of the classes. Thus, if one class is three times as wide as another class, it should be shown by a bar three times as wide. The areas of the bars should be kept proportional to class frequencies by making the heights of the bars proportional to the class frequencies divided by the class widths (Fig. 5.2).

TABLE 5.5 - FREQUENCY DISTRIBUTION WITH CLASSES OF VARYING WIDTH

Distance to nearest-neighbour town or village in km	*Number of towns and villages*	*Number divided by width of class*
2-2.9	1	1/1 = 1.00
3-5.9	19	19/3 = 6.33
6-7.9	28	28/2 = 14.00
8-8.9	11	11/1 = 11.00
9-9.9	8	8/1 = 8.00
10-11.9	6	6/2 = 3.00
12-12.9	2	2/1 = 2.00

When representing a discrete variable with a small total range, it is a good plan to use as many classes as there are separate values of the variable. The bars of the histogram are best separated slightly to emphasise the discrete nature of the variable (Fig. 5.3).

In place of a histogram, a *frequency polygon* may be used to show the frequency distribution of a continuous variable. A frequency polygon is a line graph of frequency plotted against the mid-value of each class. In Fig. 5.4A dots have been used to indicate the points on the graph where the mid-values of the various classes intersect with their respective frequencies. The lines connecting the dots form the frequency polygon.

A histogram can be transformed into a frequency polygon by drawing straight lines between the mid-points of the tops of the bars of the histogram (Fig. 5.4B).

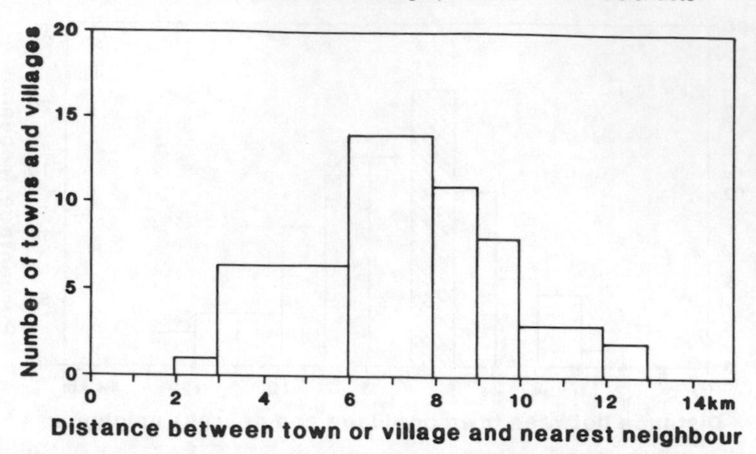

Fig. 5.2 - Histogram of the nearest-neighbour data, classes of varying width.

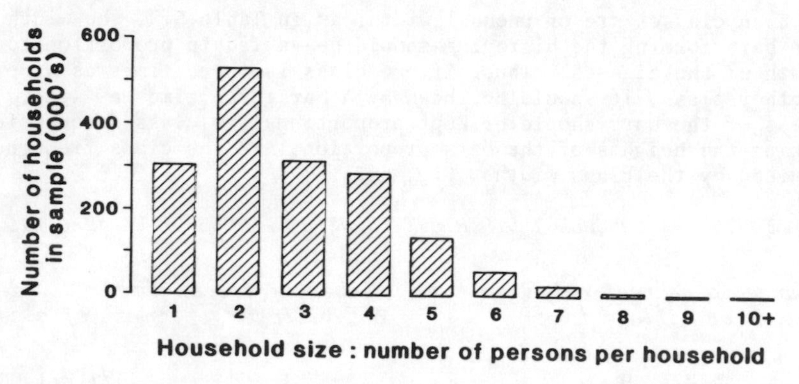

Fig. 5.3 - Variation in the size of households in England and Wales. Sample of 1,661,888 households, 1971 Census.

A frequency polygon can either be left hanging above the horizontal axis (Fig. 5.4A), or extended downwards to join the axis (Fig. 5.4B). The method of extending the polygon is to include at each end of the distribution a class of zero frequency.

A polygon extended to join the horizontal axis is more satisfying visually than one left hanging. Moreover, with the extensions the area under the polygon becomes equal in size to the area enclosed by the histogram drawn to the same scale. It is therefore a representation of the total frequency. A major disadvantage of the extensions is that they indicate a range of values greater than that actually observed. A frequency polygon must not be extended if this means including values that are clearly impossible. For example, a polygon should not be extended leftwards of zero if negative values cannot actually occur.

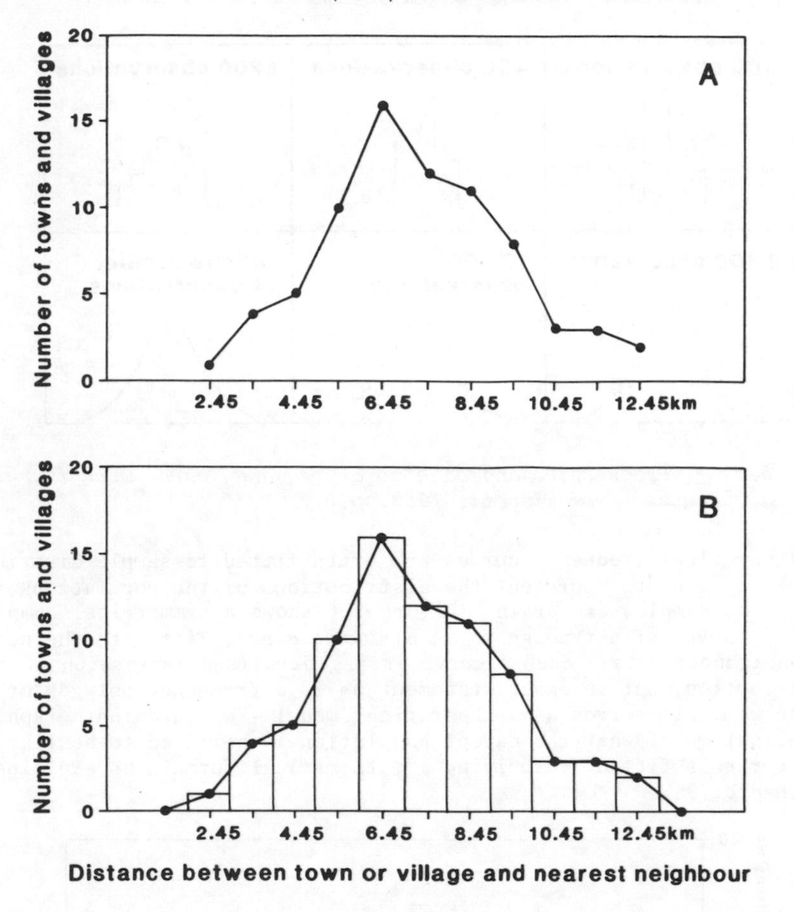

Fig. 5.4 - Frequency polygon of the nearest-neighbour distances.

The flat-topped bars of the histograms foster the impression that the observations falling within each class are evenly distributed over the whole range of the class. Frequency polygons, on the other hand, suggest that all the observations are concentrated at the mid-points of the classes. On balance, histograms are perhaps less misleading than polygons. Where two or more distributions must be shown superimposed on the same diagram, however, frequency polygons possess definite advantages over histograms. Superimposed histograms are difficult to distinguish, especially if the class limits coincide.

In the case of a continuous variable, if the number of observations is increased, and at the same time the classes forming the frequency distribution are made narrower, the resulting histogram and frequency polygon will approximate more and more closely to a smooth curve. Figure 5.5 shows that with increasing sample size the stepped outline of the histogram is gradually transformed. In the limiting case, with a sample of infinite size, both histogram and polygon become a perfectly smooth *frequency curve*.

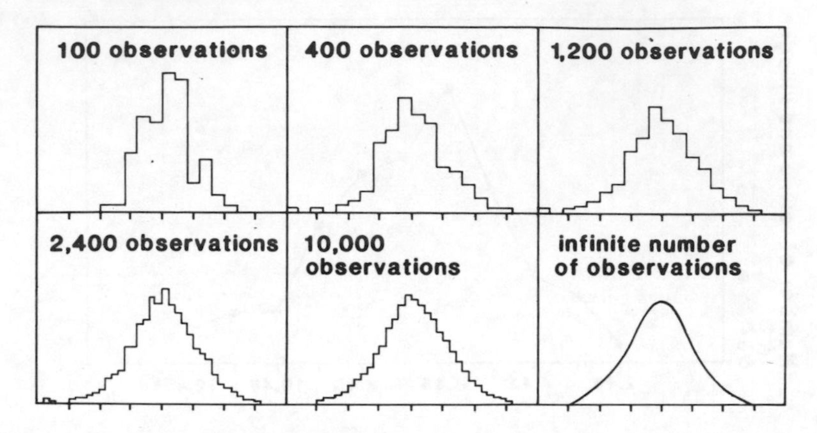

Fig. 5.5 - Progressive approach to a frequency curve with increasing sample size (from Tippett, 1952, p.5).

Theoretical frequency curves are often fitted to sample distributions in order to represent the distributions of the populations from which the samples are drawn. Figure 5.6 shows a symmetrical humpbacked curve, of a type known as a *normal curve*, fitted to the nearest-neighbour data. Such a curve is a generalised impression of the distribution, not an exact statement as is a frequency polygon or histogram. It serves as a theoretical model - a convenient graphical idealisation of what the parent population is believed to be like. The curve is fitted by applying a mathematical formula as explained in Chapter 9.

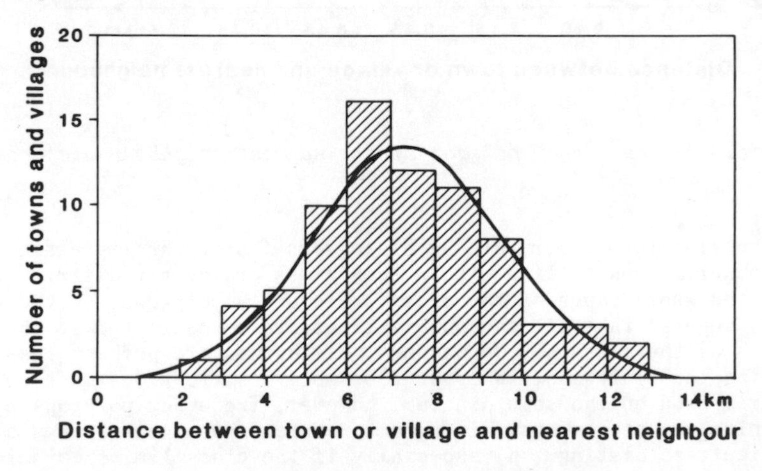

Fig. 5.6 - Frequency curve fitted to the histogram of nearest-neighbour distances.

TYPES OF FREQUENCY DISTRIBUTION

Frequency distributions assume a great variety of shapes when represented graphically. Some of the more distinctive of these shapes are illustrated in Fig. 5.7.

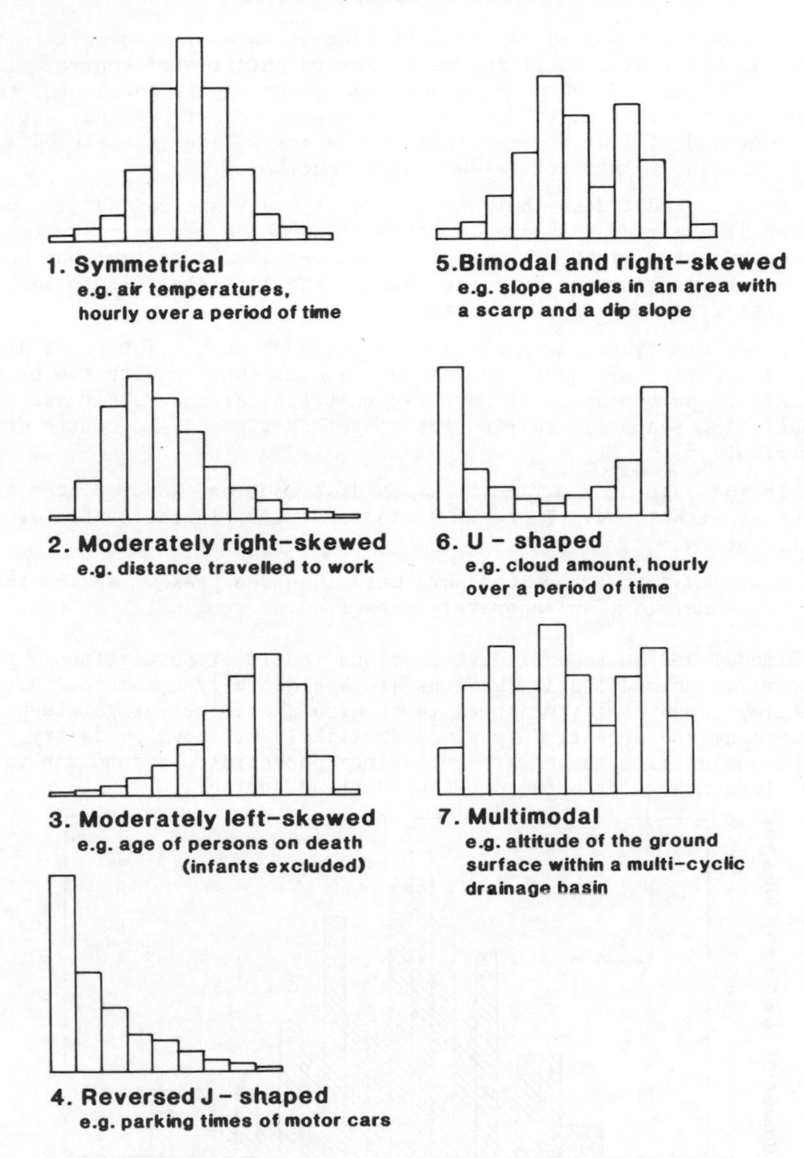

Fig. 5.7 - Typical frequency distributions.

1. Distributions that possess only one maximum or "peak" of frequency are described as *unimodal*. A number of unimodal distributions are *symmetrical* or nearly so. Frequencies decrease to zero either side of the single peak which occupies a central position. The most important symmetrical distribution is the hump-backed *normal* distribution (see Chapter 9).

2 & 3. The majority of unimodal distributions are moderately *skewed*, having single peaks lying somewhat off centre. They are said to be

right skewed (or *positively skewed*) if the longer tail occurs to the right of the peak i.e. if the peak lies to the left of centre. They are said to be *left skewed* (or *negatively skewed*) if the longer tail occurs to the left of the peak. Right-skewed distributions, e.g. the *log-normal* distribution described in Chapter 9, are encountered much more frequently than left-skewed distributions.

4. Some unimodal distributions are so strongly skewed that the peak occurs at one end. *J-shaped* distributions which are fairly rare have the peak on the right, whereas *reversed J-shaped* distributions which are rather more common have the peak on the left (that is to say, the greatest frequency is at or near zero).

5. Other distributions possess more than one peak. The peaks are not necessarily of equal prominence. A distribution with two peaks is said to be *bimodal*. It may be symmetrical or moderately skewed. Usually the peaks lie fairly close together towards the centre of the distribution.

6. In the case of the rare *U-shaped* distribution, however, the peaks occur at either end. There is a single minimum in the centre of the distribution.

7. *Multimodal* distributions have more than two peaks. Again, they may be symmetrical or moderately skewed.

Bimodal and multimodal distributions result if populations with dissimilar unimodal distributions are accidentally combined. For example, slope angles measured in an area where there is a steep escarpment and a gentle dip-slope are likely to show bimodality, the slope angles from the escarpment being concentrated around one value, the slope angles from the dip-slope around another.

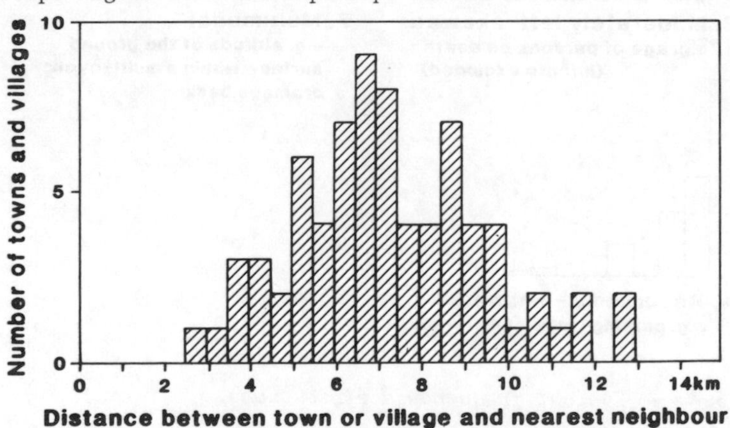

Fig. 5.8 - Histogram of the nearest-neighbour data. Class intervals of 0.5 km. Note the more pronounced asymmetry and greater irregularity of the histogram compared with that shown in Fig. 5.1.

Bimodality or multimodality can also result if too many classes are selected to form a frequency distribution. It is a useful exercise to take a distribution and vary the number of classes to see if the shape of the histogram changes appreciably. A small number of classes tends to give a unimodal distribution; many classes tend to give a number of peaks and a highly irregular histogram (Fig. 5.8).

CUMULATIVE FREQUENCY DISTRIBUTIONS

A *cumulative frequency distribution* shows how many observations are equal to or less than selected values. An example is provided by Table 5.6 which represents the nearest-neighbour distances. The cumulative frequencies shown in the table are obtained by adding together the individual frequencies shown in Table 5.3, starting from the top. Observe that the final cumulative frequency of 75 is the total frequency.

TABLE 5.6 - CUMULATIVE FREQUENCY DISTRIBUTION OF THE NEAREST-NEIGHBOUR DISTANCES SHOWN IN TABLE 5.3

Distance to nearest-neighbour town or village in km	*Cumulative frequency*
Equal to or less than 2.95	1
Equal to or less than 3.95	5
Equal to or less than 4.95	10
Equal to or less than 5.95	20
Equal to or less than 6.95	36
Equal to or less than 7.95	48
Equal to or less than 8.95	59
Equal to or less than 9.95	67
Equal to or less than 10.95	70
Equal to or less than 11.95	73
Equal to or less than 12.95	75

A cumulative frequency distribution may be represented graphically by (1) plotting dots on graph paper to represent the intersections of the "equal to or less than" figures and the corresponding cumulative frequencies, (2) joining the dots with straight lines so forming a line graph as shown in Fig. 5.9. The lines constitute a *cumulative*

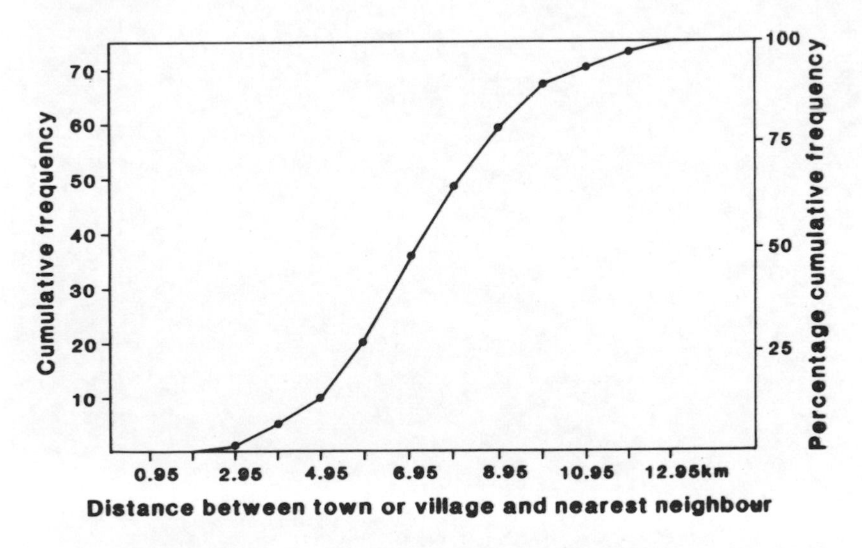

Fig. 5.9 - Graphical representation of the cumulative frequency distribution of the nearest-neighbour distances.

frequency polygon. This often approximates to an S-shaped curve, concave below and convex above. Such a curve is called an *ogive* (pronounced "oh jive").

Cumulative frequencies may be converted to *percentage cumulative frequencies* by dividing by the total frequency and multiplying by 100. Percentage cumulative frequencies can be graphed in the same way as cumulative frequencies. Figure 5.9 includes a scale of percentage cumulative frequency on the right side of the diagram.

6 Measures of Central Tendency

Statisticians use a variety of numerical measures or indices in order to summarise data in a concise yet informative manner. The best known of these measures is the *arithmetic mean*, popularly referred to as the *average*. Closely related measures include the *median* and the *mode*. All three measures define values that tend to lie centrally within a set of data arranged in order of magnitude. For this reason they are known as *measures of central tendency*.

By contrast, *measures of dispersion* seek to describe the extent of variation in data. They are discussed in the next chapter.

THE ARITHMETIC MEAN[1]

The *arithmetic mean* of a set of n numbers is obtained by adding the numbers together and dividing the total by n. As an example, consider the nine distances 5, 8, 9, 9, 10, 11, 12, 15 and 20 metres. They add up to 99 metres and so the arithmetic mean is 11 metres. Similarly, the five temperatures -6, -3, +2, +8 and +9°C add up to +10°C, giving a mean of +2°C.

In general, if X is a variable and $X_1, X_2, X_3, \ldots, X_n$ are n values of X, then the arithmetic mean of the n values is

$$\frac{X_1 + X_2 + X_3 + \ldots + X_n}{n}$$

This may be written more concisely as

$$\frac{\Sigma X}{n}$$

The symbol Σ, the Greek capital letter *sigma*, denotes the process of addition. Thus ΣX means the sum of the values of X. It is pronounced "sum X" or "sigma X".

Different symbols are usually employed for samples and populations. If the values of X form a sample, it is usual to denote their arithmetic means by the symbol $\bar{X}$, pronounced "X bar". Thus, if there are n items in the sample, each providing a value of X

$$\bar{X} = \frac{\Sigma X}{n}$$

where ΣX as before is the sum of the n values of X.

[1] It is usual to refer to the arithmetic mean simply as "the mean". Although there exist other kinds of mean, such as the geometric and harmonic means, they are rarely used, whereas the arithmetic mean is of considerable importance. Provided these other means are always referred to by their full names and not abbreviated no ambiguity can arise from referring to the arithmetic mean as "the mean".

In the case of a population the mean is usually denoted by μ, the Greek letter *mu*. Thus, if there are N items in the population

$$\mu = \frac{\Sigma X}{N}$$

ΣX referring in this instance to the sum of the N values of X.

It may seem an unnecessary complication to use symbols that differentiate between a population and a sample when the method of calculating the mean is exactly the same. Nevertheless the practice has advantages that will become apparent later on when problems of drawing inferences from samples are discussed.

ESTIMATING THE MEAN FROM GROUPED DATA

Calculating an arithmetic mean using the formula $\Sigma X/n$ or $\Sigma X/N$ can take a long time if there are many values of X to add up. If the values are already grouped so as to form a frequency distribution, the mean can be estimated from the frequencies by using a short-cut method that does not require knowledge of the individual values. Unfortunately the method is only approximate. If an exact answer is required there is no escape from the labour of adding up all the values.

The short-cut method involves multiplying the mid-value of each class by the class frequency. The resulting products are then added together and divided by the total frequency. In symbols,

$$\bar{X} = \frac{\Sigma(fm)}{n} \quad \text{or} \quad \mu = \frac{\Sigma(FM)}{N}$$

where f (or F) is the frequency in each class and m (or M) is the mid-value of the class.

The method will be illustrated using the nearest-neighbour data introduced in the last chapter.

TABLE 6.1 - METHOD OF ESTIMATING THE MEAN

Distance to nearest-neighbour town or village in km	*Number of towns and villages*	*Mid-value of class*	*Number multiplied by mid-value*
	f	*m*	*fm*
2-2.9	1	2.45	2.45
3-3.9	4	3.45	13.80
4-4.9	5	4.45	22.25
5-5.9	10	5.45	54.50
6-6.9	16	6.45	103.20
7-7.9	12	7.45	89.40
8-8.9	11	8.45	92.95
9-9.9	8	9.45	75.60
10-10.9	3	10.45	31.35
11-11.9	3	11.45	34.35
12-12.9	2	12.45	24.90
	$\Sigma f=75=n$		$\Sigma(fm)=544.75$

The estimated value of the mean $\bar{X} = \Sigma(fm)/n = 544.75/75 = 7.263'$ or 7.3 km rounded to one decimal place. The ' symbol means the last digit recurs indefinitely.

The actual value of $\bar{X}$ calculated from the individual measurements using the exact formula $\bar{X} = \Sigma X/n$ is 547.1/75 = 7.2946' or 7.3 km rounded to one decimal place. The reason for the slight discrepancy is that the formula for estimating the mean assumes that the mean of the X values falling within a class is the same as the mid-value of the class. This is only an approximation. For classes before the class of maximum frequency, the mean of the X values is usually more than the mid-value of the class; while for classes after the class of maximum frequency, the mean of the X values is usually less than the mid-value. If the distribution is symmetrical the errors may cancel out, but close agreement between the estimated mean and the true mean cannot be expected if the distribution is skewed and the class intervals are large.

An alternative method of estimating the mean is recommended if there are many classes and the mid-values run to several decimal places and are awkward to manipulate. The method is valid, however, only if the class intervals are equal. The mid-value of the largest class is selected as an *assumed mean* or *arbitrary zero*.[1] The mid-values of the remaining classes are ranked in order of their distance or deviation either side of the assumed mean. The deviations are then multiplied by the frequencies and summed. The arithmetic mean is estimated using the formulas

$$\bar{X} = X_0 + i\left[\frac{\Sigma(fd)}{n}\right] \quad \text{or} \quad \mu = X_0 + I\left[\frac{\Sigma(FD)}{N}\right]$$

where X_0 is the mid-value taken as the assumed mean, f (or F) is the frequency in each class, d (or D) is deviation of each class from the arbitrary zero, n (or N) is the number of X values, and i (or I) is the width of the classes, i.e. the class interval.

As an example, the data relating to nearest-neighbour distances will again be considered.

TABLE 6.2 - ALTERNATIVE METHOD OF ESTIMATING THE MEAN

Distance to nearest-neighbour town or village in km	*Number of towns and villages* *f*	*Position of each class relative to largest class* *d*	*fd*
2-2.9	1	-4	-4
3-3.9	4	-3	-12
4-4.9	5	-2	-10
5-5.9	10	-1	-10
6-6.9	16	0	0
7-7.9	12	+1	+12
8-8.9	11	+2	+22
9-9.9	8	+3	+24
10-10.9	3	+4	+12
11-11.9	3	+5	+15
12-12.9	2	+6	+12
	Σf=75=n		Σ(fd)=61

[1]The sole reason for selecting the mid-value of the largest class is that it minimises the subsequent arithmetic. The mid-value of any class gives the same answer in the end.

Here i, the class interval, equals 1 km. The mid-value of the class with the largest number of towns and villages is 6.45. Hence, the estimate of $\bar{X}$ = 6.45 + 1[61/75] = 7.263' or 7.3 km to one decimal place.

Note that the estimate is the same as that given by the previous method. It can be shown algebraically that the two methods are, in fact, identical. Thus there is nothing to choose between them as regards accuracy. With the second method the calculations are simpler in the sense that the numbers are kept smaller, although the steps involved may seem at first sight to be more complex.

The arithmetic mean has two properties that deserve comment:

1. *The algebraic sum of the deviations of a set of numbers from their mean is always zero, i.e.* $\Sigma(X - \bar{X}) = 0$.

For example, the mean of the distances 5, 8, 9, 9, 10, 11, 12, 15 and 20 km is 11 km. The deviations of the distances from this mean are -6, -3, -2, -2, -1, 0, +1, +4, and +9 km. The sum of these deviations is zero.

2. *The sum of the squares of the deviations about the arithmetic mean is always less than the sum of the squares of the deviations about any other number, i.e.* $\Sigma(X - \bar{X})^2$ *is a minimum.*

Continuing with the same example, the squared deviations are 36, 9, 4, 4, 1, 0, 1, 16 and 81 sq km. The sum of these squared deviations is 152 sq km. Suppose the mean had been calculated wrongly as 10 km. Then the deviations would have been -5, -2, -1, -1, 0, +1, +2, +5, and +10 km. The squares of these deviations would have been 25, 4, 1, 1, 0, 1, 4, 25 and 100 sq km. The total comes to 161 sq km, more than the sum of squares about the true mean.

The arithmetic mean can be thought of as the centre of gravity of a set of data. If the histogram corresponding to the data is cut from a piece of card, and balanced on its base, the point of balance will be found to coincide with the mean (Fig. 6.1).

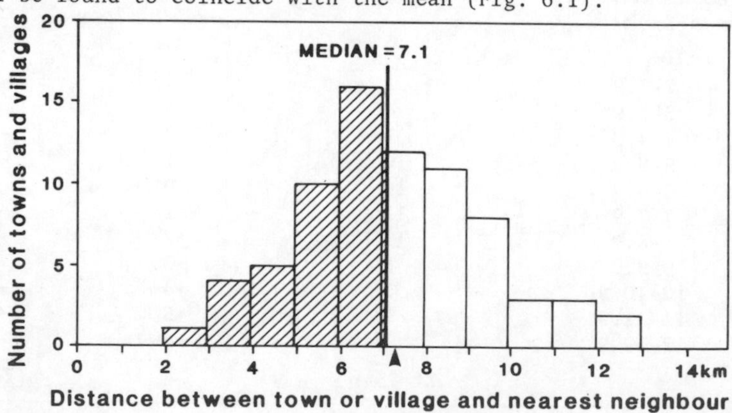

Fig. 6.1 - Histogram of the nearest-neighbour distances. The mean (indicated by a black triangle) represents the point of balance of the histogram. The median bisects the area of the histogram.

THE MEDIAN

The median of a set of n numbers arranged in order of magnitude is the number that falls mid-way i.e. divides the series so that one half or more of the numbers are equal to or less than it and one half or more of the numbers are equal to or more than it.

If n is odd, the median is the middle or $[(n/2) + (1/2)]$th number. Thus, for the 9 distances, 5, 8, 9, 9, 10, 11, 12, 15, and 20 km, the median is the 5th distance i.e. 10 km. The median of 10, 12, 13, 13, 14, 17 and 18 minutes is 13 minutes.

If n is even, the median falls mid-way between the two middle numbers i.e. it is the arithmetic mean of the $(n/2)$th number and the $[(n/2) + 1]$th number. For example, the median of 6, 7, 10, 12, 16 and 25 km is $(10 + 12)/2 = 11$ km.

In the case of a frequency distribution the median can be defined slightly differently as the value that divides the histogram area exactly in half (Fig. 6.1). The way to estimate the median is as follows

1. Calculate the cumulative frequency up to the end of each class i.e. the cumulative frequency corresponding to the upper limit of each class.

2. Determine which class contains the median, the $n/2$th observation.

3. Estimate the value of the median by substitution in the formula

$$\text{Median} = \ell + \frac{i(n/2 - c)}{f}$$

where ℓ is the lower limit of the class containing the median, c is the cumulative frequency corresponding to this limit (in other words, the number of observations with values less than ℓ), f is the number of observations, or frequency, in the class containing the median, and i is the width of the class, or class interval.

All this may sound complicated, but in fact it is easy enough. For the distances between nearest-neighbour towns and villages the cumulative frequency is set out in Table 5.6. The median is the 75/2th observation i.e. the 37.5th. It lies in the class 6.95-7.95 km. The cumulative frequency corresponding to the lower limit of this class is 36. The estimated value of the median is therefore

$$6.95 + \frac{1(37.5 - 36)}{12} = 6.95 + 0.125 = 7.075 \text{ or } 7.1 \text{ approximately.}$$

It is possible to understand how the formula works by considering this example further. There are 75 towns and villages in the sample, and hence 75 distances. By definition the median is the 75/2th or 37.5th distance in terms of length. There are 36 distances that are less than 6.95 km and 48 that are less than 7.95 km. The median must therefore have a value between 6.95 and 7.95 km. Its value may be estimated by assuming that the 12 distances with values between 6.95 and 7.95 km are distributed uniformly within the class i.e. spaced at regular intervals of $(7.95 - 6.95)/12 = 0.0833$ km. Taking the 36th distance to be 6.95 km, the 37th distance may be estimated to be 7.0333 km, the 38th distance to be 7.1167 km and so on. By interpolation, the 37.5 distance is 6.95 + (0.0833)(1.5) = 7.1 km approximately.

If preferred the value of the median can be found graphically by plotting the cumulative frequencies as shown in Fig. 6.2.

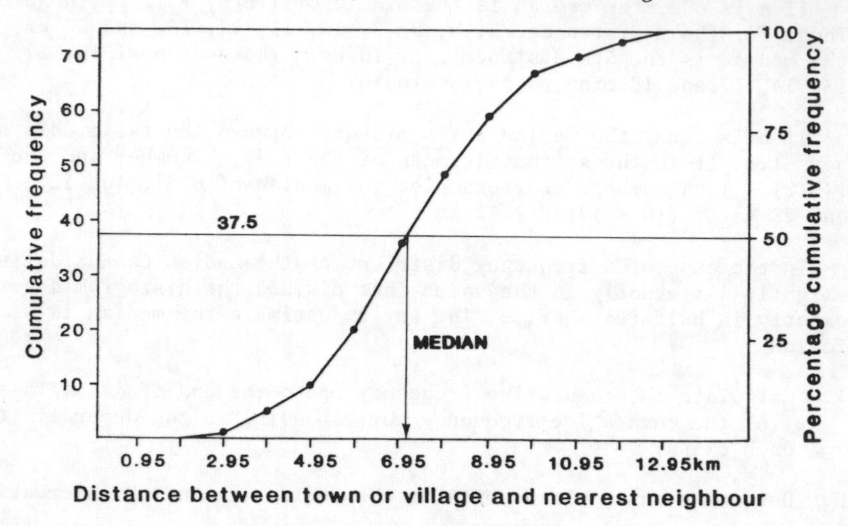

Fig. 6.2 - Graphical method of estimating the median. Data shown in Table 5.6.

The value of the median for the nearest-neighbour distances can also be found from the array set out in Table 5.2. The median is the 38th distance in order of magnitude and is equal to 7.1 km.

An interesting property possessed by the median is that *the sum of the unsigned*[1] *or "absolute" deviations from the median is less than or equal to the sum of the absolute deviations from any other value.* Put another way,

$$\Sigma|X\text{-MEDIAN}| \quad \text{is a miminum}$$

The modulus sign || is an instruction to ignore the signs of the deviations.

For example, the median of the distances 5, 8, 9, 9, 10, 11, 12, 15 and 20 km is 10 km. The absolute deviations are 5, 2, 1, 1, 0, 1, 2, 5 and 10 km. The sum of these deviations is 27 km. The absolute deviations from the mean (= 11 km) are 6, 3, 2, 2, 1, 0, 1, 4 and 9 km, making 28 km in all.

THE MODE

A further measure of central tendency is the mode. The mode of a set of numbers is the number that occurs most frequently. It can usually be determined by inspection rather than by calculation. For instance,

[1] "Unsigned" means that positive and negative signs are removed.

the mode of 5, 8, 9, 9, 10, 11, 12, 15 and 20 is clearly 9 because 9 occurs twice whereas all the other numbers occur but once. The set of numbers 5, 8, 9, 9, 10, 11, 12, 12, 15 and 20 has two modes, 9 and 12, while the set 5, 8, 9, 10, 11, 12, 15 and 20 has no mode.

The evaluation of the mode is not always straightforward. Consider the numbers 5, 8, 9, 9, 10, 10, 11, 12, 15 and 20. In a sense both 9 and 10 are modes as they each occur twice whereas the remaining numbers occur once. Because 9 and 10 are consecutive numbers, however, it makes better sense to recognise a single mode mid-way between them at 9.5. In the case of a set of numbers such as 5, 8, 9, 9, 10, 11, 12, 12, 12, 15 and 20, it is arguable whether there is a single mode, 12, or two modes, 9 and 12. Both 9 and 12 have a frequency greater than that of neighbouring numbers and so can be considered to be separate modes. On the other hand, 12 occurs more frequently than 9, and hence has some claim to be regarded as the sole mode.

To avoid these difficulties the mode of a frequency distribution is often taken to be the mid-value of the class with the greatest frequency (the *modal* class). On this definition, the mode coincides with the peak of the frequency polygon. A few statisticians, however, recommend taking into account the frequencies in the two classes lying alongside the modal class. The value of the mode is estimated using the formula

$$\text{Mode} = \ell + i\left(\frac{h_1}{h_1 + h_2}\right)$$

where ℓ is the lower limit of the modal class, h_1 is the difference between the frequency of the modal class and the frequency of the preceding class, h_2 is the difference between the frequency of the modal class and the frequency of the following class, and i is the width of the modal class. For the nearest-neighbour data, the mode on this definition is therefore

$$5.95 + 1\left(\frac{6}{6 + 4}\right) = 6.55 \text{ or } 6.6 \text{ rounded to one decimal place.}$$

Despite a greater air of precision it is doubtful whether the formula for the mode is really any improvement on the usual method of reporting the mode as the mid-point of the modal class since the modal class cannot be fixed precisely. By changing the size and number of the class intervals composing a frequency distribution, the class can be made to vary considerably. For instance, if the distances between towns and villages in Iowa are reclassified using different class intervals the modal class can be made to shift around in an alarming manner:

TABLE 6.3 - SENSITIVITY OF THE MODE TO CHANGES IN CLASS INTERVALS

Class limits	*Freq.*	*Class limits*	*Freq.*	*Class limits*	*Freq.*
2.0-4.4	8	2.5-4.9	10	2-3.4	2
4.5-6.9	28	5.0-7.4	34	3.5-4.9	8
7.0-9.4	27	7.5-9.9	23	5.0-6.4	17
9.5-11.9	10	10.0-12.4	6	6.5-7.9	21
12.0-14.4	2	12.5-14.9	2	8.0-9.4	15
				9.5-10.9	7
				11.0-12.4	3
				12.5-13.9	2
Mid-value of modal class 5.7 km		Mid-value of modal class 6.2 km		Mid-value of modal class 7.3 km	

Because the value of the mode is liable to vary according to the class intervals that are chosen, it is effective as a measure of central tendency only if the frequency distribution is strongly peaked. If no one class is clearly the biggest, the mode ceases to be a useful measure.

PROBLEMS OF CALCULATING MEANS, MEDIANS AND MODES

Various practical difficulties are encountered in calculating measures of central tendency. The difficulty of defining the mode satisfactorily has already been mentioned. Because the mode is sensitive to the class intervals used to construct a frequency distribution, it is seldom employed by statisticians, except as a crude measure of central tendency. The mean and the median are rigidly defined, but are more troublesome to obtain than the mode.

In certain situations the median has a clear advantage over the mean. Often when frequency distributions are published as tables the first and last classes are left open-ended (e.g. "10 or less", "15 or more"). The presence of such classes makes estimating the mean somewhat hazardous, since the limits of the distribution are not specified and the mid-values of the open-ended classes can only be guessed. Unlike the mean, the median can be estimated without difficulty unless it happens to fall in one of the open-ended classes, in which case no measure of central tendency has much meaning.

The median item in a sample or population can sometimes be found by mere inspection without the necessity of measuring all the other items. This can be a considerable advantage. Consider, for instance, the problem of finding the central tendency of a set of areas shown on a map, such as parishes or drainage basins. If the areas are irregular, measuring their size accurately may be a very time-consuming task. On the other hand, if they differ sufficiently in size it may well be possible to rank them just by eye. In this situation, it is obviously easier to find the median than the mean, since only one measurement is required.

In certain respects the median is a more stable measure than the mean. The value of the mean takes into account the value of every item in a sample or population. This is both a source of strength and weakness. The mean is liable to be greatly affected by the presence of a few items that are extremely large or small. Compare, for instance, the two samples

Sample A 5, 8, 9, 9, 10, 11, 12, 15 and 20

Sample B 5, 8, 9, 9, 10, 11, 12, 15 and 2001

The median is not influenced by the last item and is the same in both samples, whereas the mean jumps from 11 in the first sample to 231.1 in the second. Clearly the median is preferable as a measure of central tendency if the extreme value of 2001 is known to be highly atypical of the sampled population or is suspected of being an error.

The mean has the advantage over the median that it is readily calculated from totals. Suppose, for instance, that a comparison of maps drawn in 1900 and 1980 reveals that a sea-cliff has retreated a total of 80 m in the 80 years between observations. The mean rate of retreat can be calculated to be 1 m a year, but the median rate is indeterminate.

PROBLEMS OF INTERPRETATION

In a frequency distribution that is unimodal and symmetrical, the mean, median and mode coincide (Fig. 6.3A). It is therefore immaterial which measure is used to define central tendency.

If the frequency is multimodal (Fig. 6.3B) or skewed (Fig. 6.3C), the mean, median and mode may lie far apart. This raises the question of which measure best summarises the distribution. The answer depends upon the use to which the measure will be put.

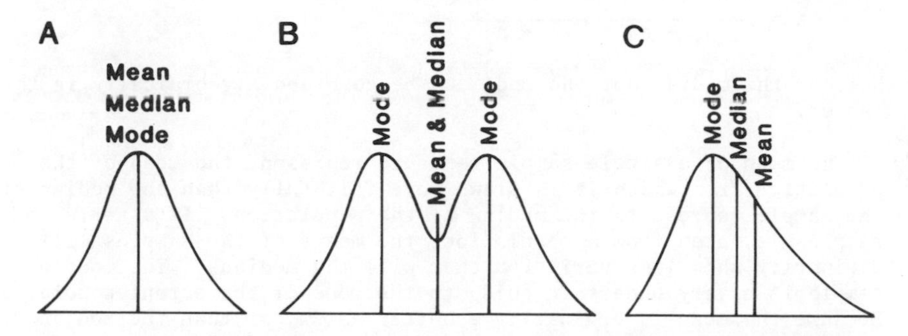

Fig. 6.3 - Location of mean, median and mode in (A) a symmetrical unimodal distribution, (B) a symmetrical bimodal distribution, and (C) a unimodal right-skewed distribution. The distributions are represented by frequency curves.

It has been shown already that the mean is very responsive to extreme values in distributions. It is therefore inappropriate in situations where extreme values are of no special interest and what is sought is a representative value. For example, if London is included in the calculation of the mean size of cities in south-east England, the result is much greater than if London is excluded. Either way the mean is liable to be misleading. The advantage of the median is that it is unaffected by a few grossly atypical values in a sample. The median is also useful because its value is more nearly like the value of all items in a skewed distribution than is the value of either the mean or the mode. The sum of the absolute deviations around a median is a minimum and consequently it provides the closest 'fit' to the data and thus in a sense is the most representative or typical value that can be found.

The mode indicates the most frequent value in a sample or population and in this respect conforms more closely to what the man in the street means by the average than does the average. As explained in the last chapter, if a sample possesses a single, clearly defined mode this is some indication that it is homogeneous. If, on the other hand, it possesses two or more modes it may have been drawn, through ignorance or carelessness, from more than one population. The advisability of dividing the sample into two or more sub-samples should be seriously considered. One must also remember that bimodality or multimodality can arise from using too many classes when constructing a frequency distribution.

One great advantage possessed by the mean is that once calculated it lends itself to further mathematical treatment. One example must suffice at this point. If several samples are combined into a single sample, the mean of the latter can be readily calculated from the means of the former, without reference to the individual measurements or observations. Thus, if there are r samples, and the mean of the first sample of n_1 measurements is $\bar{X}_1$ while the mean of the second sample of n_2 measurements is $\bar{X}_2$ and so on, the mean of the rth and final sample of n_r measurements being $\bar{X}_r$, the mean $\bar{\bar{X}}$ of the combined sample is given by the equation

$$\bar{\bar{X}} = \frac{n_1\bar{X}_1 + n_2\bar{X}_2 + \ldots + n_r\bar{X}_r}{n_1 + n_2 + \ldots + n_r}$$

Neither the median nor the mode can be combined algebraically in this manner.

The mean of a single sample tends to represent the mean of the population from which it is drawn more faithfully than the median of the sample represents the median of the population. If a series of samples is taken from a population, the means of the samples will ordinarily show less variation than will the medians. The mode of a sample is a very uncertain guide to the mode of the parent population. Because the mean is ordinarily a better *estimator* than the median or mode, it is the preferred measure of central tendency in problems involving inferences about populations from samples. The mean often features in the calculations in this book, whereas the median and mode are seldom employed.

7 Measures of Dispersion and Skewness

INTRODUCTION

Measures of central tendency are difficult to interpret unless accompanied by an indication of the variability of the data from which they derive. Average elevation above sea-level, for example, does not mean very much in an area where high mountains are dissected by equally deep valleys. It means a great deal more in a relatively flat area. A place that experiences violent extremes of temperature during the course of a year can possess the same mean annual temperature as a place where the temperature hardly changes. In a region where great estates and peasant small holdings exist side-by-side, it is merely confusing to talk about average farm size.

Measures of dispersion, as their name suggests, express the degree of variability or dispersion in sets of data. Sometimes the values are concentrated around the mean; sometimes they are so widely dispersed that little central tendency is apparent. For example, consider the two samples

Sample A 5, 10, 16, 21, 25, 29, 35, 38 47

Sample B 23, 24, 24, 25, 25, 25, 26, 26, 27

The samples have the same mean ($\bar{X} = 25$), but the spread or dispersion of the values about the mean is very different.

A crude measure of the spread is provided by the *range*; from 5 to 47 as against from 23 to 27; but this focuses on the extreme values and ignores the rest. The intermediate values have an interest equal to or greater than the extremes, which are possibly freaks or errors. A further disadvantage of the range as a measure of dispersion is that it tends to increase with the size of the sample. If the areas of 100 drainage basins are measured, the range in values is ordinarily very much greater than if only 10 basins are measured. It is not easy to be certain how much of the range of a particular sample is due to the size of the sample and how much is due to the variability inherent in the population from which it is derived.

Instead of the range, what is obviously needed is a measure of average dispersion or spread that takes into account all the values, and not merely the extremes. The average of the deviations from the mean

$$\frac{\Sigma(X - \bar{X})}{n} \text{ for a sample or } \frac{\Sigma(X - \mu)}{N} \text{ for a population}$$

is no use as a measure of spread because the positive and negative deviations always cancel out to zero, as explained in the last chapter. An obvious solution to this difficulty is to ignore the signs of the deviations when calculating their average. The resulting measure of dispersion is called the *mean deviation*. It can be represented by the formula

$$\frac{\Sigma|X - \bar{X}|}{n} \quad \text{for a sample or} \quad \frac{\Sigma|X - \mu|}{N} \quad \text{for a population.}$$

The modulus sign indicates that the absolute or unsigned deviations are required.

The mean deviation has been little used as a measure of dispersion because it has been found to have a number of undesirable mathematical properties. These stem in part from the modulus sign which is awkward to manipulate algebraically.

VARIANCE

A better measure of dispersion, called the *variance*, is obtained by squaring the deviations before averaging them. It is somewhat cumbersome to calculate, but it lends itself to further mathematical treatment in a way that the mean deviation does not.

The variance σ^2 of a population is given by the formula

$$\sigma^2 = \frac{\Sigma(X - \mu)^2}{N}$$

where $(X - \mu)$ is the deviation of each observation or X value from the mean μ, N is the total number of observations in the population i.e. its size, and σ is the lower case Greek letter *sigma*.

The variance of a sample is defined slightly differently

$$s^2 = \frac{\Sigma(X - \bar{X})^2}{n - 1}$$

where s^2 is the sample variance, $(X - \bar{X})$ is the deviation of each observation or X value from the mean $\bar{X}$, and n is the total number of observations in the sample i.e. its size.

Note that the divisor is $(n - 1)$ instead of n which would seem more natural. The reason for treating populations and samples differently is not easy to explain in simple terms. In many statistical problems it is necessary to take the variance of a sample as an estimate of the variance of the population from which it is drawn. The variance of a sample calculated using $(n - 1)$ as the divisor is just as likely to be smaller than the population variance as it is likely to be larger. In other words, s^2 is an *unbiased* estimate of σ^2. Strange as it may seem, dividing by n does not produce this desirable result. A bias is apparent. The variance tends to fall short of the population variance on average by a factor of $(n - 1)/n$. In order to remove the bias it is necessary to multiply the variance calculated using n as the divisor by the correction factor $n/(n - 1)$. By using $(n - 1)$ as the divisor the need for a correction factor is eliminated. While there can be no guarantee that the variance of any sample will be the same as that of the population from which it is taken, at least there will be no systematic tendency to under-estimate as would occur if n were used as divisor.

To find the variance of samples A and B the procedure is as follows.

TABLE 7.1 - CALCULATION OF THE VARIANCE

SAMPLE A			SAMPLE B		
X	$(X - \bar{X})$	$(X - \bar{X})^2$	X	$(X - \bar{X})$	$(X - \bar{X})^2$
5	-20	400	23	-2	4
10	-15	225	24	-1	1
16	-9	81	24	-1	1
20	-5	25	25	0	0
25	0	0	25	0	0
29	4	16	25	0	0
35	10	100	26	1	1
38	13	169	26	1	1
47	22	484	27	2	4
TOTALS	0	1500		0	12

$\Sigma(X - \bar{X})^2$ = 1500 in sample A and 12 in sample B. The variance of sample A = $\Sigma(X - \bar{X})^2/(n - 1) = 1500/8 = 187.5$. Likewise the variance of sample B = 12/8 = 1.5.

Notice that the sample with the greater spread or dispersion of values about the mean has the larger variance. The amount of variance is always strictly related to the amount by which the values in a sample (or population) differ from the mean. From a minimum of zero, variance can assume any positive value. Zero variance indicates that all the values in a sample or population are the same. Since they cannot be any less dispersed than this, negative variance is impossible.

Although the formulas for variance may look simple enough, they make for laborious calculation if the numbers involved are not integers (whole numbers). Alternative formulas can be derived that are generally more convenient for calculating variance, although they look more complicated. For a population

$$\sigma^2 = \left[\Sigma X^2 - \frac{(\Sigma X)^2}{N}\right]/N$$

while for a sample

$$s^2 = \left[\Sigma X^2 - \frac{(\Sigma X)^2}{n}\right]/(n - 1)$$

These formulas are algebraically equivalent to the basic formulas, but avoid using the mean and are therefore useful if the mean is an awkward decimal number like 24.873249. The formulas are easy to derive. Thus, for a population, by definition

$$\sigma^2 = \Sigma(X - \mu)^2/N$$

and therefore

$$N\sigma^2 = \Sigma(X - \mu)^2 = \Sigma(X^2 - 2X\mu + \mu^2)$$

Because μ is a constant, the right side of the equation can be written

$$\Sigma X^2 - 2(\Sigma X)\mu + \mu^2$$

and since

$$\mu = \Sigma X/N$$

it follows that

$$N\sigma^2 = \Sigma X^2 - 2(\Sigma X)\frac{(\Sigma X)}{N} + \frac{(\Sigma X)(\Sigma X)}{N^2} = \Sigma X^2 - \frac{(\Sigma X)^2}{N}$$

and therefore

$$\sigma^2 = \left[\Sigma X^2 - \frac{(\Sigma X)^2}{N}\right]/N$$

The formula for a sample can be derived in the same way.

For the two samples A and B the calculations using the alternative formula are set out in Table 7.2

TABLE 7.2 – ALTERNATIVE METHOD OF CALCULATING VARIANCE

SAMPLE A		SAMPLE B	
X	X^2	X	X^2
5	25	23	529
10	100	24	576
16	256	24	576
20	400	25	625
25	625	25	625
29	841	25	625
35	1225	26	676
38	1444	26	676
47	2209	27	729
$\Sigma X = 225$	$\Sigma X^2 = 7125$	$\Sigma X = 225$	$\Sigma X^2 = 5637$

Sample A:

$$s^2 = \left[7125 - \frac{(225)^2}{9}\right]/8$$

$$= [7125 - 5625]/8$$

$$= 1500/8$$

$$= 187.5$$

Sample B:

$$s^2 = \left[5637 - \frac{(225)^2}{9}\right]/8$$

$$= [5637 - 5625]/8$$

$$= 12/8$$

$$= 1.5$$

These answers are, of course, the same as before. Because the sample means are integers, no time has been saved using the alternative formula, but had the means been decimal numbers the saving might have been considerable.

Formulas that can be used to estimate the variance for a frequency distribution are

$$\sigma^2 = \frac{\Sigma[F(M - \mu)^2]}{N} \quad \text{and} \quad s^2 = \frac{\Sigma[f(m - \bar{X})^2]}{n - 1}$$

where f (or F) is the frequency in each class and m (or M) is the mid-value of the class.

Table 7.3 illustrates the estimation of the variance for the data on the spacing of towns and villages in Iowa.

TABLE 7.3 - ESTIMATE OF THE VARIANCE FOR THE NEAREST-NEIGHBOUR

Distance to nearest-neighbour town or village in km	*Number of towns and villages* f	*Mid-value of Class* m	$m - \bar{X}$	$(m - \bar{X})^2$	$f(m - \bar{X})^2$
2-2.9	1	2.45	-4.813'	23.16817'	23.16818
3-3.9	4	3.45	-3.813'	14.54151'	58.16604
4-4.9	5	4.45	-2.813'	7.91484'	39.57422
5-5.9	10	5.45	-1.813'	3.28817'	32.88178
6-6.9	16	6.45	-0.813'	0.66151'	10.58418
7-7.9	12	7.45	+0.186'	0.03484'	0.41813
8-8.9	11	8.45	+1.186'	1.40817'	15.48996
9-9.9	8	9.45	+2.186'	4.78151'	38.25209
10-10.9	3	10.45	+3.186'	10.15484'	30.46453
11-11.9	3	11.45	+4.186'	17.52817'	52.58453
12-12.9	2	12.45	+5.186'	26.90151'	53.80302
	$\Sigma f = 75 = n$				353.38666

Estimated mean, $\bar{X} = \Sigma(fm)/n = 544.75/75 = 7.263'$ km.

Estimated variance, $s^2 = \Sigma[f(m - \bar{X})^2]/(n - 1) = 353.38666/74 = 4.80252$ or 4.8 sq km to one decimal place.

As the table shows, the formulas lead to fairly heavy arithmetic. Provided the class intervals are equal, alternative formulas may be used that reduce the arithmetic substantially. The mid-value of the class with the largest frequency is selected as an assumed mean and the deviations of the mid-values of the remaining classes from this zero are recorded in terms of ranks. The formulas for estimating the variance are

$$\sigma^2 = I^2 \left[\frac{\Sigma(FD^2)}{N} - \frac{(\Sigma FD)^2}{N^2}\right] \quad \text{and} \quad s = i^2 \left[\frac{\Sigma(fd^2)}{n - 1} - \frac{(\Sigma fd)^2}{n(n - 1)}\right]$$

where d (or D) is the deviation of each class from the assumed mean and i (or I) is the class width or interval. Take care not to confuse (FD^2) with $(FD)^2$ or (fd^2) with $(fd)^2$. The first quantity in both cases is the frequency multiplied by the squared deviation;

the second quantity, which does not form part of the formulas, is the square of the product of the frequency and deviation.

To return to the nearest-neighbour data

TABLE 7.4 - RECOMMENDED METHOD OF ESTIMATING THE VARIANCE OF A FREQUENCY DISTRIBUTION

Distance to nearest-neighbour town or village in km	*Number of towns and villages*	*Position of each class relative to the assumed mean*		
	f	*d*	*fd*	*fd*^{2}
2-2.9	1	-4	-4	16
3-3.9	4	-3	-12	36
4-4.9	5	-2	-10	20
5-5.9	10	-1	-10	10
6-6.9	16	0	0	0
7-7.9	12	+1	12	12
8-8.9	11	+2	22	44
9-9.9	8	+3	24	72
10-10.9	3	+4	12	48
11-11.9	3	+5	15	75
12-12.9	2	+6	12	72
	$\Sigma f = 75 = N$		$\Sigma(fd) = 61$	$\Sigma(fd^2) = 405$

$$\text{Estimate of } s^2 = i^2 \left[\frac{\Sigma(fd^2)}{n - 1} - \frac{(\Sigma fd)^2}{n(n - 1)}\right] = (1)^2 \left[\frac{405}{74} - \frac{(61)(61)}{(75)(74)}\right]$$

$$= 5.47297 - 0.67045$$

$$= 4.80252 \text{ or } 4.8 \text{ sq km to one decimal place.}$$

The estimate is the same as that obtained previously. This is because the two methods of calculation are equivalent algebraically, although the formulas at first sight may look very different.

It is an unfortunate fact that variances estimated from frequency distributions generally tend to be slightly larger than the true variances calculated for the raw data. The present example is no exception. While the estimated variance is 4.80252 sq km, the true variance calculated using the formula $s^2 = \Sigma(X - \bar{X})/(n - 1)$ works out at 4.76754 sq km. Even for symmetrical distributions, estimates of sample variances are generally slightly too large, whereas estimates of the means are without bias. In estimating the mean and the variance for a frequency distribution the mean value of the items in any class is assumed to equal the mid-value of the class. In point of fact, with a symmetrical distribution that tapers off to zero in both directions, the mean value of each class is located not at the mid-value but at a value lying closer to the mean of the distribution. In estimating the mean of the distribution no overall error results from taking the mid-values of the classes as the means of the classes because the errors on one side of the mean of the distribution offset those on the other. Unfortunately, estimating the variance involves squaring the deviations of the mid-values of the classes, and as a result the errors on either side of the mean add together instead of cancelling out. The effect of the errors is to inflate the estimate of the variance.

A correction factor is sometimes applied to try to compensate for the errors so as to bring the estimated variance closer to the true value. It is known as *Sheppard's correction*. Because of doubts about its reliability, it is not presented here. The interested reader should consult Yule and Kendall (1950, p.133) for further information.

STANDARD DEVIATION

Standard deviation is simply the square root of the variance and is denoted by the symbols σ for a population and s for a sample. The definitional formulas are therefore:

$$\sigma = \sqrt{\frac{\Sigma(X - \mu)^2}{N}} \quad \text{and} \quad s = \sqrt{\frac{\Sigma(X - \bar{X})^2}{n - 1}}$$

Standard deviation is in some ways a more natural and straightforward measure of dispersion than is variance. It is expressed in the same units of measurement as the raw data, instead of in squared units as is variance.

On the other hand standard deviation does have one slight disadvantage compared with variance. As already explained the variance of a sample is an unbiased estimate of the variance of the population from which it is drawn (provided n - 1 is used as the divisor in the formula for s^2). Strange as it may seem, if s is used to estimate σ, a slight bias creeps in. For hump-backed unimodal distributions of the type shown in Fig. 5.6 (the so-called *normal* distribution - see Chapter 9) the value of s must be multiplied by a correction factor, $(4n - 3)/(4n - 4)$, in order to obtain an unbiased estimate of σ. The value of n must exceed 10 for the correction factor to be valid (Dixon and Massey, 1957, p.136). In the case of other distributions different correction factors apply.

It must be stressed that the amount of bias in s as an estimator of σ is seldom large and can ordinarily be ignored.

Beginners sometimes find the the concept of standard deviation holds little meaning for them. They understand how to calculate a standard deviation but fail to grasp its significance. Yet in essence the concept is simple. A sample or population that is very variable has a large standard deviation; one that has all the values clustered closely around the mean has a low standard deviation. If all the values are the same, the standard deviation is zero.

The standard deviation of a single sample or population is often of little interest if studied in isolation. The same is true of the arithmetic mean. Usually the purpose of calculating a standard deviation (or a mean) is to compare it with that of another sample or population. For example, if an area is found to have a standard deviation of altitudes twice that of another area, this suggests it is twice as rugged. An investigator who makes repeated measurements of the length of a road shown on a map and gets a standard deviation four times that of another investigator can be said to be four times as careless. Comparing standard deviations is often a more meaningful exercise than considering them in isolation.

Geographical variations in standard deviation are often of interest. Figure 7.1 depicts the variation of annual rainfall over Zambia as measured by the standard deviation. Variability is seen to be least in the south where the mean annual rainfall is low and highest in the north where the mean annual rainfall is greatest. The value of the standard deviation is about one quarter that of the annual rainfall throughout Zambia. Knowing the annual rainfall at a particular locality, one can roughly predict the variability.

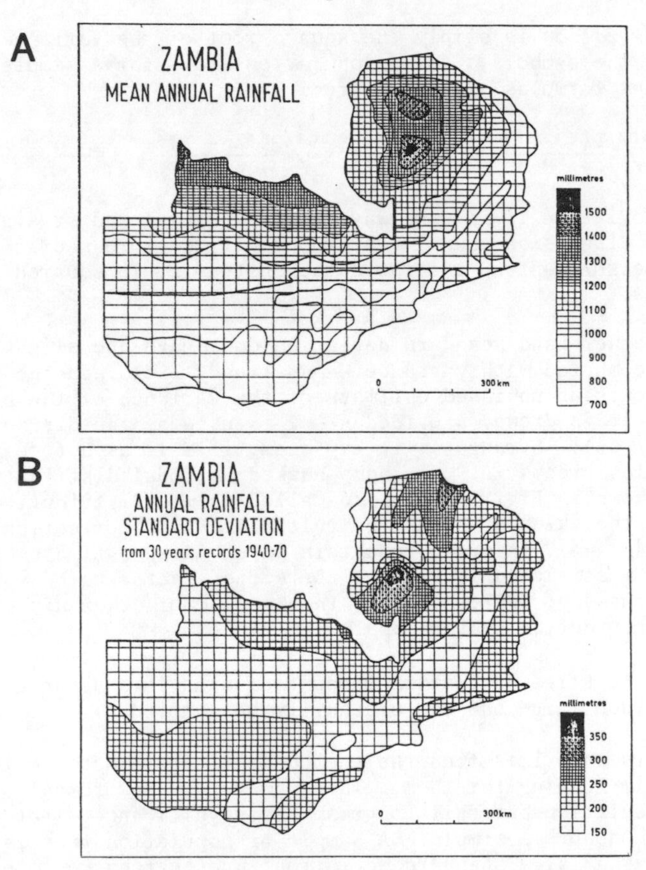

Fig. 7.1 - (A) Mean annual rainfall in Zambia and (B) its standard deviation. From Hutchinson (1974). Amounts of rainfall in mm.

Many matters that arise in everyday life can be expressed in terms of standard deviation. Social egalitarians who want all incomes to be the same seek in effect to reduce the standard deviation of existing incomes to zero. Those trade unionists who are concerned to protect differentials want to leave the standard deviation unchanged. In areas of the world where the standard deviation of monthly or annual rainfall is high, farming tends to be a very risky operation. In areas like the British Isles where the standard deviation is relatively low, farming is less risky and more profitable. One of the major obstacles to making railways pay is the standard deviation of the number of passengers per hour of the day. Great numbers of passengers want to travel during the rush hours of early morning and late afternoon while few want to travel at mid-day. Expensive rolling stock is used only for short periods and stands idle for the rest of the day.

Sometimes it is useful to replace the original units of measurement on a graph or histogram with units of standard deviation. In Fig. 7.2, for example, the distances between nearest-neighbour towns and villages in Iowa are indicated in units of standard deviation (or standard deviations) from the mean. It is immediately clear not only that all distances are less than three standard deviations, but also that the great majority are less than 2 standard deviations. The greatest distance listed in Table 5.2 is 12.7 km or 12.7 - 7.3 = 5.4 km more than the mean, 7.3 km. This is equivalent to 5.4/2.18 = 2.48 standard deviations from the mean, one standard deviation being 2.18 km. The shortest distance in Table 5.2 is 2.7 km or 7.3 - 2.7 = 4.6 km less than the mean. This is 4.6/2.18 = 2.11 standard deviations from the mean.

Standard deviation is therefore not just a measure of the variability of a sample or population. It is also a measure of the degree of aberrance or peculiarity of individual items in a sample or population.

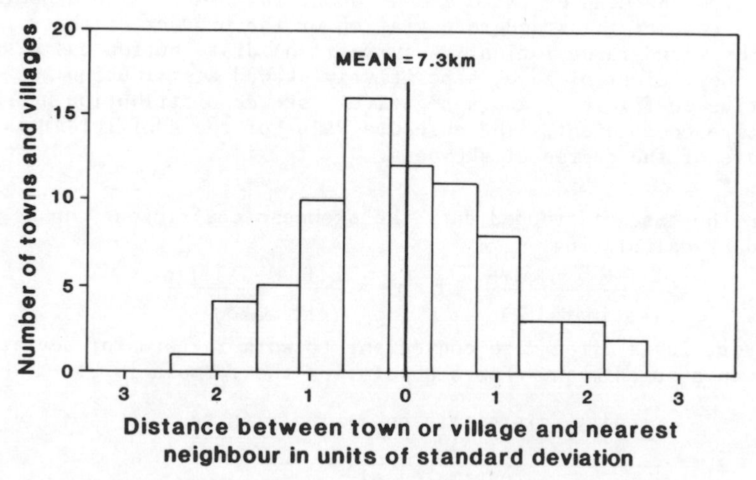

Fig. 7.2 - Distances separating towns and villages in Iowa. The distances are shown in units of standard deviation from the mean. One standard deviation = √4.76754 = 2.18 km.

COEFFICIENT OF VARIATION

The *coefficient of variation* is merely the standard deviation expressed as a percentage of the mean. The coefficient V of a sample is thus given by the formula

$$V = \frac{100\ s}{\bar{X}}$$

where s is the standard deviation, and $\bar{X}$ the mean.

The coefficient of variation possesses no great advantage over the standard deviation, and is not often used as a measure of variability.

The next section should be omitted on first reading. The formulas to be given are of minor importance and will not be used in subsequent chapters.

*SKEWNESS

Although some frequency distributions are symmetrical or approximately symmetrical, the majority are demonstrably skewed. The direction and degree of skew can be measured by calculating the skewness coefficient, β_1 or b_1. The definitional formulas for ungrouped data are

$$\beta_1 = \frac{[\Sigma(X - \mu)^3]/N}{\sigma^3} \text{ for a population,}$$

and $$b_1 = \frac{[\bar{\Sigma}(X - \bar{X})^3]/n}{s^3} \text{ for a sample.}$$

The numerator of the skewness coefficient is the average cubed deviation from the mean or *third moment about the mean*.[1] The denominator is the cube of the standard deviation or the product of the variance and the standard deviation. A symmetrical distribution has a skewness coefficient of zero, a positively skewed distribution has a positive coefficient, and a negatively skewed distribution has a negative coefficient. The absolute value of the coefficient is a measure of the degree of skewness.

In the case of grouped data the skewness coefficient can be estimated by calculating

$$\beta_1 = \frac{\Sigma[F(M - \mu)^3]/N}{(\text{estimated } \sigma)^3} \quad \text{or} \quad b_1 = \frac{\Sigma[f(m - \bar{X})^3]/n}{(\text{estimated s})^3}$$

However, it is often more convenient to work in terms of deviations from an assumed mean. For a population the formula is

$$\beta_1 = \frac{\frac{\Sigma(FD^3)}{N} - \frac{3[\Sigma(FD)][\Sigma(FD^2)]}{N^2} + \frac{2[\Sigma(FD)]^3}{N^3}}{(\text{estimated } \sigma)^3}$$

---

[1] The average squared deviation from the mean, $\Sigma(X - \mu)^2/N$ or $\Sigma(X - \bar{X})^2/n$, is called the *second moment about the mean*. In the case of a population it is equal to the variance; in the case of a sample it is the variance multiplied by $(n - 1)/n$. The *first moment*, $\Sigma(X - \mu)/N$ or $\Sigma(X - \bar{X})/n$, is always zero. The *fourth moment*, $\Sigma(X - \mu)^4/N$ or $\Sigma(X - \bar{X})^4/n$, forms the numerator in a formula for the degree of "peakedness" (*kurtosis*) of the hump-backed unimodal distributions.

where F is the frequency in each class and D is the deviation of the class from the assumed mean. Note carefully that $\Sigma(FD^3)$ is the sum of the products of F and D^3, which is not the same as $\Sigma(FD)^3$, the sum of the cubes of F multiplied by D. $\Sigma(FD)^3$ does not appear in the formula for β_1.

The computational formula for a sample is the same as the formula for a population except for the customary differences in notation

$$b_1 = \frac{\frac{\Sigma(fd^3)}{n} - \frac{3[\Sigma(fd)][\Sigma(fd^2)]}{n^2} + \frac{2[\Sigma(fd)]^3}{n^3}}{(\text{estimated } s)^3}$$

The values of fd^3 for the nearest-neighbour data are -64 for the lowest class (2-2.9 km), -108 for the next class (3-3.9 km), and -40, -10, 0, 12, 88, 216, 192, 375 and 432 for the following classes. The sum of these values, $\Sigma(fd^3)$, is 1082. The value of $\Sigma(fd)$ is 61 and therefore $[\Sigma(fd]^3$ is $61^3 = 226{,}981$. The remaining quantities in the formula for b_1 have already been calculated. The skewness is given by

$$b_1 = \frac{\frac{1082}{75} - \frac{3(61)(405)}{(75)^2} + \frac{2(226981)}{(75)^3}}{(2.19147)^3} = \frac{2.32672}{10.52456} = 0.22$$

to two decimal places. This is a very trivial amount of positive skew. For most practical purposes the distribution of nearest-neighbour distances can be regarded as symmetrical.

8 The Binomial and Poisson Distributions

This chapter considers two discrete distributions that are formulas constructed purely from theory. Each distribution represents a random process operating in a strictly defined manner, chosen to be as simple and as general as possible so that each distribution models a wide variety of real world phenomena. Many other discrete distributions have been constructed, but the two discussed here have proved particularly useful.

THE BINOMIAL DISTRIBUTION

As explained in Chapter 4, if an unbiased coin is tossed K times, the probability of obtaining X heads is

$$\frac{K!}{X!(K-X)!}(1/2)^K$$

The formula is correct only if the coin is unbiased, in other words if the probability of obtaining a head on any one toss is the same as the probability of obtaining a tail.

With a biased coin the probability of obtaining X heads in K tosses can be shown to be

$$\frac{K!}{X!(K-X)!}p^X p^{K-X}$$

where p is the probability of obtaining a head on any toss and q is the probability of obtaining a tail, i.e. not obtaining a head. Under the addition rule, $p + q = 1$.

In this modified form the formula can be used to calculate probabilities associated with repeated throws of an unbiased die. The probability of obtaining, say, a 6 on a single throw is 1/6 whereas the probability of not obtaining a 6 is 5/6. Referring to the formula, the probability of obtaining X sixes (or any other face) in K throws is

$$\frac{K!}{X!(K-X)!}(1/6)^X(5/6)^{K-X}$$

The formula for the biased coin can, in fact, be used in a great many contexts. To generalise: suppose there exists a process or experiment that has two possible outcomes. Suppose further that one of the outcomes can be classified as an *occurrence* and the other as a *non-occurrence* (an alternative classification might be *success* or *failure*). Let p be the probability of an occurrence and q the probability of a non-occurrence. Then, if the process or experiment is repeated K times, the probability of X occurrences is the formula already given

$$\frac{K!}{X!(K - X)!} p^X q^{K-X}$$

Each repetition of the process or experiment is defined as a *trial*. A technical requirement is that the trials be *independent*. In other words, the outcome of one trial must not affect the outcome of any other. The probabilities of an occurrence and a non-occurrence must remain the same throughout the trials.

As an illustration, suppose that five trials are made. Only six values of X are possible, namely the integers 0, 1, 2, 3, 4 and 5. The probabilities associated with each of these values are as follows

Probability of no occurrence $(X = 0) = \frac{(5!)}{(0!)(5!)} p^0 q^5 = q^5$

Probability of one occurrence $(X = 1) = \frac{(5!)}{(1!)(4!)} p^1 q^4 = 5pq^4$

Probability of two occurrences $(X = 2) = \frac{(5!)}{(2!)(3!)} p^2 q^3 = 10p^2 q^3$

Probability of three occurrences $(X = 3) = \frac{(5!)}{(3!)(2!)} p^3 q^2 = 10p^3 q^2$

Probability of four occurrences $(X = 4) = \frac{(5!)}{(4!)(1!)} p^4 q^1 = 5p^4 q$

Probability of five occurrences $(X = 5) = \frac{(5!)}{(5!)(0!)} p^5 q^0 = p^5$

These probabilities correspond to the successive terms in the expansion $(q + p)^5$

$$(q + p)^5 = (q + p)(q + p)(q + p)(q + p)(q + p)$$

$$= q^5 + 5pq^4 + 10p^2q^3 + 10p^3q^2 + 5p^4q + p^5$$

The correspondence is not fortuitous. In general, if there are k trials, the probabilities attached to X = 0, 1, 2, 3,...,k are given by the successive terms of the expansion of (q + p) raised to the kth power. According to the binomial expansion, $(q + p)^k$ is equal to

$$q^k + kpq^{k-1} + \frac{k(k - 1)}{2!} p^2 q^{k-2} + \frac{k(k - 1)(k - 2)}{3!} p^3 q^{k-3}$$
$$+ \frac{k(k - 1)(k - 2)(k - 3)}{4!} p^4 q^{k-4} + \ldots + p^k$$

The first of these terms is the probability that X = 0, the second the probability that X = 1, the third that X = 2, and so on. Observe that the (X - 1)th term in the binomial expansion, i.e. the term involving p^X, is

$$\frac{k(k-1)(k-2)\ldots(k-X+1)}{X!}p^X q^{k-X}$$

Rewritten, this becomes

$$\frac{k!}{X!(k-X)!}p^X q^{k-X}$$

which is the formula given earlier.

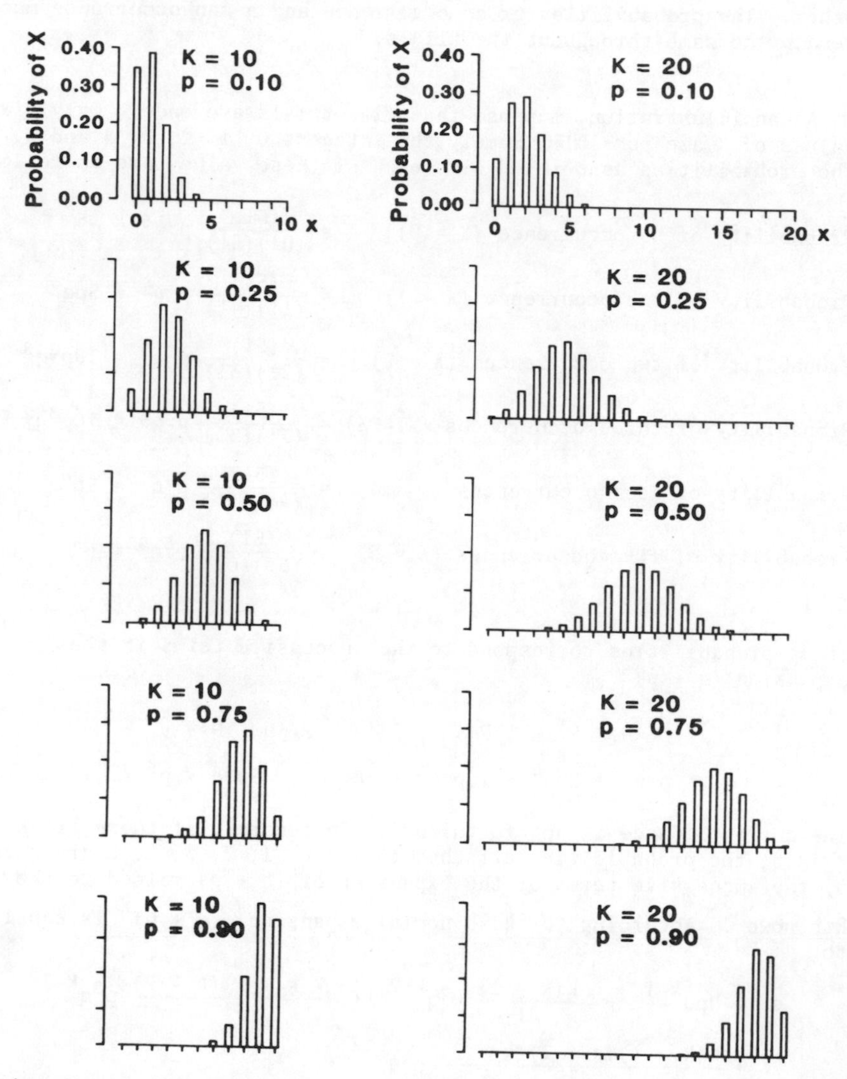

Fig. 8.1 - Examples of the binomial distribution for various values of k and p.

The sum of the successive terms of the binomial expansion of $(q + p)^k$ is clearly 1 because (q + p) equals 1 and therefore $(q + p)^k$ equals 1. This is as it should be since all possible values of X are represented and in k independent trials one of the values is bound to occur with probability equal to 1.

The probabilities associated with the binomial expansion constitute a *probability distribution* in the sense that the total probability (= 1) is distributed among the various possible values of X. The distribution is called the *binomial distribution* or less commonly the *Bernoulli distribution* and is a major cornerstone in the theory of probability.

Although determined by a single formula, the binomial distribution is in a sense not a single distribution but a whole family of distributions having very different graphical forms, depending on the values assigned to k and p (Fig. 8.1). If p = q = 1/2 the histogram is symmetrical irrespective of the value of k. If p is larger than q, the distribution is skew with the peak displaced to the left of centre (right-skew). If p is smaller than q, the distribution is skew in the opposite sense. The amount of skewness decreases with increasing k. If k is larger than 50 the distribution is approximately symmetrical unless p or q is extremely small.

FITTING THE BINOMIAL DISTRIBUTION TO AN ACTUAL DISTRIBUTION

The binomial distribution is a theoretical model, a product of mathematics rather than empirical data. Nevertheless it reproduces successfully the appearance of a number of actual distributions that arise from random processes. Figure 8.2 shows the number of days per week in 1973 when measurable rain fell at Balcombe in Sussex. Also shown for comparison is a binomial distribution with the same k and mean as the observed distribution. In this example the agreement between the binomial and observed frequencies is fairly good, but by no means exact.

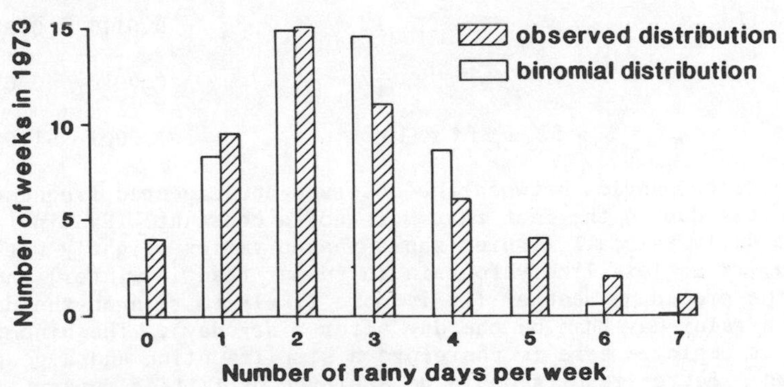

Fig. 8.2 - Number of rainy days per week in 1973, Balcombe, Sussex, with fitted binomial distribution.

Use of the binomial model requires that values be assigned to p and k. The value of k is usually self evident, if it exists at all. The value of p is sometimes specified by theory, but generally it is necessary to estimate p from the data, making use of the fact that the mean is the product of k and p. Thus in the case of the rainfall data represented in Fig. 8.2 and in Table 8.1 the value of k is 7, the maximum possible number of rainy days in a week. The empirical value of p can be obtained by dividing the mean number of rainy days per week by k. During the 52 weeks there were 135 days with rain. Hence μ, the mean number of rainy days per week, = 135/52 = 2.596. The probability of a rainy day, p, = μ/k = 2.596/7 = 0.3709. Conversely, the probability of a dry day, q, = 1 - p = 0.6291. The binomial expansion of $(q + p)^7$ is $q^7 + 7pq^6 + 21p^2q^5 + 35p^3q^4 + 35p^4q^3 + 21p^5q^2 + 7p^6q + p^7$. Substituting for p and q gives the binomial probabilities listed in the table. Multiplying each of the probabilities by the total number of weeks (52) gives the expected frequencies corresponding to the binomial distribution.

TABLE 8.1 - NUMBER OF RAINY DAYS PER WEEK IN 1973 AT BALCOMBE, SUSSEX

Number of days per week with measurable rain *X*	*Number of weeks in year* *f*	*Number of rainy days* *fX*	*Binomial probabilities*	*Expected frequencies*
0	4	5	q^7 = 0.0390	2.028
1	9	9	$7pq^6$ = 0.1610	8.370
2	15	30	$21p^2q^5$ = 0.2847	14.803
3	11	33	$35p^3q^4$ = 0.2797	14.545
4	6	24	$35p^4q^3$ = 0.1649	8.574
5	4	20	$21p^5q^2$ = 0.0583	3.033
6	2	12	$7p^6q$ = 0.0115	0.596
7	1	7	p^7 = 0.0010	0.050
Totals	$\Sigma f = N = 52$	$\Sigma fX = 135$	1.0001	51.999

The discrepancies between the observed and expected frequencies are partly due to the fact that p is not a constant. Examination of the daily rainfall figures shows that p varies slightly with the season (it is less likely to rain in summer than in winter) and with the preceding weather (it is more likely to rain on the day after a rainy day than on the day after a dry day). The binomial model as employed here is therefore a simplification and a generalisation. Better results could be achieved by fitting separate binomial distributions to each of the seasons, or by restricting the analysis to days of heavy rain which occur less frequently and more randomly.

DISCUSSION

The binomial distribution is useful in situations where only two outcomes need to be considered e.g. rain or no rain, success or failure, yes or no, north or south, woodland or non-woodland. If more than two outcomes need to be considered, the multinomial distribution is applicable (see Chapter 15) rather than the binomial.

The binomial model requires that the observed variable (X) be limited to zero and positive whole numbers. Negative or fractional values cannot be accommodated and still less can observations or measurements based upon a continuous scale of magnitude. The variable must also have a definite upper limit, i.e. a finite value of k. In Fig. 8.2 for example, the variable cannot exceed a value of 7, the number of days in a week. Unfortunately, in many geographical situations it is not possible to specify an upper limit, and this restricts the applicability of the binomial model.

The model assumes that the probability of an occurrence (p) is fixed. One occurrence does not make it any more or less likely that there will be another. This assumption is not particularly realistic since in practice occurrences are often linked to one another. The more people catch an infectious illness, the more effectively it spreads. Where there is one house, there are likely to be others. If an area is flooded once, it is likely to be flooded again. This tendency of one occurrence to be followed by another means that the binomial model often fails to give a satisfactory representation of empirical data. The observed frequencies are found to differ substantially from the expected frequencies calculated according to the binomial model.

The assumption that p is fixed is violated by sampling without replacement because the composition of the population alters as each member is withdrawn to form the sample. Suppose, for instance, that a ball is to be drawn at random from a box containing ten blue and fifteen green balls. The chance of drawing a blue ball is $p = 0.4$, while the chance of drawing a green ball is $p = 0.6$. Suppose that a blue ball is selected on the first draw and is not replaced. Only nine blue balls will be left in the box as against fifteen green balls. On a second drawing, therefore, the chance of obtaining a blue ball is less than before ($p = 0.375$) while the chance of obtaining a green ball is increased ($p = 0.625$).

If the population is large relative to the sample, the probability p will remain fairly constant even though the sampling is without replacement. The binomial model can therefore be employed as a good approximation. Only if the sample is greater than about 10 per cent of the population will p vary so much as to exclude the use of the binomial. In that event the hypergeometric distribution should be used as explained by Hays (1973).

THE POISSON DISTRIBUTION

The *Poisson distribution* is used to describe random phenomena that have a low probability of occurrence. It is named after the French mathematician Siméon-Denis Poisson, who developed it to investigate the probabilities of incorrect judgements in law courts. Later writers have found that the distribution has applications in a wide variety of problems in the social and physical sciences.

The first few terms of the Poisson distribution are

$$e^{-\lambda},\ \lambda e^{-\lambda},\ \frac{\lambda^2}{2!}e^{-\lambda},\ \frac{\lambda^3}{3!}e^{-\lambda},\ \frac{\lambda^4}{4!}e^{-\lambda},\ldots,\ \frac{\lambda^X}{X!}e^{-\lambda}$$

In this formula e = 2.71828..., the base of natural logarithms[1] and λ, the Greek lower case letter *lambda*, can be any positive number. The first term in the series gives the probability of 0 occurrences, the second term gives the probability of 1 occurrence, the third term the probability of 2 occurrences, and so on. The last (or general) term gives the probability of X occurrences where X can be any positive integer (0, 1, 2, 3 etc.) and is therefore a discrete variable. Note that $e^{-\lambda} = 1/e^{\lambda}$.

As an example, suppose that λ happens to equal 2. The value of $e^{-\lambda}$ is given by

$$\frac{1}{e^2} = \frac{1}{(2.71828\ldots)^2} = \frac{1}{(7.38905\ldots)} = 0.13533\ldots$$

and hence the Poisson distribution is

0.13533..., 0.27067..., 0.27067..., 0.18044..., 0.00902..., etc.

The probability of no occurrences is 0.13533..., the probability of 1 occurrence is 0.27067..., as is the probability of 2 occurrences, the probability of 3 occurrences is 0.18044..., the probability of 4 occurrences is 0.00902..., and so on. The probabilities become neglibibly small if the series is continued beyond a certain point.

In practical applications of the Poisson series λ is not usually an integer, but an inconvenient decimal number. The value of $e^{-\lambda}$ can be found either by using logarithms or a pocket calculator that has a power key, or by referring to published tables of $e^{-\lambda}$.

The binomial distribution closely resembles the Poisson if p is small. In the limiting case when p approaches zero and k approaches infinity, the binomial becomes identical to the Poisson with $\lambda = kp$. A proof is given in Yule and Kendall (1950, pp.189-191). Table 8.2 demonstrates that agreement between the two distributions is good even if k is as low as 10, provided that p is kept very small.

---

[1] e is an example of what is called an *irrational* number - it has an endless non-repeating series of digits after the decimal point. Another irrational number is π, the circumference of a circle divided by its diameter = 3.14159...

e plays an important part in the calculus. It may be defined as the limit of the function $(1 + X)^{1/X}$ as X decreases and approaches zero. Thus when X = 0.1, $(1 + X)^{1/X} = 2.594$; when X = 0.01, $(1 + X)^{1/X} = 2.706$; when X = 0.001, $(1 + X)^{1/X} = 2.717$; when X = 0.0001, $(1 + X)^{1/X} = 2.718$; and so on.

TABLE 8.2 - EXAMPLES OF THE BINOMIAL AND POISSON DISTRIBUTIONS

X	*Binomial*	*Poisson*	*Binomial*	*Poisson*
	k=10, p=0.01	λ=k, p=0.01	k=10, p=0.05	λ=k, p=0.05
0	0.9044	0.9048	0.5987	0.6065
1	0.0914	0.0905	0.3151	0.3033
2	0.0042	0.0045	0.0746	0.0758
3	0.0001	0.0002	0.0105	0.0126
4	0.0000	0.0000	0.0010	0.0016
5	0.0000	0.0000	0.0001	0.0002
6	0.0000	0.0000	0.0000	0.0000

Figure 8.3 shows how the graphical form of the Poisson distribution varies according to the value of λ. The distribution is strongly skewed if λ is small, becoming less so if λ is increased. When λ is greater than about 20, the distribution is for all practical purposes symmetrical. Figure 8.3 should be compared carefully with Fig. 8.1.

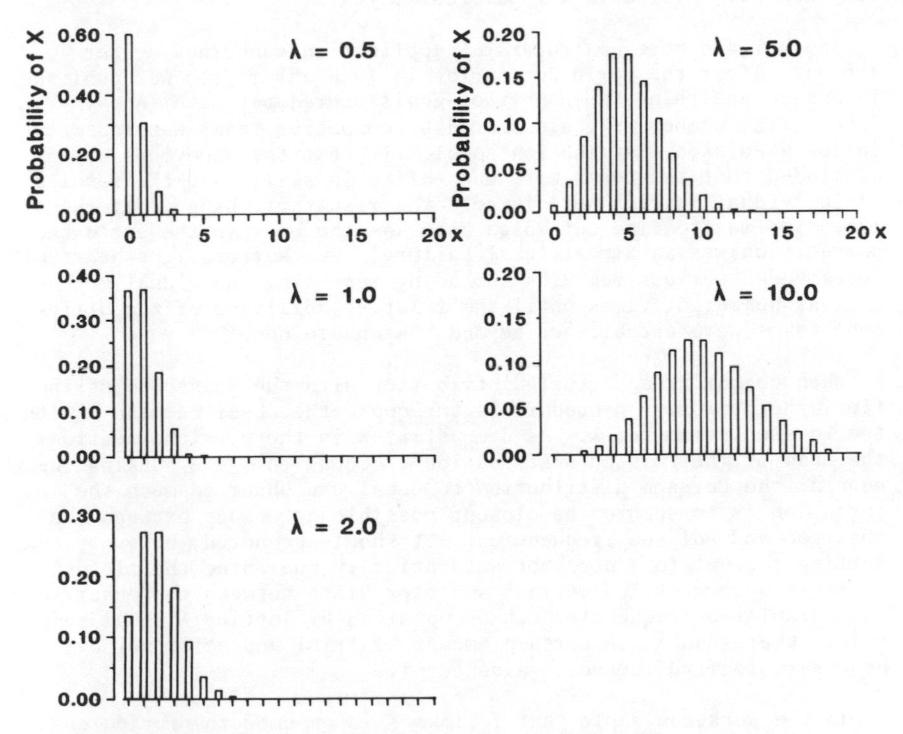

Fig. 8.3 - Examples of the Poisson distribution for various values of λ.

Despite the variation in graphical form, the mean of the Poisson distribution is always equal to λ, as is also the variance. The equality of mean and variance arises because the probability of one or more occurrences is low. The mean of the binomial distribution, it will be recalled, is kp, and the variance kpq. If p is made increasingly small, q = 1 - p approaches closer and closer to 1. The variance of the distribution therefore tends to approach the mean in value.

The mathematical theory of the Poisson distribution is a little complicated and will not be considered here. The interested reader will find full details in Parzen (1960) and other standard texts on probability.

PRACTICAL APPLICATIONS OF THE POISSON DISTRIBUTION

The Poisson distribution is useful in the analysis of what may be called accidents: infrequent events that arise through random chance. One of its first and most famous applications concerned the number of men in the Prussian Army who were killed each year as a result of the kick of a horse (Bortkiewicz, 1898). It was found that the numbers of men killed each year agreed very closely with the numbers given by a Poisson distribution. Apparently chance determined how many men were killed in any particular year.

Another and more controversial application concerned soccer scores. After the World Cup series in 1966 an article was published in *Nature* analysing the number of goals scored per math (Anon, 1966). The number of goals scored by competing teams was found to follow a Poisson distribution so closely that the anonymous author concluded that the teams were much alike in skill, and the results of individual matches were largely the result of chance. If the intention was to find out which team was the best in the world the competition was an almost total failure! The author light-heartedly recommended various remedies including repeating individual matches a great number of times until the relative abilities of the different teams were established beyond reasonable doubt.

When comparing an actual distribution with the Poisson distribution, the customary procedure is to equate the observed mean X with the Poisson parameter λ. As demonstrated in the previous section, the mean of the Poisson distribution is equal to λ. By taking the mean of the Poisson distribution to equal the observed mean the intention is to secure the closest possible agreement between the observed and Poisson frequencies. It should be noted, however, that setting λ equal to X does not automatically guarantee the closest possible agreement. Sometimes a better "fit" between the observed and calculated frequencies can be obtained by letting λ equal some value other than X. A certain amount of trial and error may be necessary to find the best value for λ.

In the worked example that follows X is assumed to provide the best value of λ in order to save time and no attempt is made to establish whether a better fit might be obtained by adopting a different value of λ.

Table 8.3 records the incidence of heavy rains at Balcombe in Sussex during the years 1909 to 1973 inclusive. It would be permissible to fit a binomial distribution of the form $(p + q)^{365}$ to the data. Because p is small, however, the Poisson distribution provides a very close approximation.[1] The total number of days with 30 mm or more rainfall recorded during the 65 years is (12)(0) + (23)(1) + (19)(2) + (7)(3) + (3)(4) + (1)(5) = 99. The mean number of days of heavy rainfall per year is therefore 99/65 = 1.52308... Taking this to be the value of λ, $e^{-\lambda} = 0.21804$ approximately.

The Poisson probabilities $e^{-\lambda}$, $\lambda e^{-\lambda}$, $\lambda^2 e^{-\lambda}/2!$ etc. are set out in the third column of the table. These probabilities multiplied by 65, the total number of years, yield the frequencies that would be expected given a perfect Poisson distribution of heavy rains (fourth column).

TABLE 8.3 - NUMBER OF DAYS WITH HEAVY RAIN AT BALCOMBE, SUSSEX, 1909 - 1973

Number of days per year with 30 mm or more rainfall	*Number of years*	*Poisson probabilities*		*Expected number of years*
0	12	$e^{-\lambda}$	= 0.21804	14.173 ≃ 14
1	23	$\lambda e^{-\lambda}$	= 0.33209	21.586 ≃ 22
2	19	$\lambda^2 e^{-\lambda}/2!$	= 0.25290	16.439 ≃ 16
3	7	$\lambda^3 e^{-\lambda}/3!$	= 0.12840	8.346 ≃ 8
4	3	$\lambda^4 e^{-\lambda}/4!$	= 0.04889	3.178 ≃ 3
5	1	$\lambda^5 e^{-\lambda}/5!$	= 0.01489	0.968 ≃ 1
6	0	$\lambda^6 e^{-\lambda}/6!$	= 0.00378	0.246 ≃ 0
7	0	$\lambda^7 e^{-\lambda}/7!$	= 0.00082	0.053 ≃ 0
8	0	$\lambda^8 e^{-\lambda}/8!$	= 0.00016	0.010 ≃ 0
Totals	65		0.99997	64.999

The agreement between the observed and expected frequencies is so close that there is little need to investigate whether another value of λ might secure even closer agreement. The agreement is already sufficient to suggest strongly that heavy rains are random and independent events. The differences between the observed and expected frequencies are so slight they could easily be the result of chance.

It must be noted, however, that the Poisson model makes the assumption that λ is a constant over time. The model does not allow for the possibility that the average time between heavy rains is increasing or decreasing. For instance, the number of heavy rains

[1]At first glance the binomial might seem to be a more appropriate distribution to fit than the Poisson because it explicitly limits the number of days per year to 365. However, the Poisson distribution predicts a vanishingly small number of years with 366, 367, etc. days of heavy rain. The Poisson predictions are accurate enough for all practical purposes and the labour of calculation is much less than in the case of the binomial.

per year might have been zero in the years 1909 - 1920, one in the 1921 - 1943, two in the years 1944 - 1962, three in the years 1963 to 1969, four in the years 1970 - 1972, and five in the year 1973. Such a distribution would accord with that shown in Table 8.3, yet the rains would not be occurring randomly in time. The intervals between successive rains would be getting shorter and shorter. Some meteorologists and hydrologists (e.g. Rodda, 1970), are convinced that heavy rains in Britain are becoming more frequent and more intense at least at certain sites, but their views are not universally accepted. As far as Balcombe is concerned, the number of heavy rains appears to vary randomly from year to year (Fig. 8.4A). No pronounced long term trend is apparent. Taking the years in random order (Fig. 8.4B) does not introduce any greater disorder, i.e. does not alter the year to year variation appreciably. It therefore seems reasonable to take the value of λ as constant, and to employ the Poisson model to describe the frequency distribution.

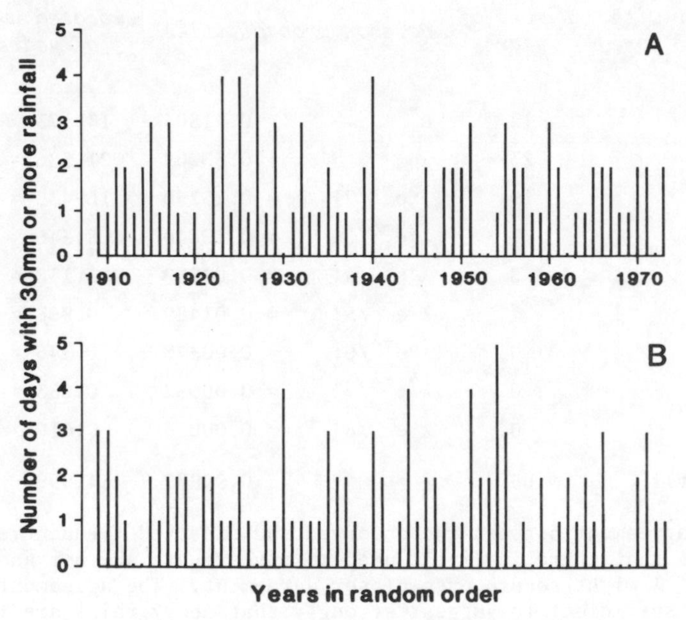

Fig. 8.4 - Number of days per year with heavy rain, Balcombe, Sussex.

A: actual data, B. years rearranged in random order.

The Poisson distribution is not restricted to the analysis of events over time, and can also be used to analyse events that occur randomly within an area, as was first shown by the Japanese geographer Isamu Matui in 1932. Using a map, Matui investigated the distribution of houses in a rural area of central Japan (Fig. 8.5A). He divided the map evenly with a grid composed of 1200 small squares, each representing 0.01 sq km and then counted the number of houses in each square (Fig. 8.5B). The number of squares containing 0, 1, 2, 3 etc. houses is entered in Table 8.4.

A binomial distribution cannot be fitted to the data since there is no definite upper limit to the number of houses that can be cram-

TABLE 8.4 - NUMBER OF HOUSES PER GRID SQUARE IN AN ARBITRARILY CHOSEN AREA OF THE TONAMI PLAIN, CENTRAL JAPAN

Number of houses per grid square	*Observed number of grid squares*	*Expected number of grid squares (Poisson distribution)*
0	584	562
1	398	426
2	168	162
3	35	41
4	9	8
5	4	1
6	0	0
7	1	0
8	0	0
9	1	0
Total	1200	1200

med on to a grid square. The value of k is undefined, in contrast to the last example where it equalled 365. The Poisson distribution is less restricted in its applications because it does not require a value of k. According to the Poisson distribution, if N points (houses, farms, settlements or whatever) are distributed randomly within an area, subdivided into Z squares equal in size, the numbers of squares containing 0, 1, 2, 3, etc. points will be on average

$$Ze^{-\lambda},\ Z\lambda e^{-\lambda},\ \frac{Z\lambda^2 e^{-\lambda}}{2!},\ \frac{Z\lambda^3 e^{-\lambda}}{3!},\ \text{etc.}$$

where $\lambda = N/Z$, the mean number of points per square, i.e. the density of the pattern.

The total number of houses in the area studied by Matui was 911, giving a value for λ of 911/1200 = 0.7592. $e^{-\lambda} = 0.4680$. The expected frequencies for a Poisson distribution with this value of $e^{-\lambda}$ are listed in Table 8.4. They differ slightly from those calculated by Matui who took $e^{-\lambda}$ to be 0.4730, a value which secures marginally greater correspondence with the observed frequencies.

Even leaving $e^{-\lambda}$ equal to 0.4680 the degree of correspondence between the observed and expected frequencies is impressive. The fact that the Poisson model describes the observed distribution so well suggests that the houses are located at least partly at random.

The analysis is insufficient as it stands because it does not consider the spatial distribution of the frequencies. Table 8.4 would be no different if all the squares with few or no houses were concentrated in one part of the area and all the squares with many houses were in a different part. Moreover the frequencies obtained are dependent on the size of the grid squares. If squares of different size were selected they might not yield the same close correspondence between the observed and Poisson frequencies. Complete proof that the pattern of houses is random or near random is therefore lacking. However, the analysis is perfectly sound as far as it goes, and it can be elaborated and improved upon if required in several ways. Dacey (1964), for example, suggests a method of combining the grid squares into successively larger squares in order to test for independence in the frequencies of adjacent cells. It is also possible to subject the data in Table 8.4 to a chi-square test as explained in the companion volume to this book. The test suggests that the houses may not be distributed perfectly randomly, and further analysis suggests that a very small amount of spatial autocorrelation (clustering) is present in the data.

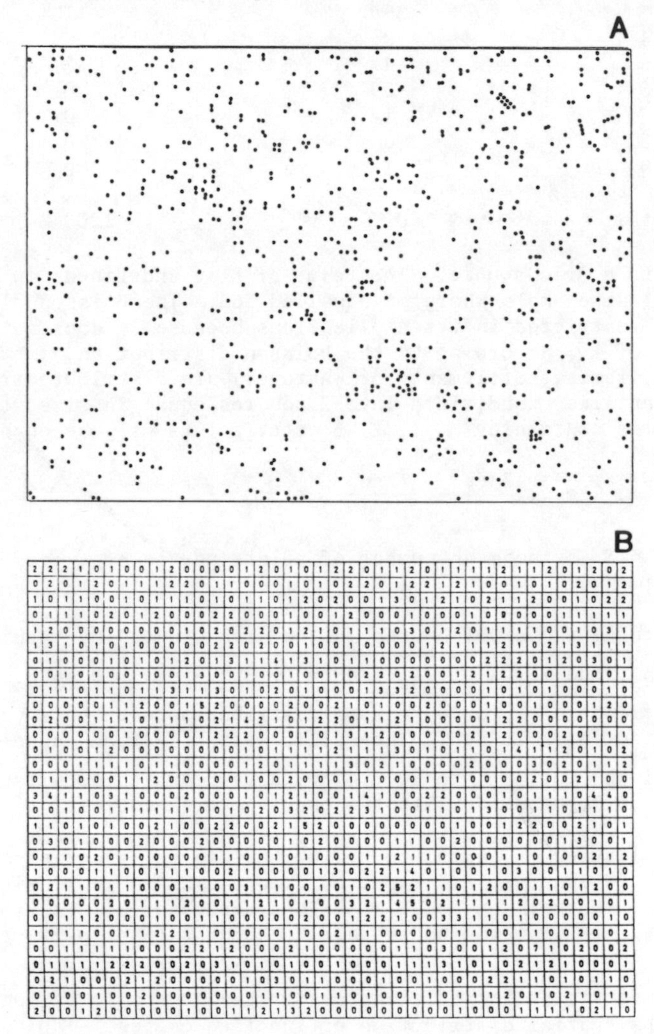

Fig. 8.5 - Distribution of houses in an area 0.4 by 0.3 km, Toyama Prefecture, Japan.

A. Dot map showing the individual houses. B. Number of houses falling in each grid square.

9 The Normal and Log-Normal Distributions

The binomial and Poisson distributions restrict the variable X to integer values. Two theoretical distributions that allow it to assume fractional values as well are the *normal* and *log-normal distributions*. Each is applicable to a wide variety of random variables.

THE NORMAL DISTRIBUTION

The normal distribution is symmetrical with a convex central peak and concave "tails" stretching away on either side (Fig. 9.1). The smooth curve which outlines this shape is called the *normal curve*. Its equation is somewhat daunting at first glance

$$Y = \frac{N\ e^{-(X - \mu)^2/2\sigma^2}}{\sigma\sqrt{(2\pi)}}$$

where Y is the height of the curve corresponding to a given value of X (X being measured along the horizontal axis and Y along the vertical), while μ, the Greek lower-case letter mu, is the mean value of X, and σ is the standard deviation of X. N is the total number of items in the population, i.e. the size of the population or total number of X values.[1] π and e are the familiar constants 3.14159... and 2.71828... respectively.

The normal curve attains its greatest height above the mean, μ. With distance from the mean, the curve descends closer and closer to the horizontal axis but never actually reaches it. The curve can be visualised as extending to plus infinity in one direction and minus infinity in the other. Although the curve has an unlimited range, the area between the curve and the horizontal axis is finite. If the X and Y scales are made equal, the area is N square units of X (or Y), the exact size of the population. If the X and Y scales are made unequal, the area is Nr square units of X, where r is the fraction or multiple of the X scale represented by the Y scale. For example, if the Y scale is half the X scale, the area is N/2 square units.

The reason why the area under the curve is finite even though the curve extends to infinity in both directions can be seen from Fig. 9.2. The area under the curve increases with increasing distance either side of the mean, but at an ever decreasing rate. As the distance approaches plus and minus infinity so the rate of increase of area approaches zero.

---

[1]The symbols μ, σ and N are employed in the above equation rather than $\bar{X}$, s and n because the normal curve is used mainly as an idealisation or model to represent sampled populations. If the curve is used to represent a sample, $\bar{X}$, s and n are the appropriate symbols.

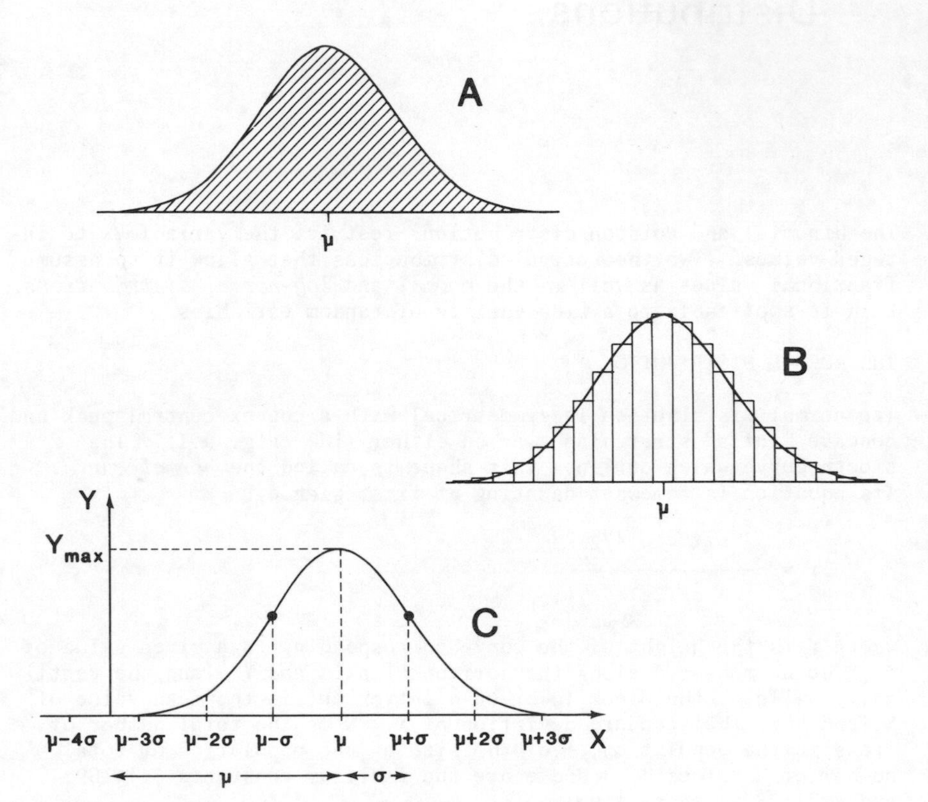

Fig. 9.1 -

A: Normal distribution bounded above by a normal curve.

B: Histogram approximating to a normal distribution with fitted normal curve.

C: Normal curve showing co-ordinate system. The origin (0,0) lies μ *units of X to the left of the mean. The dots indicate points of inflection.*

The *points of inflection* on the normal curve where it changes in slope from convex to concave lie at a distance of one standard deviation either side of the mean, i.e. at $(\mu + \sigma)$ and $(\mu - \sigma)$. The value of σ determines the extent to which the curve spreads out from the mean and also its height. The three curves shown in Fig. 9.3, for instance, represent populations of equal size with the same mean but different standard deviations. The areas under all three curves are the same, but the dimensions of the curves differ. The smaller the value of σ, the higher is the peak.

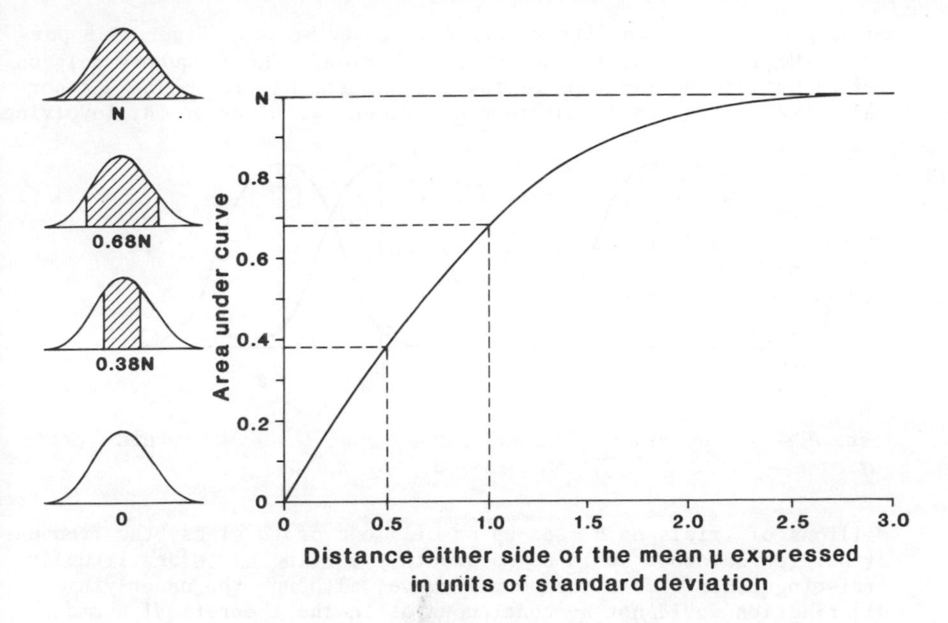

Fig. 9.2 - Area under the normal curve at different distances from the mean.

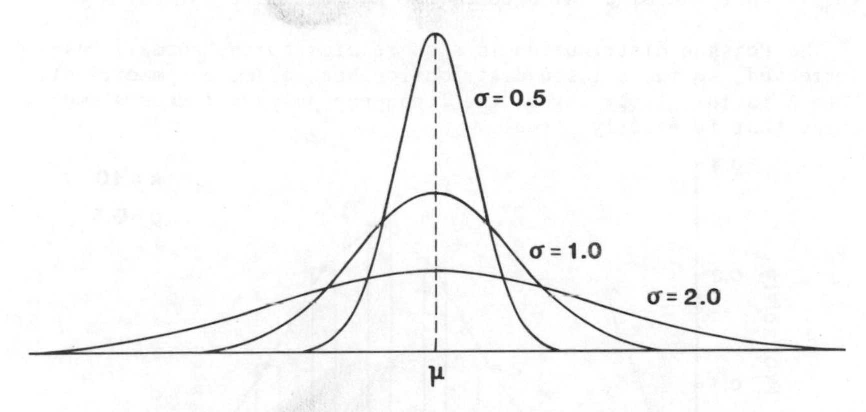

Fig. 9.3 - Examples of normal curves with the same mean but different standard deviations.

Altering the value of μ does not affect the shape of the normal curve in any way, but does affect its location along the horizontal axis as Fig. 9.4 illustrates.

The normal distribution is closely related to the binomial, although this is a discrete distribution. The binomial approaches the normal in form when neither p nor q is close to zero and k is very large. For instance, if an unbiased coin is tossed 10 times, the most probable result is 5 heads and 5 tails (probability = 0.2461...). In a long series of trials, each involving tossing the coin 10 times, 5 heads would be expected to occur most frequently, 4 or 6 heads a little less often (probability = 0.2051...), 3 or 7 heads less frequently again (probability = 0.1172...), and so on. Figure 9.5 portrays the theoretical frequency distribution. The frequency polygon formed by joining the tops of the columns roughly resembles the normal curve. If a really marathon experiment were performed, involving

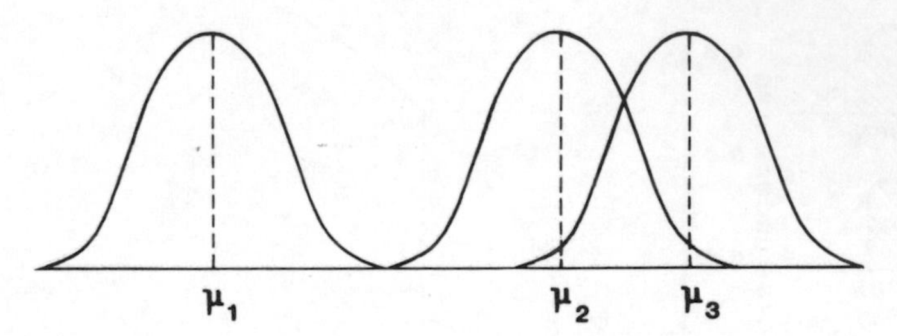

Fig. 9.4 - Examples of normal curves with the same standard deviation σ = 0.5 but different means μ_1, μ_2 and μ_3.

millions of trials each made up of millions of tossings, the frequency polygon obtained would be so smoothly curving as to be virtually indistinguishable from the normal curve, although the underlying distribution would not be continuous as in the theoretical normal distribution. In the limiting case, however, when k is infinite, the discrete variable can be visualised as becoming continuous with the result that the binomial becomes the normal distribution.

The Poisson distribution is also related to the normal. As λ is increased, so the Poisson distribution becomes more symmetrical. When λ is infinitely large, the frequency polygon traces a smooth curve that is exactly normal.

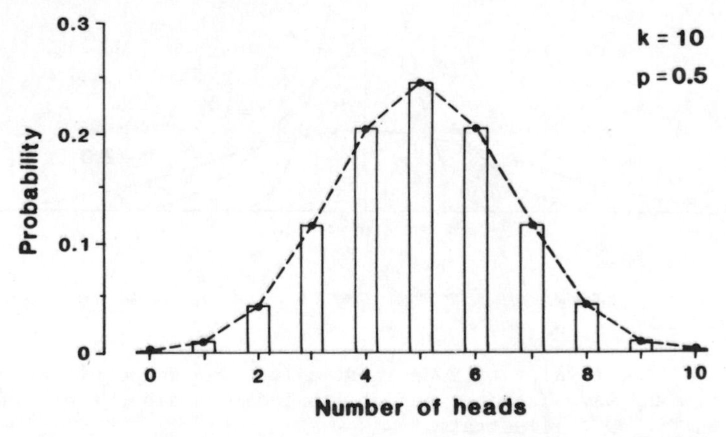

Fig. 9.5 - Symmetrical binomial distribution representing the probabilities of obtaining 0, 1, 2, 3 etc. heads in 10 tossings of an unbiased coin.

THEORETICAL BACKGROUND AND PRACTICAL APPLICATIONS

The equation of the normal curve can be derived by making the following assumptions

1. Each value of the variable X results from the operation of an infinite number of chance factors.

2. Each factor operates independently of all the other factors.

3. Each factor increases or decreases the value of X by a fixed amount which is infinitely small.

4. Chance determines whether an increase or decrease occurs.

The mathematical operations needed to extract the equation from these assumptions are somewhat complicated and will not be given here. The interested reader is referred to Feller (1957) for an extended discussion. For the present, an illustrative rather than mathematical approach must suffice.

The assumptions listed above are given tangible expression in the mechanical device shown in Fig. 9.6. A stream of steel balls, glass

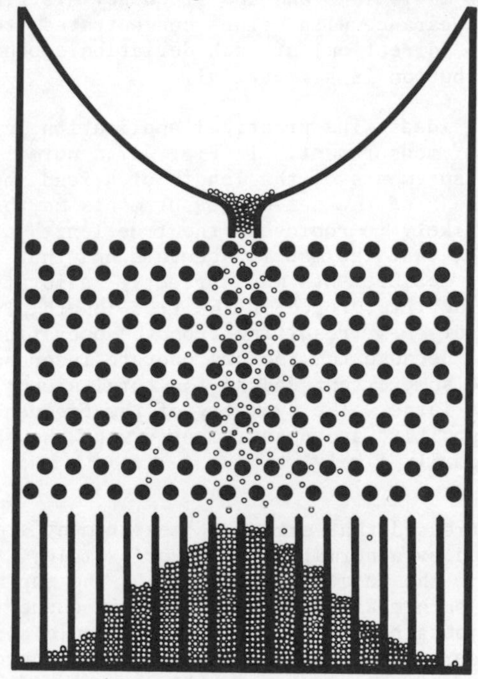

Fig. 9.6 - Apparatus for demonstrating the symmetrical binomial and normal distributions. Modified from Galton (1899).

beads or lead shot is allowed to descend a board past an array of pins of circular cross-section. The balls hit the pins and are deflected either left or right at random. Below the array of pins are compartments in which the balls collect. In the absence of the pins, all the balls would roll straight down the board and arrive in the two central compartments. However, as a result of the deflections caused by the pins, the balls distribute themselves between the various compartments to give a symmetrical distribution that resembles the normal.

Each ball arriving in a compartment can be thought of as an individual value of X, the pins it encounters on the way being chance factors which help to determine its value. The distance of the compartment from the central compartments represents the amount of deviation from the mean value. Thus each ball can be thought of as representing a fixed value (the mean) together with a variable amount of deviation. In symbols,

$$X = \mu + d$$

where d, the amount of deviation, may be either positive or negative.

By observing the way the balls descend the board, it is possible to gain an intuitive understanding of why the normal distribution develops. The deflections produced by successive pins cancel out to some extent, the balls being deflected first one way and then the other. Consequently, small deviations from the mean are more probable than large deviations and the frequency distribution takes on a hump-backed appearance with values concentrated around the mean. Because the sign (direction) of each deviation depends on chance, the frequency distribution is symmetrical.

The foregoing ideas find practical application in the interpretation of errors of measurement. In Fig. 9.7 a normal curve has been fitted to 75 measurements of the length of a road shown on a map. The arithmetic mean of the set of measurements may be thought of as the value most likely to represent the true length of the road. (Provided, of course, that the map is accurate and that the instrument used to make the measurements was correctly calibrated.) Each measurement can be interpreted as the true length plus or minus a variable amount of *error* resulting from the operation of a great many chance factors. Because the factors operate independently they tend to cancel out to some extent and so the total amount of error is more often small than large. The symmetry of the frequency distribution arises because positive and negative errors of the same numerical magnitude are equally likely.

It must be stressed that errors of measurement are not always closely described by a normal curve. For example, if the instrument used to determine the length of the road on the map had become more inaccurate with each measurement, a skewed frequency distribution would have been obtained. The distribution would also have been skewed if the user of the instrument had become increasingly careless. Non-random errors of this sort are called *systematic errors*. If systematic errors are present, the distribution need not be approximately normal in form, although it may be if the errors are constant. Thus, if the road was inaccurately shown on the map, the errors of measurement would still tend to be normally distributed, but the mean would not represent the true length of the road.

Normal curves find many other uses in depicting data. A wide range of variables exhibit distributions that are approximately normal, including the lengths of ears of corn, the heights of men or women, their intelligence quotients, and the velocities of molecules in a gas. However, it is important to realise that no naturally

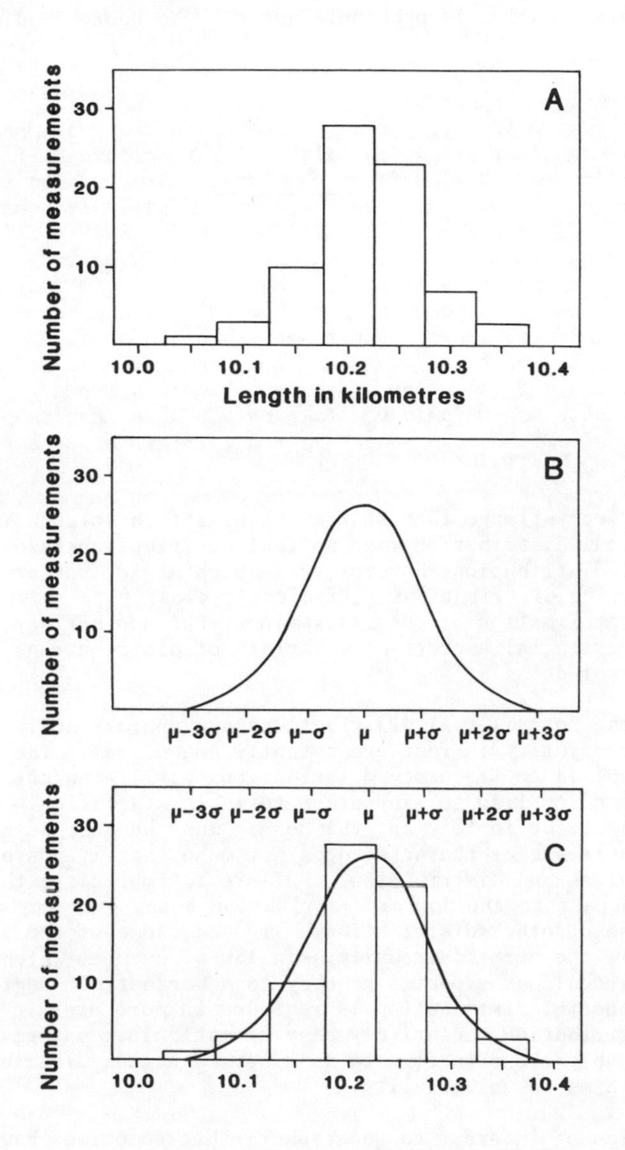

Fig. 9.7 - Diagram to illustrate errors of measurement.

A: Histogram representing 75 measurements of the length of a road (a stretch of the A275 between Lewes and Newhaven in Sussex).

B: Normal curve with the same mean (μ), standard deviation (σ) and area as the histogram.

C: Normal curve superimposed on the histogram.

occurring distribution is precisely normal. To quote Bradley (1976, p. 384):

"The normal distribution is a mathematical fiction. It is impossible, not just unlikely, for any variable in the real world to have an exactly normal distribution, although for certain variables the bulk of their distribution is approximately normal. The impossibility is ensured by the fact that the normal distribution extends from $-\infty$ to $+\infty$. It therefore treats all values in that infinite range as possible values that the variable could actually assume on some occasions, however rare they might be. But of course no real-world variable has an unlimited range of possible values, and for most the range is relatively short. Distributions of heights, or weights, of people are often cited as examples of a normal distribution. Within the interval $\mu \pm 2\sigma$ they are often approximately normal. But since neither heights nor weights can take values less than zero, they certainly cannot extend to $-\infty$ as required if their distributions are to be exactly normal."

It may seem strange that statisticians attach so much importance to the normal distribution when no real distribution is exactly normal The normal distribution, however, resembles a wide variety of naturally occurring distributions sufficiently closely to serve as a convenient approximation or generalisation. Put another way, it reproduces the essential aspects of a variety of distributions while leaving out minor details.

Of course not all real distributions are approximately normal in form. The majority in fact are patently non-normal. The term "normal" as applied to frequency distributions is ill-chosen and invites misunderstanding like so many other terms in statistics. Normal distributions are definitely not the norm. Just because something is of normal occurrence or character does not mean that it possesses a normal or near normal distribution. Failure to appreciate this point led in the past to the normal distribution being over-emphasised at the expense of other distributions. Indeed, some of the first statisticians saw the normal virtually as a law of nature, which most distributions could be expected to obey to a perceptible degree. Nowadays, the normal distribution is regarded in more prosaic terms as a useful distribution, relatively easy to manipulate mathematically, and reasonably close in form to a number of actual distributions, but without claims to universality.

Variables of interest to geographers that sometimes have normal or near normal distributions include

1. Maximum angles of valley side slopes within an area of uniform lithology, soil, vegetation, etc.

2. Altitudes (heights of the ground surface) within a drainage basin that has undergone extensive erosion during the course of one cycle of erosion.

3. Daily temperature extremes during the summer, hourly temperatures per month, annual rainfall totals and various other meteorological phenomena.

4. Annual runoff totals and various other hydrological phenomena.

5. Yields of grain from plots or fields of equal size and fertility.

6. Distance between settlements and their nearest neighbours in certain areas (e.g. Fig. 5.6).

7. Local variations in the prices of certain goods and services.

8. Population sizes of certain administrative areas, e.g. Parliamentary constituencies in the United Kingdom.

It is important to realise that these phenomena do not always display distributions that are near normal. The examples shown in Fig. 9.8 have been carefully selected to be as convincing as possible. Contrary examples could easily be produced. Thus while it is true that maximum angles of slope are sometimes distributed in a strikingly normal fashion as Fig. 9.8 demonstrates, there are so many exceptions that no general law can be claimed. Quite the reverse, for it would seem that slope angles generally follow a right-skewed distribution (Speight, 1971). In southern England, for instance, lowland areas and cuestas both tend to be characterised by distributions of slope angles that are strongly skewed to the right.

Geographers and geologists have been somewhat incautious in claiming altitudes to be approximately normally distributed. The examples that can be cited (e.g. Fig. 9.8) seem to be the exception rather than the rule. The majority of drainage basins exhibit a distribution of heights that is strongly right-skewed or multimodal.

Distances between settlements and their nearest neighbours are often anything but normal. According to Curry (1964) settlements located entirely at random would exhibit an exactly normal distribution of distances, but this conflicts with the theoretical findings of Thompson (1956) and would seem to be incorrect. Thus, although the spacing of settlements in Iowa (Fig. 5.6) can be adequately described by the normal distribution, the settlements are not randomly located, but are spaced fairly evenly.

The population sizes of English constituencies are distributed approximately normally because of careful design rather than random chance. The boundaries of the constituencies are so drawn that on the whole each constituency has the same number of voters and is a recognisable community with its own distinct character, problems and traditions (Rowley, 1975). The emphasis on population equality serves to distinguish Parliamentary Constituencies from other types of administrative areas which tend to have populations that are markedly unequal and distributions that are right-skewed. As ever, the normal distribution must be recognised as something of a special case.

Despite the fact that few variables are distributed approximately normally, statisticians continue to pay a great deal of attention to the normal distribution. One reason is that the normal provides a convenient starting point from which to derive other theoretical distributions. A second and more cogent reason is that non-normality results only from the way variables are measured. Any distribution can be converted to normality by a change in the scale of measurement. For example, moderately right-skewed distributions can be transformed into normal or near normal distributions by taking the logarithms of the measurements instead of the measurements themselves (Fig. 9.9). Other methods of converting distributions to the normal form include

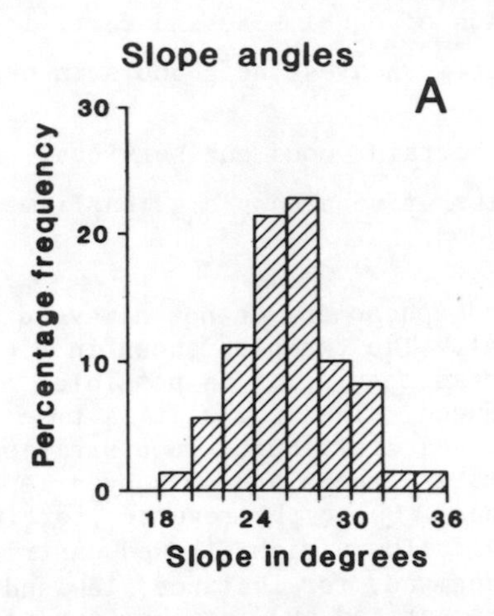
Slope angles
A
Percentage frequency
30
20
10
0
18
24
30
36
Slope in degrees

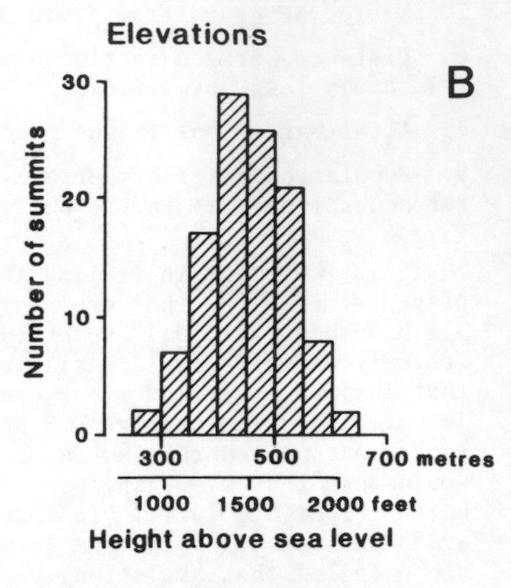
Elevations
B
Number of summits
30
20
10
0
300
500
700 metres
1000
1500
2000 feet
Height above sea level

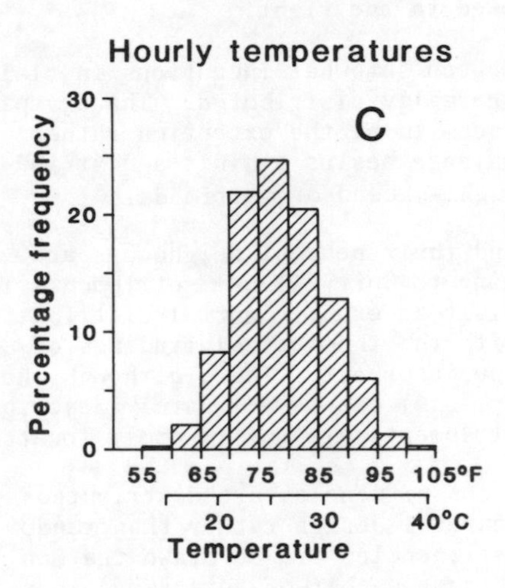
Hourly temperatures
C
Percentage frequency
30
20
10
0
55
65
75
85
95
105°F
20
30
40°C
Temperature

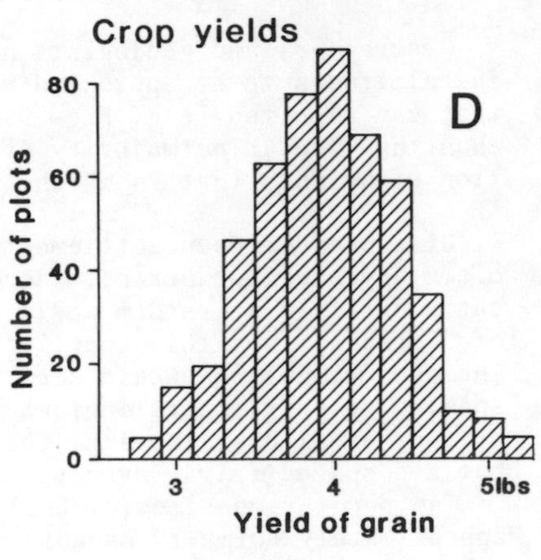
Crop yields
D
Number of plots
80
60
40
20
0
3
4
5lbs
Yield of grain

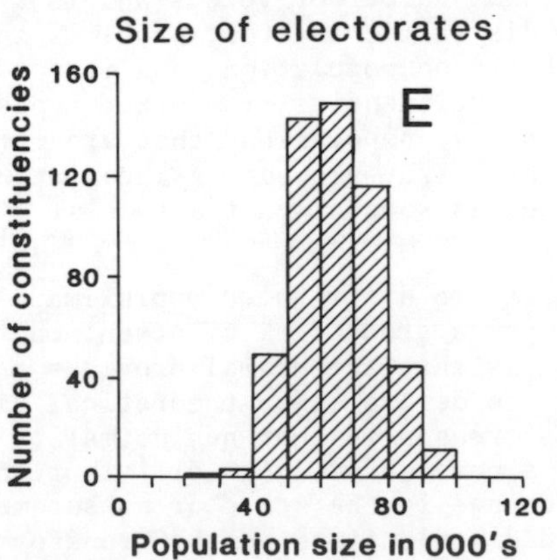
Size of electorates
E
Number of constituencies
160
120
80
40
0
0
40
80
120
Population size in 000's

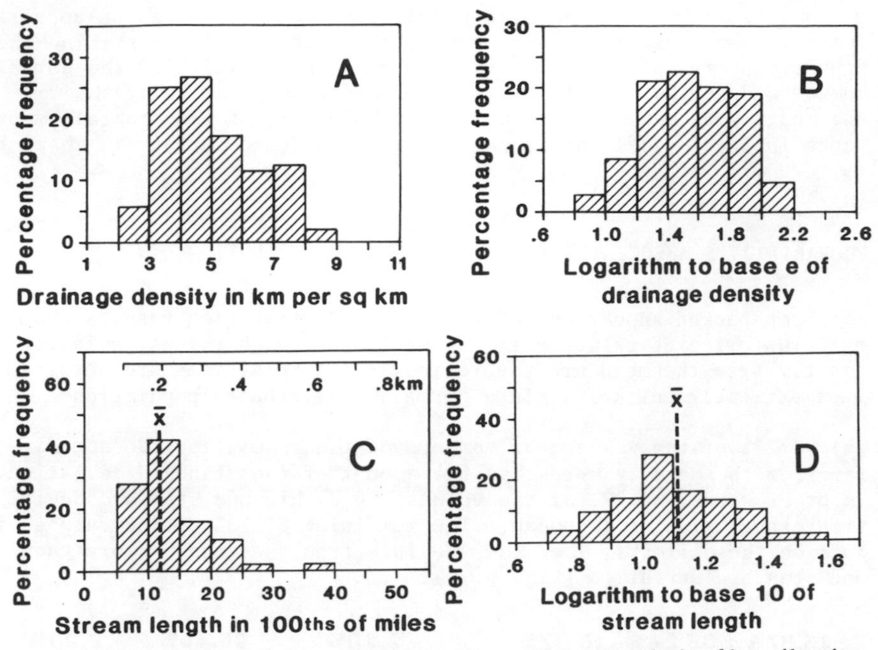

Fig. 9.9 - Comparison of arithmetic and logarithmic distributions.

A: Histogram of drainage density for 105 drainage basins in the New Forest (Data: D.A. Robinson).

B: Histogram of the logarithms of drainage density.

C: Histogram of lengths of first order streams developed on the Copper Ridge Dolomite, Virginia (Data from Miller, 1953).

D: Histogram of the logarithms of the lengths (from Strahler, 1954).

replacing the measurements by their square roots, or by angles whose sines are proportional to the square roots. The logarithmic transformation, however, is easily the most useful. Variables whose logarithms tend to be distributed normally are encountered very frequently in geographical studies. Because of the importance of such "log-normal" distributions, as they are called, they are considered separately later in this chapter. A further reason why the normal distribution plays a central role in statistical theory is that the means

---

Fig. 9.8 (opposite) - Examples of near normal distributions.

A: Maximum valley-side slopes, Sante Fe formation, Bernalillo, New Mexico. From Strahler (1950).

B: Heights of summits, Dartmoor. From Gerrard (1974).

C: Hourly temperatures in July, Washington, D.C. (1946-55). From

D: Yields of wheat in pounds from plots of 0.002 acres. From Mercer and Hall (1911).

E: Population sizes of English Parliamentary constituencies, 1974. From Rowley (1975).

of samples taken from the same population tend to plot as an approximately normal distribution, even if the measurements comprising the population are not normally distributed. This ability of the normal curve to describe the distribution of means, regardless of the way the individual measurements are distributed, is of fundamental importance in statistical inference, and is considered further in Chapter 11.

PROBABILITIES ASSOCIATED WITH THE NORMAL DISTRIBUTION

The hump-backed appearance of the normal distribution results from the clustering of values near the mean. Although values deviating greatly from the mean are theoretically possible, they are nevertheless extremely unlikely. In a normally distributed population

(a) 68.27% of the values (or members of the population) do not differ from the mean by more than the amount of the standard deviation. In other words, 68.27% of the values lie within one standard deviation either side of the mean. The remaining 31.73% of the values lie outside these limits, i.e. they deviate from the mean by more than one standard deviation (Fig. 9.10).

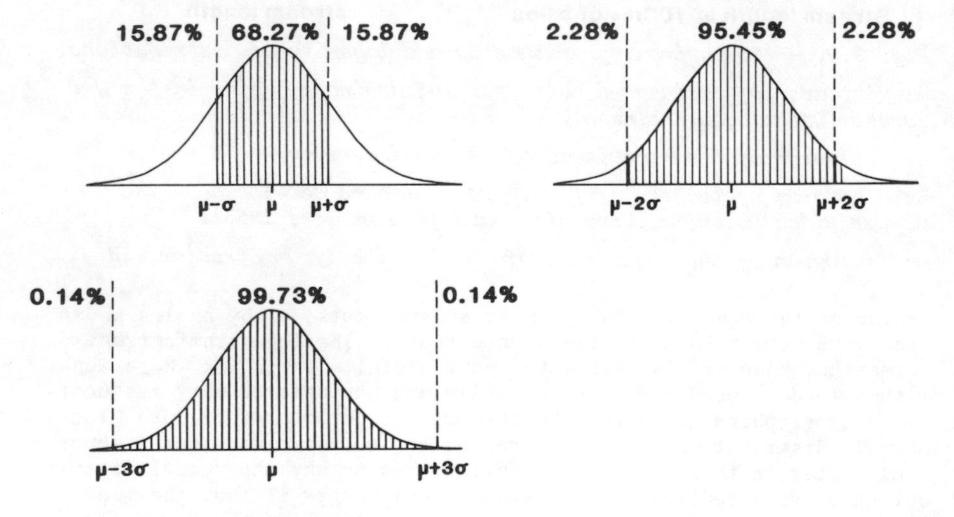

Fig. 9.10 - Areas under the normal curve within 1, 2 and 3 standard deviations of the mean.

(b) 95.45% of the values do not differ from the mean by more than twice the amount of the standard deviation, i.e. 95.45% of the values lie within two standard deviations either side of the mean. 4.55% of the values differ from the mean by more than two standard deviations.

(c) 99.73% of the values fall within 3 standard deviations of the mean, while only 0.27% of the values fall outside these limits.

Note also that 50% of the values in a normally distributed population lie within a distance 0.6745 times the standard deviation either side of the mean. 95% of the values lie within a distance 1.96 times the standard deviation.

The percentages given above are the percentage areas under the relevant portions of the normal curve. Thus, the area under the curve within one standard deviation of the mean is 68.27% of the total area. The area forming the tail ends of the distribution is 31.73% of the total area. Because the curve is symmetrical about the mean, the area in the left tail is the same as the area in the right viz. 15.87%. The area between the mean and the left tail is 34.13%, as is also the area between the mean and the right tail.

Dividing by 100 converts the percentages to probabilities. For instance, the probability that a single member drawn from the population will deviate from the mean by more than the standard deviation is 0.3173, the probability that it will deviate by more than two standard deviations is 0.0455, and the probability that it will deviate by over three standard deviations is 0.0027.

Table B at the back of the book sets out these and other probabilities relating to the two tails of the normal distribution. It is worth taking some trouble to understand how to use Table B as many problems in statistics to be discussed later cannot be solved without reference to it. In the table deviations from the mean are expressed as decimal fractions or multiples of the standard deviation and are denoted by the letter "Z". The value of Z can be ascertained as follows

$$Z = \frac{|X - \mu|}{\sigma}$$

where X is the value of the variable corresponding to the inner boundary of the tail, μ is the mean of the distribution, and σ the standard deviation. Put another way, Z is the distance of the inner boundary of the tail measured in units of standard deviation.

The figures down the left side of the table are values of Z to one decimal place. Thus the eleventh entry from the top (1.0) refers to a deviation of one standard deviation from the mean, the twelfth entry (1.1) refers to a deviation of 1.1 standard deviations (i.e. a deviation equivalent to 1.1 times the standard deviation) and so on. The figures across the top of the table give the second decimal place of Z. In the body of the table are set out the probabilities corresponding to the respective values of Z. Thus row twenty-one (labelled "Z = 2.0"), column one (headed "0.00") gives the probability (0.0455) that a member of a normally distributed population will differ in value from the mean by more than 2.00 standard deviations (Z = 2.00). Similarly row twenty (labelled "Z = 1.9"), column seven (headed "0.06") gives the probability (0.0500) that a member of the population will differ in value from the mean by more than 1.96 standard deviations (Z = 1.96).

The probabilities given in the table can of course be converted into percentages by multiplying by 100. For example, the percentage of members in a normal population that differ in value from the mean by over 1.96 standard deviations is 5%. Also the area included in the two tails of the normal distribution lying 1.96 standard deviations from the mean is 5% of the total area.

By subtracting the probabilities given in the table from 1.0000, the probabilities relating to the central portion of the normal distribution can be found. Thus, the probability that a member of the population will deviate from the mean by less than a Z value of 1.96 (i.e. by less than 1.96 standard deviations) is 1.0000 - 0.0500 = 0.9500. The probability that it will deviate by less than a Z value of 1.0 is 1 - 0.6827 = 0.3173.

Table C lists the probabilities associated with half the central portion of the normal distribution, i.e. the interval between the mean and specified values of Z. The probabilities are therefore equivalent to those of Table B subtracted from 1.000 and then halved.

FITTING A NORMAL CURVE TO A HISTOGRAM

The curve needs to have the same mean and standard deviation as the data in order to fit as closely as possible. The area under the curve must be the same as the area enclosed by the histogram representing the data. The equation of the normal curve employed up to now limits the area under the curve to Nr square units, where r is the fraction or multiple of the X scale represented by the Y scale. For curve fitting it is necessary to modify the equation so that it takes into account the size of the class intervals used to construct the histogram. Assuming the class intervals are equal

$$Y = \frac{Ni}{\sigma\sqrt{(2\pi)}} e^{-(X - \mu)^2/2\mu^2}$$

where N is the total number of observations or measurements (the total number of X values) and i is the class interval - in other words the width of the classes that make up the histogram. The area under the curve is Nri and this is also the area of the histogram since each of the classes making up the histogram is represented by a bar or column whose height is proportional to the number of observations and whose width is the class interval. The total area of the histogram in square units of X = area of bar A + area of bar B + area of bar C + ... = (height of bar A in units of X) i + (height of bar B in units of X) i + (height of bar C in units of X) i + ... = (total height of all the bars in units of X) i = Nri.

When fitting a normal curve to a histogram use the following procedure

(1) Calculate the mean μ (or $\bar{X}$) and standard deviation σ (or s) of the data. If fitting a normal curve to a population, remember to use N as the divisor in the formula for standard deviation. If fitting a normal curve to a sample, use (n - 1) in order that the curve may represent the normal population most likely to have been the source of the sample.

(2) Calculate the area of the histogram, Ni (or ni).

(3) Determine the highest point on the normal curve. The curve is highest at the mean, μ. Substituting μ for X in the equation of the normal curve, the maximum height of the curve is given by

$$Y_{max} = \frac{Ni}{\sigma\sqrt{(2\pi)}} e^{-(\mu-\mu)^2/2\sigma^2} = \frac{Ni}{\sigma\sqrt{(2\pi)}} e^0 = \frac{Ni}{\sigma\sqrt{(2\pi)}} \simeq \frac{0.39894228\ Ni}{\sigma}$$

(4) Determine additional points on the curve at selected distances either side of the mean. Refer to Table D which gives the heights (ordinates) of the normal curve relative to the maximum height (designated 1.0000) at selected distances from the mean. The distances (Z) are expressed in terms of the standard deviation. Thus Z = 0.5 indicates a distance from the mean equal to half the standard deviation, while Z = 2.0 denotes a distance equal to two standard deviations. By multiplying the relative heights given in the table by the maximum height, the heights of the curve at selected distances from the mean can be readily determined.

Use of Table D obviates the need to evaluate the equation for the normal curve.

(5) Superimpose the points on the histogram and draw in the normal curve by interpolation. Enough points should be plotted to enable the curve to be drawn to whatever accuracy is required.

As an illustration, consider the sample of the nearest-neighbour distances between towns and villages in Iowa. The calculations involved in fitting a normal curve to the histogram shown in Fig. 5.1 are as follows

1. Mean distance, $\bar{X}$ = 7.2947 km (see Chapter 6). Standard deviation, $s = \sqrt{4.76754} = 2.18347$ km (see Chapter 7). These are the best estimates of the mean and standard deviation of the sampled population.
2. n = 75, i = 1 km.
3. $Y_{max} = \frac{ni}{s\sqrt{(2\pi)}} = \frac{(0.39894)(75)}{2.18347} = 13.70327$
4. See Table 9.1.
5. The fitted normal curve is shown in Fig. 5.6.

TABLE 9.1 - CALCULATION OF THE ORDINATES OF THE NORMAL CURVE

Selected distances from the mean, $\bar{X}$ = 7.2947	Z	*Height of curve at Z as a proportion of its maximum height. Values from Table D*	*Height of curve to scale, i.e. maximum height of curve multiplied by proportional height*
$\bar{X}$ + 3.0s = 13.8451	3.0	0.0111	0.15
$\bar{X}$ + 2.5s = 12.7534	2.5	0.0439	0.60
$\bar{X}$ + 2.0s = 11.6616	2.0	0.1353	1.85
$\bar{X}$ + 1.5s = 10.5699	1.5	0.3247	4.45
$\bar{X}$ + 1.0s = 9.4782	1.0	0.6065	8.31
$\bar{X}$ + 0.5s = 8.3864	0.5	0.8825	12.09
$\bar{X}$ = 7.2947	0.0	1.0000	13.70
$\bar{X}$ - 0.5s = 6.2030	0.5	0.8825	12.09
$\bar{X}$ - 1.0s = 5.1112	1.0	0.6065	8.31
$\bar{X}$ - 1.5s = 4.0195	1.5	0.3247	4.45
$\bar{X}$ - 2.0s = 2.9278	2.0	0.1353	1.85
$\bar{X}$ - 2.5s = 1.8360	2.5	0.0439	0.60
$\bar{X}$ - 3.0s = 0.7443	3.0	0.0111	0.15

CALCULATING EXPECTED FREQUENCIES

The frequencies corresponding to a normal distribution can be found by first drawing the distribution and then estimating what proportion of the total area is included in each class. The proportions, when multiplied by N or n, the total number of observations, yield the required frequencies. The trouble with this procedure is that it is tedious and inexact. Since tables are available setting out the areas under different portions of the normal curve, there is in fact no need to resort to drawing the distribution. The recommended procedure for obtaining the expected frequencies is as follows

1. Calculate the mean and standard deviation of the data.

2. Calculate the Z values for the class limits. The Z values are the distances of the class limits from the mean converted into units of standard deviation, that is

$$Z = \frac{|X - \mu|}{\sigma}$$

where X is a class limit, μ is the mean of the data and σ is the standard deviation. The two verticals || indicate that the difference should always be taken as positive. Negative signs, in other words, should be ignored.

3. Refer to Table C to find what proportion of the total area under the normal curve lies between the mean and each class limit, i.e. between Z = 0 and the Z value for the class limit.

4. Determine what proportion of the total area under the normal curve lies within each class i.e. between the values of Z assigned to the upper and lower limits in step 3 above. For classes other than the one containing the mean the proportions determined in step 3 must be subtracted as shown in Fig. 9.11A. For the class which contains

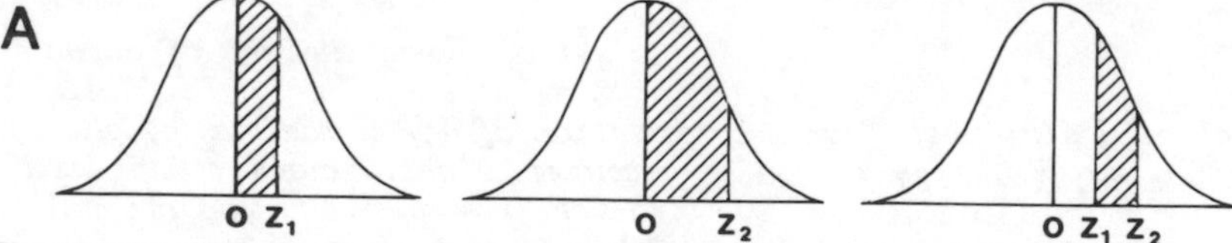

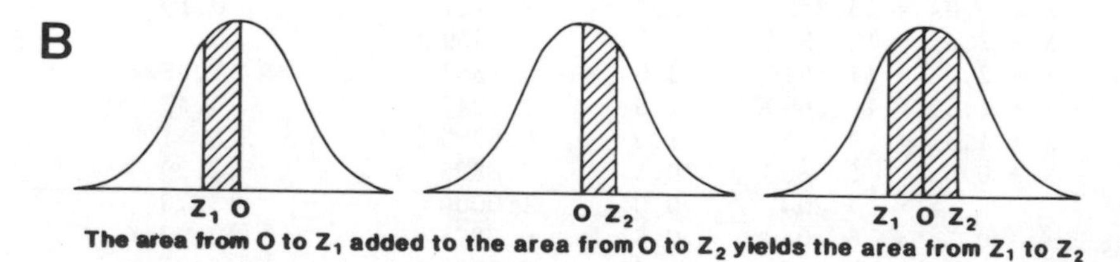

Fig. 9.11 - Areas under the normal curve. For explanation see text.

the mean, the proportion must be added and not subtracted (Fig. 9.11B). Multiply the proportion of the total area lying within each class by N, the total number of observations. The products are the frequencies predicted by the normal distribution. They should be compared with the actual or observed frequencies.

A worked example may serve to make the procedure clear. Table 9.2 demonstrates the various steps in the calculation of the expected frequencies for the nearest-neighbour distances. The expected frequencies represent a normal distribution whose size (n = 75), mean ($\bar{X}$ = 7.2947 km), and standard deviation (s = 2.1835 km) are the same as the actual distribution. The observed frequencies are shown on the right side of the table. Note that the normal distribution fits the data very closely, the expected frequencies agreeing closely with the observed frequencies. What differences exist can be reasonably attributed to chance. A sample, by definition, represents only part of the population from which it is drawn, and therefore it cannot be expected to possess precisely the same characteristics as the population.

TABLE 9.2 - CALCULATION OF EXPECTED FREQUENCIES

Class limit X	*Deviation from mean* $\lvert X-\bar{X}\rvert$	*Z for class limit* $\frac{\lvert X-\bar{X}\rvert}{s}$	*Area under normal curve between mean and X (Table C)*	*Area within each class* A	*Expected frequency* nA	*Observed frequency*
-∞	-∞	-∞	0.5000 \			
				subtract → 0.0018	0.135 ≈ 0	0
0.95	6.3446'	2.906	0.4982 /			
				subtract → 0.0053	0.398 ≈ 0	0
1.95	5.3446'	2.448	0.4929 \			
				subtract → 0.0162	1.215 ≈ 1	1
2.95	4.3446'	1.990	0.4767 /			
				subtract → 0.0395	2.962 ≈ 3	4
3.95	3.3446'	1.532	0.4372 \			
				subtract → 0.0786	5.895 ≈ 6	5
4.95	2.3446'	1.074	0.3586 /			
				subtract → 0.1275	9.562 ≈ 10	10
5.95	1.3466'	0.616	0.2311 \			
				subtract → 0.1683	12.662 ≈ 13	16
6.95	0.3445'	0.158	0.0628 /			
				add → 0.1807	13.552 ≈ 14	12
7.95	0.6533'	0.300	0.1179 \			
				subtract → 0.1579	11.842 ≈ 12	11
8.95	1.6553'	0.758	0.2758 /			
				subtract → 0.1122	8.415 ≈ 8	8
9.95	2.6553'	1.216	0.3880 \			
				subtract → 0.0649	4.868 ≈ 5	3
10.95	3.6553'	1.674	0.4529 /			
				subtract → 0.0306	2.295 ≈ 2	3
11.95	4.6553'	2.132	0.4835 \			
				subtract → 0.0117	0.878 ≈ 1	2
12.95	5.6553'	2.590	0.4952 /			
				subtract → 0.0037	0.278 ≈ 0	0
13.95	6.6553'	3.048	0.4989 \			
				subtract → 0.0011	0.082 ≈ 0	0
+∞	+∞	+∞	0.5000 /			
				Total ≈ 1.0000	74.999 ≈ 75	75

ARITHMETIC PROBABILITY PAPER

The frequencies of a normal distribution can be cumulated and plotted on ordinary graph paper to form an s-shaped curve or ogive (Fig. 9.12). Alternatively, the cumulative frequencies can be plotted on special graph paper known as *arithmetic* (or *normal*) *probability paper* (Fig. 9.13). The vertical scale of arithmetic probability paper is increasingly stretched out towards the top and bottom so that the cumulative frequency distribution appears not as an s-shaped curve, but as a straight line. The scale is so designed that the slope of the line is inversely proportional to the standard deviation. The greater the slope the smaller is the standard deviation.

The value of the median can be found by locating the 50% cumulative frequency on the vertical scale and extending a line horizontally to meet the sloping straight line representing the cumulative normal distribution. The value of the variable on the horizontal scale directly below the point of intersection of the two lines is the value of the median. It is also of course the value of the mean and the mode since the normal distribution is symmetrical. Remem-

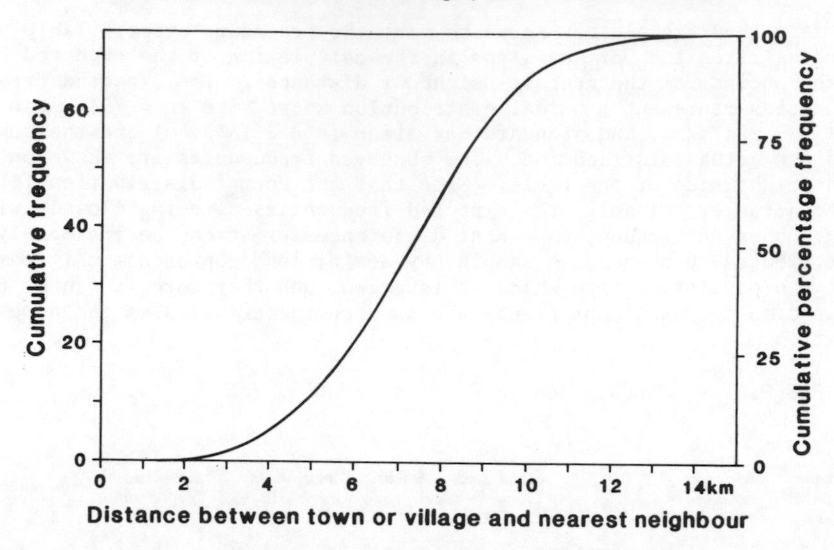

Fig. 9.12 - Cumulative distribution corresponding to the expected frequencies calculated in Table 9.2.

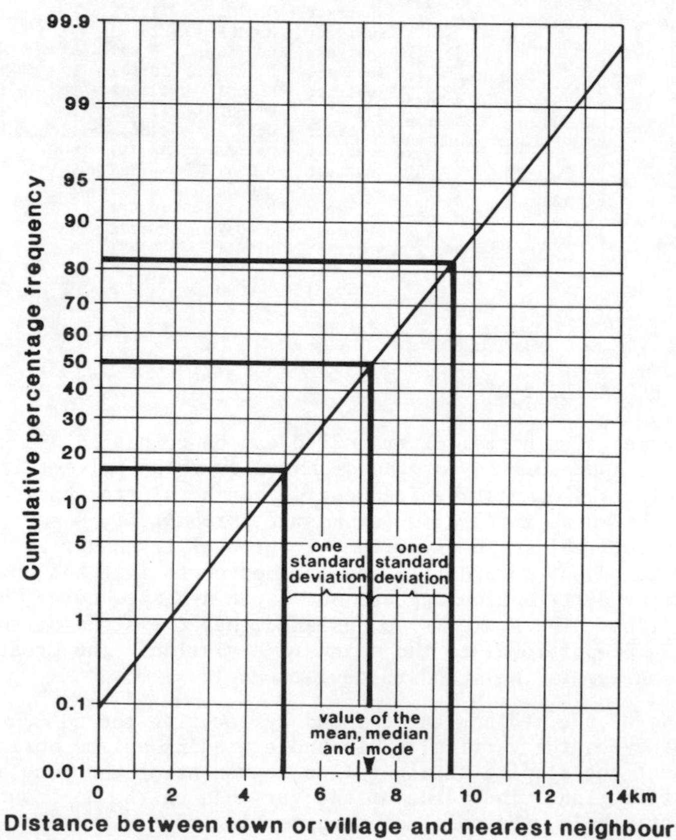

Fig. 9.13 - Cumulative distribution plotted on arithmetic probability paper.

bering that in a normal distribution 68.27% of the items (or members of the population) lie within one standard deviation of the mean, the value of the standard deviation can be found by subtracting the value of X corresponding to a cumulative frequency of 50 - 34.13 = 15.87% from the value corresponding to a cumulative frequency of 50 + 34.13 = 84.13%. In Fig. 9.13 the mean is approximately 7.3 km and the standard deviation is 9.7 - 4.9 = 4.8 km.

Arithmetic probability paper is useful in that it allows for a simple visual test of whether a given sample is normal or approximately normal. Fig. 9.14 shows the nearest-neighbour distances plotted as cumulative percentage frequencies on arithmetic probability paper. The fact that the points lie close to a straight line indicates that the nearest-neighbour distances are satisfactorily approximated by a normal distribution. The linearity of the points also allows the values of the mean and the standard deviation to be estimated quickly.

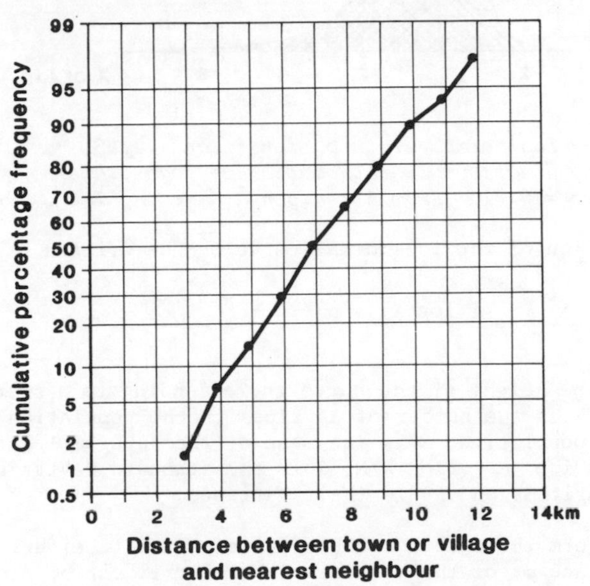

Fig. 9.14 - Arithmetic probability plot of the cumulative frequency distribution of nearest-neighbour distances. Compare with Fig. 5.9.

*THE LOG-NORMAL DISTRIBUTION

Closely allied to the normal distribution is the log-normal distribution which is also unimodal but differs in being moderately skewed to the right. If each observation or measurement is replaced by its logarithm, the distribution becomes normal (Fig. 9.15). Logarithms to any base can be employed to correct the skew and render the distribution normal, but logarithms to base 10 (*common logarithms*) have the practical advantage that tables are generally available, while logarithms to base e (*natural logarithms*) are mathematically more convenient. Common logarithms can be converted to natural logarithms by multiplying by $\log_e 10 = 2.302585...$ Thus the common logarithm of 50 is 1.69897... and the natural logarithm is therefore 1.69897... × 2.302585... = 3.91202... In the discussion that follows, natural logarithms are used exclusively.

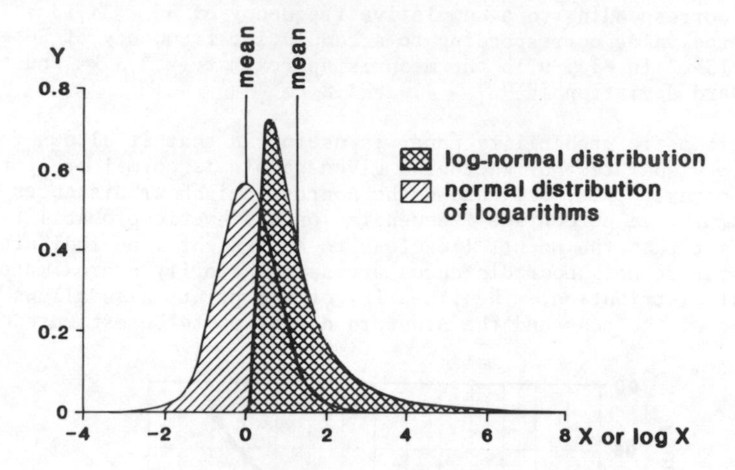

Fig. 9.15 - Log-normal distribution (mean = 1.28, variance = 1.07) with corresponding normal distribution of logarithms to base e (mean = 0, variance = 0.5). From Aitchison and Brown, 1957.

The equation of the log-normal curve can be written as

$$Y = \frac{N}{X\beta\sqrt{(2\pi)}} e^{-[(\log X) - \alpha]^2/2}$$

where Y is the height of the curve corresponding to a specified value of X; N is the number of X values in the population, i.e. the size of the population; α is the mean of the natural logarithms of the X values i.e. $\alpha = \Sigma \log X/N$; β is the standard deviation of the natural logarithms and β^2 is their variance.

In this form the equation makes no reference to either the mean μ or the variance σ^2 of the X values. However it can be proved that

$$\alpha = \frac{1}{2} \log \left(\frac{\mu^4}{\mu^2 + \sigma^2}\right)$$

and

$$\beta^2 = \log \left(\frac{\sigma^2 + \mu^2}{\mu^2}\right)$$

If desired, the right-hand expressions can be substituted for α and β^2 in the equation of the log-normal curve given above.

Log X is free to assume any value from $-\infty$ to $+\infty$. This means that X can assume only positive values because the logarithm of 0 is $-\infty$. From X = 0, the log-normal curve ascends to a single mode at $X = e^{\alpha-\beta^2}$, then descends to meet the X axis asymptotically at $X = +\infty$. The mean of the distribution is located to the right of the mode at $X = \mu = e^{\alpha+\frac{1}{2}\beta^2}$. The median is located between the mean and the mode at $X = e^{\alpha}$. The variance of the X values is given by

$$\sigma^2 = e^{2\alpha+\beta^2} \; (e^{\beta^2} - 1) = \mu^2 \; (e^{\beta^2} - 1)$$

The log-normal distribution varies somewhat in appearance according to the values assumed by α and β^2. Figure 9.16 provides some idea of the extent of variation. Note that β^2 not only measures the spread of values around the mean α, but also measures the degree of skewness. The greater the value of β^2, the greater is the skewness.

The equation of the log-normal distribution can be derived by slightly altering the assumptions used to derive the equation of the normal distribution. Each value of X is assumed to result from the operation of innumerable chance factors, but the factors are seen as multiplicative in their effects, rather than additive. The mathematics are too complex to be discussed here, but details are included in Aitchison and Brown's monograph (1957) on the log-normal distribution.

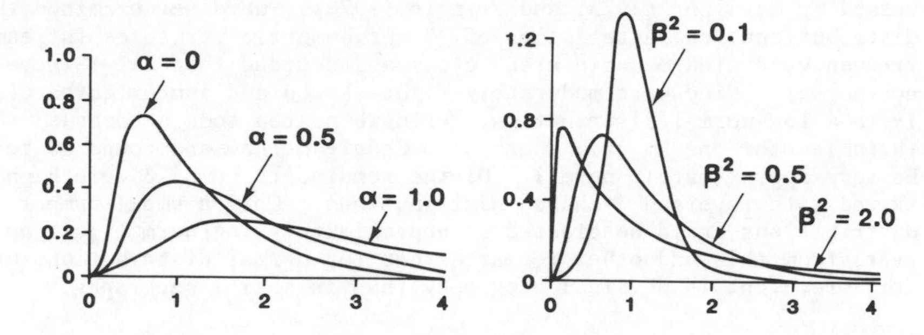

Fig. 9.16 - The log-normal distribution for selected values of α and β^2, the mean and variance respectively of the logarithms to base e. From Aitchison and Brown, 1957.

*PRACTICAL APPLICATIONS OF THE LOG-NORMAL DISTRIBUTION

No real variable possesses a distribution that is perfectly log-normal, although a great many variables possess right-skewed distributions that are approximately log-normal. Variables of interest to geographers include

1. The length of streams of a given order in a drainage basin and their mean channel gradients (Schumm, 1956; Chorley, 1957; Maxwell, 1967).

2. Drainage density i.e. the length of streams in a drainage basin divided by the area of the basin (Schumm, 1956; Chorley, 1957; Maxwell, 1967).

3. Slope angles in areas where there are many low-angle slopes and a smaller proportion of steep ones (Speight, 1971).

4. Sizes of sedimentary particles, e.g. the diameters of sand grains (Hatch and Choate, 1929; Blench, 1952).

5. Maximum yearly rainfall (McIllwraith, 1953).

6. Flood discharges, and many other hydrological variables (Chow, 1954; Flood Studies Report, 1975).

7. Distances travelled to work within a given area.

8. Populations of settlements in a given region (Berry, 1961; Parr and Suzuki, 1973).[1]

9. Numbers of employed persons in industries and firms in a given region (Aitchison and Brown, 1957).

10. Numbers of persons resident in households in a given region (Aitchison and Brown, 1957) and household incomes (Mogridge, 1969).

Distributions that are moderately right-skewed are most nearly log-normal. Distributions that are strongly right-skewed retain some of their skewness when the measurements are converted to logarithms. Distributions that are slightly right-skewed are rendered left-skewed by taking logarithms.

The normality and log-normality of geographic data have been discussed by Gardiner (1973) and Pringle (1976). Gardiner examined the distributional characteristics of 15 morphometric variables (stream frequency, drainage basin area, etc.) and decided that all 15 were non-normal. Nine were moderately right-skewed and approximated closely to a log-normal distribution. Pringle tested some 64 census variables for one km grid squares in County Durham and found 32 to be very approximately normal. Of the remainder, some 23 were highly skewed with reversed j-shaped distributions. Only a small number of distributions could be classed as approximately log-normal. It appears from this and other research that log-normal distributions are more prevalent in physical geography than in social geography.

ESTIMATING μ AND σ^2 IN A LOG-NORMAL POPULATION

It rarely happens that an entire population is available for study. Usually only a sample is available, and the problem then arises of estimating the population mean μ and variance σ^2 from the sample data.

---

[1] Although geographers have repeatedly claimed that populations of settlements are distributed approximately log-normally, no convincing evidence has ever been produced. Population sizes of towns and cities almost always plot as a reversed j-shaped distribution (see Davis, 1969, for world data). Were smaller settlements to be included (villages, hamlets, etc.) the distribution might conceivably become log-normal, but it is difficult to obtain satisfactory data on the population sizes of small settlements in order to test the hypothesis of log-normality. Census reports for most countries do not distinguish between the population sizes of small settlements and the population sizes of the surrounding areas. In Britain, for example, there are no population figures for individual villages, but only for the parishes in which the villages are located. While it is possible that population sizes of small settlements form the left-hand side of a log-normal distribution, the right-hand side of which has the reversed j-shape seen in the distribution of towns and cities, it is equally possible that the small settlements merely extend the reversed j-shape further upwards to the left. Just looking at a map, one gets the strong impression that hamlets are more common than small villages, and small villages more common that large villages, which is consistent with the notion that the total distribution is reversed j-shaped and not log-normal.

If the population is distributed perfectly normally, μ is best estimated by the sample mean $\bar{X}$. For instance, if the sample mean is 35, the population mean is most likely to be 35. The sample mean provides an *unbiased* estimate of μ in the sense that it is equally likely to be too large as too small. Moreover, no other measure can be devised that lies closer on average to μ. The sample mean is a *fully efficient estimator* or *minimum variance estimator* of μ. In the same way, the sample variance s^2 (with $n - 1$ as the denominator) is an unbiased estimator of σ^2. It is also fully efficient in that the variance of the estimate of σ^2 is as small as possible.

If the population is not normally distributed, but is skewed, $\bar{X}$ fails to provide a fully efficient estimate of μ. Although $\bar{X}$ is unbiased as an estimator of μ, alternative estimators of μ can be devised that have smaller variances. In addition, s^2 is not a fully efficient estimate of σ^2 if the population is skew, although it is an unbiased estimate.

In the case of a population that is perfectly log-normal, Finney (1941) has shown that a fully efficient estimate of μ is given by

$$\Psi e^{\bar{X}_{log}}$$

where $\bar{X}_{log}$ is the mean value of the natural logarithms of the X values forming the sample, $e^{\bar{X}_{log}}$ is the natural antilogarithm of $\bar{X}_{log}$, and Ψ (the Greek upper-case letter *psi*) is defined by the relation

$$\Psi = 1 + \left[\frac{n-1}{n}\right]\left[\frac{1}{2}\, s^2_{log}\right] + \frac{1}{2!}\left[\frac{(n-1)^3}{n^2(n+1)}\right]\left[\left(\frac{1}{2}\, s^2_{log}\right)^2\right]$$

$$+ \frac{1}{3!}\left[\frac{(n-1)^5}{n^3(n+1)(n+3)}\right]\left[\left(\frac{1}{2}\, s^2_{log}\right)^3\right] + \ldots$$

n being the sample size.

Finney (1941) has also shown that the population variance can be estimated with full efficiency by calculating

$$\Phi \bar{X}_{log}$$

where Φ (the Greek capital letter *phi*) is defined by the relation

$$\Phi = s^2 + \frac{(n-1)[4(n-1)^2 - (n-2)^2]}{(2!)(n^2)(n+1)}\,(s^2)^2$$

$$+ \frac{(n-1)^2[8(n-1)^3 - (n-2)^3]}{(3!)(n^3)(n+1)(n+3)}\,(s^2)^3$$

$$+ \frac{(n-1)^3[16(n-1)^4 - (n-2)^4]}{(4!)(n^4)(n+1)(n+3)(n+5)}\,(s^2)^4 + \ldots$$

These estimates of μ and σ^2 are both unbiased and efficient. They are termed *maximum likelihood estimates*.

Tables of ψ and Φ are available (see, for example, Aitchison and Brown, 1957; Koch and Link, 1970) which considerably reduce the labour of calculating μ and σ^2.

The method of estimating μ and σ^2 will be illustrated using some data on the variation of drainage density over the outcrop of the Dartmoor granite (Table 9.3 and Fig. 9.17). Seventy five of the kilometre squares of the National Grid were selected at random, representing approximately 10% of the granite outcrop. The total length of the drainage network in each square was measured using 1:25,000 topographic maps. Contour crenulations were used to delimit the drainage network as well as the blue lines shown on the map. Analysis of the data (see below) suggests a value of 2.87 for μ and a value of 1.53 for σ^2. Note that these estimates differ only slightly from the sample mean ($\bar{X}$ = 2.86) and variance (s^2 = 1.43). The differences are slight because the sample is fairly large and the variance σ^2 is small. Had the sample been smaller or the variance larger, much bigger differences might easily have occurred.

TABLE 9.3 - DRAINAGE DENSITIES OF 75 GRID SQUARES, DARTMOOR, DEVON. MEASUREMENTS IN KILOMETRES PER SQUARE KILOMETRE

0.7, 1.2, 1.3, 1.4, 1.4, 1.4, 1.5, 1.5, 1.5, 1.6, 1.6, 1.8,
1.8, 1.8, 1.9, 2.0, 2.0, 2.1, 2.1, 2.1, 2.2, 2.2, 2.2, 2.2,
2.2, 2.2, 2.3, 2.3, 2.4, 2.4, 2.4, 2.4, 2.5, 2.5, 2.6, 2.6,
2.7, 2.7, 2.7, 2.8, 2.8, 2.8, 2.8, 2.8, 2.9, 3.0, 3.0, 3.0,
3.0, 3.1, 3.1, 3.1, 3.1, 3.1, 3.2, 3.3, 3.3, 3.4, 3.4, 3.6,
3.7, 3.9, 4.2, 4.2, 4.3, 4.3, 4.5, 4.5, 4.7, 4.7, 5.3, 5.3,
5.4, 5.4, 7.4

The sample size, n, = 75. The sum of the natural logarithms of the sample observations, $\Sigma \log X$, = 72.70626. The mean value of the logarithms, $\bar{X}_{\log}$, = $\Sigma \log X/n$ = 0.96942. $(\Sigma \log X)^2$ = 5286.20024. $(\Sigma \log X)^2/n$ = 70.48267. The sum of the squares of the natural logarithms, $\Sigma (\log X)^2$, = 83.22345.
The variance $s^2_{\log} = [\Sigma(\log X)^2 - (\Sigma \log X)^2/n]/(n - 1)$

$$= [83.22354 - 70.48267]/74$$

$$= 0.17217$$

$$\Psi = 1 + \frac{74(0.08609)}{75} + \frac{(74)^3(0.08609)^2}{(2)(75)^2(76)} + \frac{(74)^5(0.08609)^3}{(6)(75)^3(76)(78)} + \frac{(74)^7(0.08609)^4}{(24)(75)^4(76)(78)(80)} + \ldots$$

$$= 1 + 0.084942 + 0.003513 + 0.000094 + 0.000002 + \ldots$$

$$= 1.08855 \text{ approximately}$$

Hence an unbiased and efficient estimate of μ is $(e^{0\cdot 96942})(1.08855)$ = 2.87 approximately. The sample mean, $\bar{X}$, = $\Sigma X/n$ = 214.8/75 = 2.86 approximately.

$$\Phi = 0.17217 + \frac{74(16575)}{(2)(75)^2(76)}(0.17217)^2 + \frac{(74)^2(2852775)}{(6)(75)^3(76)(78)}(0.17217)^3 + \frac{(74)^3(451386975)}{(24)(75)^4(76)(78)(80)}(0.17217)^4$$

$$= 0.22048 \text{ approximately.}$$

Hence an unbiased and efficient estimate of σ^2 is $(e^{1.93884})(0.22048)$ = 1.53 approximately.

The sample variance, $s^2 = [\Sigma X^2 - (\Sigma X)^2/n]/(n - 1)$

$$= [721.34 - 46139.04/75]/74$$
$$= [721.34 - 615.18720]/74$$
$$= 1.43 \text{ approximately}$$

ESTIMATING α AND β^2 IN A LOG-NORMAL POPULATION

The best estimate of α is provided by the mean value of the natural logarithms of the sample observations

$$\bar{X}_{log} = \frac{\Sigma \log X}{n}$$

Similarly, the best estimate of β^2 is the variance of the natural logarithms of the observations

$$s^2_{log} = \left[(\Sigma \log X)^2 - \frac{(\Sigma \log X)^2}{n}\right] / (n - 1)$$

These estimates of α and β^2 are called maximum likelihood estimates. They are unbiased and fully efficient since the distribution of the logarithms is normal.

A less efficient method of estimating α and β^2 is to take the mean $\bar{X}$ and variance s^2 of the actual observations

$$\bar{X} = \frac{\Sigma X}{n} \text{ and } s^2 = \left[\Sigma X^2 - \frac{(\Sigma X)^2}{n}\right] / (n - 1)$$

and to substitute $\bar{X}$ for μ and s^2 for σ^2 in the equations

$$\alpha = \frac{1}{2} \log \left(\frac{\mu^4}{\mu^2 + \sigma^2}\right) \text{ and } \beta^2 = \log \left(\frac{\mu^2 + \sigma^2}{\mu^2}\right)$$

This method of estimating α and β^2 is known as the *method of moments*. In general, it does not yield such accurate results as the method of maximum likelihood and therefore it is not recommended.

In the case of the drainage density data, the maximum likelihood method estimates α to be 0.96942 ($= \bar{X}_{log}$) and β^2 to be 0.17217 ($= s^2_{log}$). The sample mean $\bar{X}$ is 2.86400 and the sample variance s^2 is 1.43450. According to the method of moments

$$\alpha = \frac{1}{2} \log \left[\frac{\bar{X}^4}{\bar{X}^2 + s^2}\right] = \frac{1}{2} \log \left[\frac{67.28094}{9.63699}\right] = \frac{1}{2} \log 6.98153$$
$$= 0.97163$$

$$\beta^2 = \log \left[\frac{\bar{X}^2 + s^2}{\bar{X}^2}\right] = \log \left[\frac{9.63699}{8.20250}\right] = \log 1.17489 = 0.16117$$

These estimates are likely to be less accurate than those obtained by the maximum likelihood method.

FITTING A LOG-NORMAL CURVE TO A HISTOGRAM

A variable that is log-normally distributed can have either (1) a normal curve fitted to a histogram of the logarithms or (2) a log-normal curve fitted to a histogram of the actual values (Fig. 9.17). Opinions differ as to which method of representation is the more effective.

1. Fitting a normal curve to the logarithms involves treating the logarithms as if they were actual values. The method of fitting the curve is the same as that discussed earlier in this chapter. The first step is to calculate the mean $\bar{X}_{log}$ and the standard deviation s_{log}, $\bar{X}_{log}$ being an estimate of α and s_{log} being an estimate of β. The height of the curve at the mean is given by

$$Y_{max} = \frac{Ni}{s_{log}\sqrt{(2\pi)}}$$

where i is the class interval, that is, the width of the classes used to construct the histogram. Put another way, i is simply the arithmetical difference between the log value forming the upper limit of a class and the log value forming the lower limit. Heights of the curve at selected distances either side of the mean are obtainable by multiplying the relevant entries in Table D by the height of the curve at the mean.

2. The method of fitting a log-normal curve to a histogram of the actual values is slightly more complicated. The equation of the curve can be written

$$Y = \frac{Ni}{Xs_{log}\sqrt{(2\pi)}} e^{-[(\log X) - \bar{X}_{log}]^2/2s^2_{log}}$$

where Y is the height of the curve corresponding to a given value of X, and i is the class interval (the arithmetical value of the interval, not the log). The presence of the term Ni on the right-hand side of the equation ensures that the area under the curve is equal to Ni, the area under the histogram.

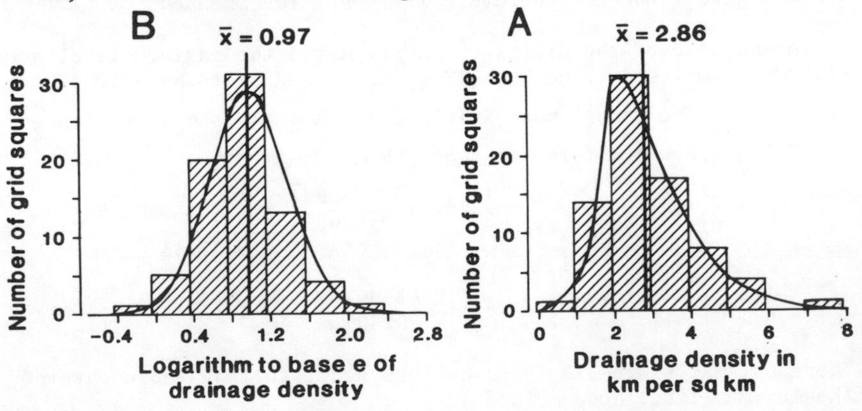

Fig. 9.17 - Drainage densities of 75 one km grid squares, Dartmoor, Devon

A: Distribution of logarithms with fitted normal curve.

B: Distribution of actual values with fitted log-normal curve.

Sufficient values of X should be selected, and the corresponding values of Y calculated, to permit the log-normal curve to be drawn smoothly and accurately.

TABLE 9.4 - HEIGHTS (Y) OF THE LOG-NORMAL CURVE FITTED TO THE DRAINAGE DENSITY DATA. LOGARITHMS TO BASE e.

X	$(\log X)$	$(\log X)-\bar{X}_{log}$	$\frac{[(\log X)-\bar{X}_{log}]^2}{2s^2_{log}}$	$e^{-\frac{[(\log X)-\bar{X}_{log}]^2}{2s^2_{log}}}$	Y
0	$-\infty$	$-\infty$	∞	0.0000	0.000
1	0.00000	-0.96942	2.72914	0.0653	4.707
2	0.69315	-0.27627	0.22166	0.8012	28.886
3	1.09861	+0.12919	0.04847	0.9527	22.899
4	1.38629	+0.41687	0.50468	0.6037	10.883
5	1.60944	+0.64002	1.18956	0.3044	4.389
6	1.79176	+0.82234	1.96383	0.1403	1.686
7	1.94591	+0.97649	2.76910	0.0627	0.646

Table 9.4 illustrates the calculations for the drainage density data. The equation of the curve is

$$Y = \frac{(75)(1)}{X(0.41494)(2.50663)} e^{-[(\log X) - 0.96942]^2/(2)(0.17217)}$$

$$= \frac{72.10869}{X} e^{-[(\log X) - 0.96942]^2/0.34435}$$

The mode falls at $X = e^{\alpha - \beta^2} = e^{0 \cdot 79725} = 2.219$. The corresponding value of Y is

$$\frac{Ni}{(e^{\alpha - \beta^2})\beta\sqrt{(2\pi)}} e^{-\beta^2/2} = \frac{72.10869}{(e^{0 \cdot 79725})} e^{-0 \cdot 08609} = 29.809$$

Instead of evaluating

$$e^{-[(\log X) - \bar{X}_{\log}]^2/2s^2_{\log}}$$

directly as in the above table, it is possible to make use of the entries in Table D which are values of $e^{-\frac{1}{2}Z^2}$ or $e^{-(Z - \mu)^2/2}$ where $Z = |\log X - \bar{X}_{\log}|/s_{\log}$. For example, if $X = 2$, $Z = |\log_e X - \bar{X}_{\log}|\ s_{\log} = 0.27627/\sqrt{0.17217} = 0.6658$. Entering Table D with $Z = 0.6658$, the height of the curve is seen to be approximately 0.8012, the value calculated in Table 9.4.

The log-normal curve defined by the X and Y values shown in Table 9.4 is fitted to the histogram in Fig. 9.17A. For comparison a normal curve is also shown fitted to the logarithms of the measurements (Fig. 9.17B). It will be observed that both curves provide useful representations of their respective histograms.

CALCULATING EXPECTED FREQUENCIES FOR A LOG-NORMAL DISTRIBUTION

The method is largely the same as that for the normal distribution discussed previously. The mean $\bar{X}_{log}$ and standard deviation s_{log} of the logarithms are first calculated, then the Z values for the class limits

$$Z = \frac{|\log X - \bar{X}_{log}|}{s_{log}}$$

where X is the class limit. The remaining steps in the calculation are the same as those for the normal distribution. Table 9.5 sets out the calculation for the drainage density data.

The expected frequencies are seen to be closely in agreement with the observed frequencies. This fact suggests that the spatial variation of drainage density in the Dartmoor region may be considered to be approximately log-normal. Using the chi-square test discussed in Chapter 15 it can be shown that the differences between the observed and expected frequencies are no greater than the differences that could be ordinarily expected to occur with chance sampling of a log-normal distribution.

TABLE 9.5 - CALCULATION OF EXPECTED FREQUENCIES FOR A LOG-NORMAL CURVE FITTED TO THE DRAINAGE DENSITY DATA

Class Boundary X	*log X*	$\frac{\lvert \log X - X_{log}\rvert}{s_{log}}$	*Area under normal curve between mean and class limit (Table C)*		*Area within each class A*	*Expected frequency nA*	*Observed frequency*
0	- ∞	∞	0.5000 \				
				Subtract →	0.0069	0.518 ≈ 1	1
0.95	-0.05129	2.45991	0.4931 /				
				Subtract →	0.2268	17.010 ≈ 17	14
1.95	0.66783	0.72683	0.2663 \				
				Add →	0.3732	27.990 ≈ 28	30
2.95	1.08181	0.27085	0.1069 /				
				Subtract →	0.2282	17.115 ≈ 17	17
3.95	1.37372	0.97435	0.3351 \				
				Subtract →	0.1004	7.53 ≈ 8	8
4.95	1.59939	1.51822	0.4355 /				
				Subtract →	0.0396	2.97 ≈ 3	4
5.95	1.78339	1.96167	0.4751 \				
				Subtract →	0.0152	1.14 ≈ 1	0
6.95	1.93874	2.33606	0.4903 /				
				Subtract →	0.0058	0.435 ≈ 0	1
7.95	2.07317	2.66004	0.4961 \				
				Subtract →	0.0039	0.292 ≈ 0	0
+ ∞	∞	∞	0.5000 /				

*MODIFIED LOG-NORMAL DISTRIBUTIONS

A variable that is distributed log-normally is effectively restricted to positive values. It cannot take on a value of 0 or less because log 0 is - ∞. This restriction has important practical consequences. Variables are often encountered in geographical studies that are moderately right skewed, but which include some zero or negative values. On the face of it, the log-normal distribution might seem inapplicable. However, the problem of the zero or negative values can be overcome by adding a positive constant to all the values in the observed distribution so as to displace the zero point to the left edge of the distribution.

Sometimes distributions are encountered that are truncated on the left side at a certain value, although in every other respect they can be considered log-normal (Fig. 9.18). The point of truncation represents a threshold which the variable in question must attain or exceed in order to be taken into account. A possible example is the distribution of cities by size of population. Cities, as ordinarily defined, have populations above a certain minimum size. While the size distribution of all the villages, towns and cities in a region may be approximately log-normal, the distribution of cities alone is inevitably restricted to that portion of the distribution beyond the minimum size adopted in the definition. Because the distribution of cities is truncated it has a reversed j-shape.

The mathematical procedures for dealing with truncated distributions are somewhat complex and will not be discussed here. Full details are given in Aitchison and Brown's monograph.

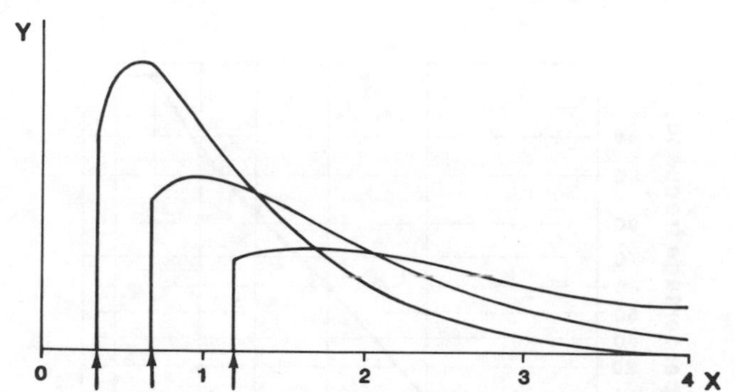

Fig. 9.18 - Examples of truncated log-normal distributions. Arrows indicate point of truncation.

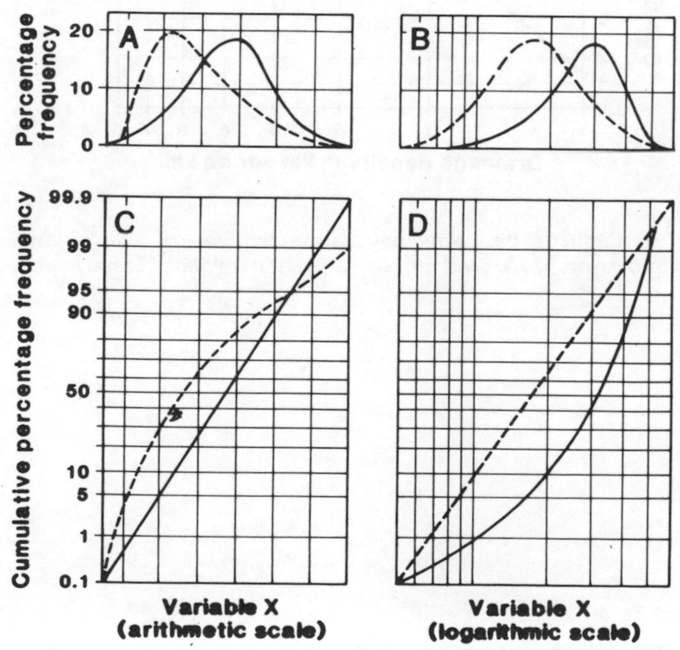

Fig. 9.19 - Normal (solid line) and log-normal (pecked line) distributions referred to (A) arithmetic and (B) logarithmic scales. The same distributions are also plotted in cumulative form on (C) arithmetic and (D) logarithmic probability paper. From Chorley (1958).

LOGARITHMIC PROBABILITY PAPER

Log-normal distributions if plotted as cumulative percentage frequencies on arithmetic probability paper appear as curves (Fig. 9.18). However, the distributions plot as straight lines on *logarithmic probability paper*, which uses a logarithmic scale on the horizontal axis in place of an arithmetic scale. They also plot as straight lines on arithmetic probability paper if the arithmetic values of the variable in question are first converted to logarithmic values. Figure 9.20 is a logarithmic probability plot of the drainage density measurements. It can be seen that the measurements closely follow a normal distribution.

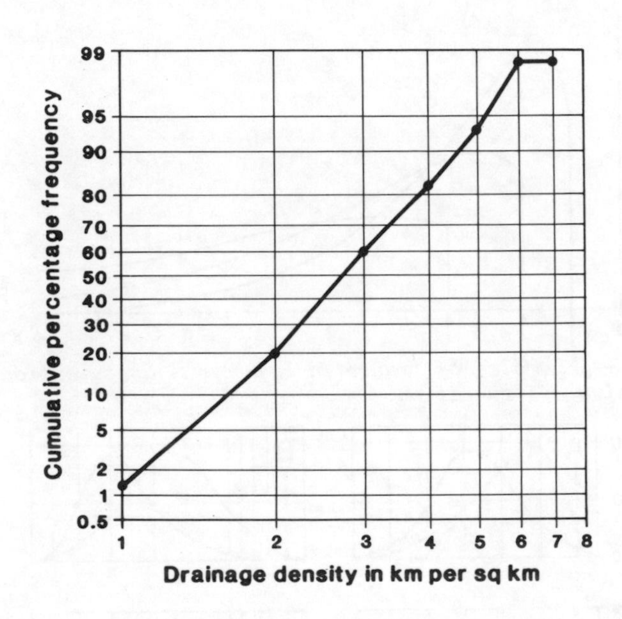

Fig. 9.20 - Cumulative frequency distribution of the 75 drainage density measurements plotted on logarithmic probability paper.

10 Hypothesis Testing: An Introduction

Sampling is undertaken in order to secure information about a population that for reasons of time or otherwise cannot be studied in full. From the sample inferences are made concerning the nature of the population. Thus, if the sample mean is 35 units, the inference might be drawn that the population mean is equal to 35 units. If the variance of the sample is 5 units, the variance of the population might be inferred to be also 5 units.

Unfortunately inferences drawn from samples are subject to uncertainty since no sample, however carefully devised, can be relied upon to represent exactly the population from which it is drawn. By definition, a sample includes only certain members of the population and there is a danger that they will not be representative. This danger exists even if the sample is drawn at random. For example, only 20 out of the 2000 settlements in an area might be cities, yet a random sample of 20 settlements might by chance include them all. The probability of such an extreme circumstance arising is very low, but nevertheless with any sample some degree of misrepresentation is almost inevitable. How much error has to be expected depends on the size of the sample. The smaller the sample, the more likely it is to differ from the population from which it is drawn. The larger the sample, the less important are the effects of chance. For instance, a sample of 100 items will tend to give a more reliable picture of the population from which it is drawn than a sample of 25 items. A sample of only 5 items will tend to show considerable differences from the population.

The failure of samples to reproduce all the characteristics of their parent populations is known as *sampling error*. The existence of sampling error makes it difficult to know what significance to attach to samples. Suppose, for instance, that a sample survey gives results differing from those of a full survey, or census, carried out some years previously. Is the difference due to changed conditions, or is it an accident of sampling, Again, if the mean or variance of a sample is different from that suggested by theory, can the difference be dismissed as sampling error, or does it indicate that the theory is incorrect? If two samples yield different results, is it because of chance sampling or because the samples represent two different populations? Problems like these are the concern of *inferential statistics*. They are handled by using what are called *tests of significance*. Statisticians have devised a great number of tests, each designed to deal with a different type of problem. All the tests, however, have certain elements in common. Here is a very simple example known as the *Z test*. The facts are imaginary.

A complete survey or census of all the arable farms in a region reveals that their sizes form a normal distribution (Fig. 10.1). The mean size of the farms is 500 hectares and their standard deviation is 50 hectares.

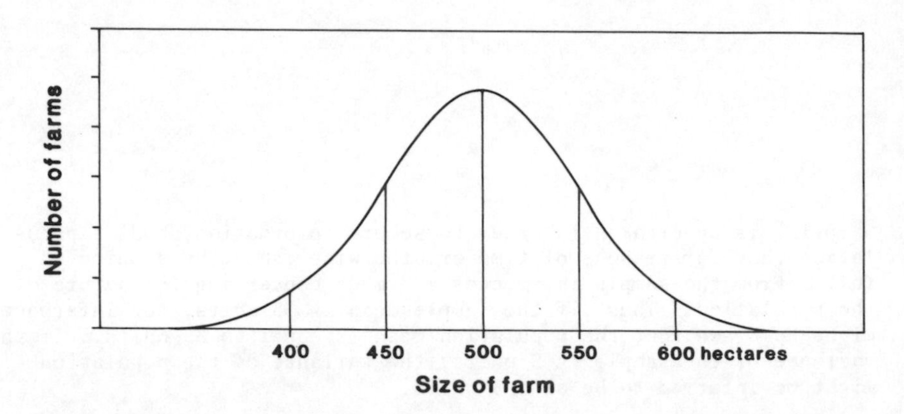

Fig. 10.1 - Hypothetical frequency distribution of the sizes of arable farms in a region. The number of farms is presumed to be so large that the stepped outline of the histogram of farm sizes can be approximated by a smooth curve. Although a normal curve is shown for the sake of simplicity, a log-normal curve would probably have been more realistic.

The survey does not include livestock farms although these are also present in the region. Suppose it is of interest to establish whether the livestock farms differ in size from the arable farms. Suppose further that the size of one livestock farm selected at random is found to be 375 hectares. Since the vast majority of the arable farms are larger than 375 hectares, is it reasonable to assume that the livestock and arable farms differ in size?

As a starting point consider the hypothesis that the livestock farms are in fact no different in size from the arable farms. A complete survey of all the livestock farms would show that their mean size is 500 hectares and their standard deviation is 50 hectares, the same as the arable farms. The size distribution, moreover, is normal like the arable farms. Figure 10.1 on this hypothesis represents livestock farms as well as arable farms.

Note that the hypothesis is deliberately one of caution. It assumes that there is nothing special about the sizes of livestock farms. The sample farm happens to be especially small, but this is solely the result of chance (sampling error). Although the farm is 125 hectares smaller than the mean, it might with equal probability have been 125 hectares larger.

The hypothesis under consideration has the virtue that it is easily testable. If a farm is selected at random, the probability P that it will deviate from the mean by 125 hectares or more is only 0.0124. This probability can be obtained by calculating Z as explained in the last chapter

$$Z = \frac{|X - \mu|}{\sigma} = \frac{125}{50} = 2.5$$

The probability corresponding to this value of Z is 0.0124 according to Table B. Note carefully that P is the probability relating to both the left and the right tails of the normal distribution of farm sizes (Fig. 10.2), so it means the probability of selecting either (a) a farm of 375 hectares or less or (b) a farm of 625 hectares or more. Because the normal distribution is symmetrical the probability of (a) is the same as that of (b), i.e. half 0.0124, or 0.0062.

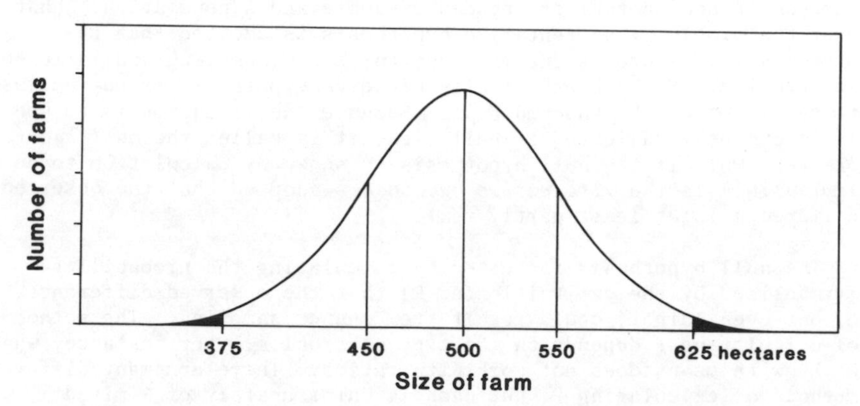

Fig. 10.2 - Two tails of the frequency distribution of the sizes of livestock farms. The two tails are shown in black.

The probability P is calculated on the assumption that the livestock and arable farms are the same size. The calculation shows that only 1.24% of the livestock farms differ from the mean by 125 hectares or an even greater amount. If a series of livestock farms were to be selected at random, only 1.24% (on average) would be as unrepresentative as the sample farm, or even less representative. Because the probability of obtaining the sample farm, or one even more divergent from the mean, is so low, serious doubt is cast on the hypothesis that livestock farms are not different in size from arable farms. The hypothesis is not disproven, but it is plainly so unreasonable as to be best discarded. In its place the alternative hypothesis can be adopted that livestock farms are different in size from arable ones.[1]

The alternative hypothesis does not state what form the difference takes, because there is not enough information. Possibly the livestock farms are smaller on average than the arable farms. Possibly the mean size is the same, but the livestock farms are more variable than are the arable farms. Possibly the livestock farms are not normally distributed like the arable farms. Possibly all these

---

[1] The foregoing reasoning can perhaps be clarified by focusing on sampling error. According to the hypothesis under investigation the small size of the sample farm is due to sampling error. However, the probability of obtaining the amount of sampling error in question (125 hectares), or a greater amount, is only 0.0124. Hence the hypothesis is scarcely credible and can be discounted.

differences exist. With a sample consisting of only one farm, the alternative hypothesis cannot be made precise.

Of course, samples are usually made up of a number of items, not just one. The reason is that larger samples tend to be more reliable than small ones. But although the problem just discussed is somewhat unrealistic, it nevertheless serves to demonstrate the basic elements of hypothesis testing. Every problem in inferential statistics is solved using reasoning similar to that employed in this example. There is always a difference to consider: either a difference between one sample and another or between a sample and a population. Whatever the problem, the tentative hypothesis is adopted that the observed difference is entirely due to chance sampling and is in no way genuine. Such an essentially negative hypothesis is put up just to see if it can be knocked down. Because the intention is to see if it can be invalidated or nullified, it is called the *null hypothesis*. Only if the null hypothesis is shown by calculation to be implausible is the *alternative hypothesis* adopted that the observed difference is at least partly real.

The null hypothesis is tested by calculating the probability (symbolised by the capital letter P) that the observed difference, or one even larger, could result from chance sampling. The method of calculating P depends on the type of problem. For instance, what works with means does not work with ratios. There are many different methods of calculating P, but each is unfortunately of limited applicability. A correct method must be chosen for the particular problem. An additional consideration is time. All too often the amount of arithmetic needed to calculate P exactly turns out to be prohibitive, and one must make do with an approximate method which, though not entirely accurate, has the advantage of being relatively easy to use.

If the value of P is very small, the observed difference is unlikely to be entirely a matter of chance. The null hypothesis cannot be regarded as satisfactory, and has to be abandoned in favour of the alternative hypothesis that the difference is at least partly real. The difference is described as *statistically significant*. This use of the word 'significant' is perhaps misleading. A difference that is statistically significant is not necessarily of practical or substantive significance. Although by definition it is likely to be real, it may or may not be sufficiently important to bother about. If a sample is big enough, even a tiny difference will show up as statistically significant, but it may have little or no practical consequence. On the other hand, it may be enormously important. Everything depends on the nature of the problem.

Although the null hypothesis is rendered implausible if P is very small, it is not totally disproved (unless, of course, $P = 0$). For instance, suppose P works out at 0.03. This means that there are only 3 chances in 100 of getting the observed difference, or one greater, as a result of chance alone. If the alternative hypothesis is accepted in place of the null, there is therefore a slight risk of being wrong. Assuming the null hypothesis is correct, the risk amounts to 0.03 or 3%. Consequently the alternative hypothesis cannot be regarded as fully proven and must be looked upon as somewhat provisional. Although justified by the existing data, it may conceivably prove to be wrong if additional data become available.

If the value of P is not small, the null hypothesis cannot be rejected without appreciable risk of error, and must therefore be retained. A significant difference is not demonstrated.[1] The fact that the null hypothesis is retained does not mean that it is necessarily true and the alternative hypothesis is wrong. While the data do not provide convincing evidence that the hypothesis is faulty, they do not prove it to be correct. The test of the null hypothesis is therefore inconclusive.

In most problems interest centres on the alternative hypothesis rather than on the null. If the observed results derive entirely from chance they are simply not worth knowing about. The standard procedure of examining the null rather than the alternative hypothesis may seem therefore to be putting the cart before the horse, but in fact it is the only way of establishing the likelihood of an alternative hypothesis, indirect though it may be. In inferential statistics one is dealing with probabilities, not absolute truth. While it is possible to discredit a hypothesis by showing it to be improbable, it is not possible to demonstrate its correctness except indirectly by discrediting whatever rival hypothesis exists. Hence the strategy of attacking the null hypothesis in order to support the alternative hypothesis.

An advantage of this strategy is that it errs on the side of caution. In most research investigations it is less disastrous if a real effect is mistaken for chance than if a chance effect is mistaken for real. By assuming tentatively that the observed difference is due to chance, the test is designed to "fail-safe". If the mistake is made of assuming that something that is chance is real, the assumption may later come to be accepted as a proven fact; explanations may be constructed to justify it, and perhaps no one will think of trying to disprove it. If on the other hand the mistake is made of assuming that an effect that is perfectly real is pure chance, the investigation will have failed to make progress, but the possibility is left open that someone else will try at a later date to establish that the effect is real.

An analogous strategy is employed in courts of law. There, defendants on criminal charges are assumed to be innocent unless proved guilty beyond reasonable doubt. The burden of proof is placed on the prosecution, it being deemed better that a guilty man should go free than that an innocent man should be committed to jail. Neither mistake is desirable, but the first is felt to be preferable to the second.

Courts of law experience difficulty in deciding what constitutes a "reasonable doubt". In inferential statistics similar difficulty is encountered in deciding whether the case against the null hypothesis is established beyond reasonable doubt. At what level of P is rejection of the null hypothesis justified?

---

[1] The usual practice of reporting the difference as "not statistically significant" has little to commend it except brevity. The difference may be "statistically significant", i.e., not due to chance, although the test fails to demonstrate this beyond reasonable doubt.

Although there are, unfortunately, no totally objective rules for deciding whether a given value of P is low enough to permit rejection of the null hypothesis, the custom has grown up over the years of using 0.05 (5%) as an upper limit. If P is less than 5%, the null hypothesis is rejected as untenable and the observed difference is reported as statistically significant or probably significant. If P is greater than 5% judgement is reserved. This figure of 5% is, of course, entirely arbitrary. It appears to have originated with the statistician R.A. Fisher, who used it extensively in his analyses of agricultural experiments. Subsequently it was adopted by other scientists, mainly in the interests of standardising procedure.

It must be stressed that the figure of 5% is appropriate only if the consequences of wrongly rejecting the null hypothesis are not serious. If great expense is involved, or human lives are at risk, a lower figure of 1% (0.01) or even 0.1% (0.001) is adopted as a matter of prudence. The effect is to make it more difficult to reject the null hypothesis of no real difference.

The figures of 5%, 1% and 0.1% are referred to as *levels of significance*. Which level is adopted in a given problem depends on the nature of the problem and the use to which the results will be put. If there is any uncertainty, two levels of significance should be used. Thus a P value of 0.03 could be reported as "significant at the 5% level, but not at the 1% level. In view of this result the null hypothesis is regarded as scarcely tenable and is accordingly rejected". Such a form of words makes clear the subjective nature of the argument, and at the same time allows others to judge for themselves.

Although the practice will not be followed here, some statisticians denote levels of significance by asterisks. A single asterisk indicates a probability between 0.05 and 0.01, a pair indicates probabilities equal to or less than 0.01 but greater than 0.001. Three asterisks indicate a probability equal to or less than 0.001. For example, Z = 3.0** means that P is significant at the 1% level, but not at the 0.1% level; Z = 0.67 without an asterisk means that P does not reach the 5% level.

ONE-TAILED AND TWO-TAILED TESTS

In the case of the livestock farms referred to earlier the value of P was the probability of selecting a farm of 375 hectares or less plus the probability of selecting a farm of 625 hectares or more. P was the sum of these two probabilities because the problem was essentially two-sided. Even though livestock farms might be different from arable farms, before drawing the sample there was nothing to indicate the direction in which the difference lay. If the livestock and arable farms differed in means, for instance, the difference might be that the mean size of livestock farms was smaller than that of arable farms, or that the mean size of livestock farms was larger. Because of this uncertainty the test of significance took both tails of the normal distribution into account, instead of just one tail. A test of this type is called a *two-tailed test*.

A *one-tailed test* would have been appropriate if it had been clear from the outset that any real difference between livestock and arable farms could only be in a certain direction, for instance if the mean size of livestock farms could be smaller but not larger than that of arable farms. P would have been calculated using only one tail of the normal distribution and so its value would have been exactly half that calculated for the two-tailed test, i.e. 0.0062 instead of 0.0124. A one-tailed test can be used only in a lop-sided situation where a real difference, if it exists, can only be in a certain direction. If, after sampling, the difference turns out to be in the opposite direction, there must be complete certainty that it is due entirely to chance.

A one-tailed test would be justified in connection with the following problem (as before the data are hypothetical). A Railway Company keeps a record of the number of passengers travelling each day on a certain branch line. To try to increase the number of passengers the Company decides to reduce the fares for an experimental period. The first day of reduced fares is a Monday, and 625 passengers travel on the line. Since the last change in fares, the mean number of passengers on Mondays has been 500, and the standard deviation has been 50. The number of passengers follows a normal distribution. Can the Company conclude that the reduction in fares has had the desired effect?

In no circumstances could the reduction in fares have decreased the number of passengers. Nobody ever minds being charged less than before. If the reduction has had any effect at all, it has been to increase the number of passengers. But is the increase in passengers over and above the mean due merely to chance? Despite the reduction the fares are possibly still too high compared with the fares charged by competing forms of transport, for instance the local Bus Company, to tempt more travellers to use the railway. Possibly, the fares are already so low that everybody who wishes to travel by the railway is already doing so. Clearly, the reduction in fares does not automatically guarantee an increase in the number of passengers.

Had the numbers of passengers on the first day been less than 500 (the previous mean) the difference could instantly have been dismissed as a chance fluctuation. A test of significance would not be necessary because the difference could not conceivably be the result of the reduction in fares. But because the number of passengers is more than 500, a test of significance is needed to decide if the difference is due entirely to chance or to the reduction in fares.

The problem is thus one-sided. P is calculated using the same arguments as in the problem of the farm sizes except that only the right tail of the normal distribution is considered (625 or more). The value of P is 0.0062 rather than 0.0124 as previously. Because the value of P is so low the increase in the number of passengers cannot reasonably be explained as a chance fluctuation. Before the fares were reduced, 625 passengers or more travelled by train on only about 6 Mondays in every thousand. On the assumption that the first Monday the fares are reduced is not one of these altogether exceptional Mondays, the increase in passengers has to be explained as due to the reduction in fares.

Such clear-cut problems as the above are rather rare. Usually one has a hunch that, barring chance, the observed difference will be in a certain direction, but no absolute certainty. A two-tailed test has then to be used. One-tailed tests require knowledge that the difference, if real, will lie in one direction only. This knowledge must be derived independently of the sample. It is illegitimate to argue, for instance, that a one-tailed test is applicable simply because the sample result lies in a certain direction. Unless it is clear that an opposite result could not be obtained except by chance, the test of significance needs to be two-tailed.

11 One-Sample Tests Based on Z and t

Suppose that a series of random samples, all the same size, are drawn from a population with a mean of μ. Suppose further that each sample is returned to the population before the next sample is drawn. The population is therefore the same for each sample and never becomes exhausted however many samples are drawn.

Each sample will tend to contain different members of the population and so will tend to have a different mean. If there are k samples there will be k numerical values of $\bar{X}$, possibly none of them the same. Let the sample means be $\bar{X}_1, \bar{X}_2, \bar{X}_3, \ldots, \bar{X}_k$. The means can be grouped into classes to form a frequency distribution just as the X values of each individual sample can be grouped to form a frequency distribution of the sample (Fig. 11.1). The mean of the sample means or grand mean $\bar{\bar{X}}$ is given by

$$\bar{\bar{X}} = \frac{\bar{X}_1 + \bar{X}_2 + \bar{X}_3 + \ldots + \bar{X}_k}{k} = \frac{\Sigma \bar{X}}{k}$$

The larger the value of k, the more closely $\bar{\bar{X}}$ will approximate to the population mean μ. If there are an infinite number of samples, $\bar{\bar{X}}$ will be exactly equal to μ.

Just as it is possible to calculate the mean of the sample means so it is possible to calculate their standard deviation. For k samples, the standard deviation of the sample means is given by

$$\frac{\Sigma(\bar{X} - \bar{\bar{X}})^2}{k - 1}$$

This is the usual formula for standard deviation but with a change in notation to make clear that it is not the standard deviation of a single sample that is referred to but the standard deviation of k values of $\bar{X}$.

The standard deviation of the means of an infinite number of samples is referred to as the *standard error of the mean*. A simple relation exists between the standard error and the standard deviation of the population yielding the samples.[1] In symbols

---

[1]Take care not to confuse the various standard deviations: the standard deviation (standard error) of the set of sample means, the standard deviation of the population, and the standard deviations of the individual samples.

$$\sigma_{\bar{X}} = \frac{\sigma}{\sqrt{n}}$$

where $\sigma_{\bar{X}}$ (pronounced "sigma sub X bar") symbolises the standard error, σ is the standard deviation of the population, and n is the size of the samples (not their number, which is infinite). Note that $\sigma_{\bar{X}}$ is one symbol - it does not mean σ multiplied by $\bar{X}$.

If each sample is drawn without individual replacements,[1] the standard error of the mean is given by the equation

$$\sigma_{\bar{X}} = \frac{\sigma}{\sqrt{n}} \left(\sqrt{\frac{N - n}{N - 1}} \right)$$

where N is the size of the population. The quantity in parentheses is referred to as the *finite population correction or FPC*. If N is infinite, it takes the value of 1, and the equation becomes the same as for sampling with replacements.

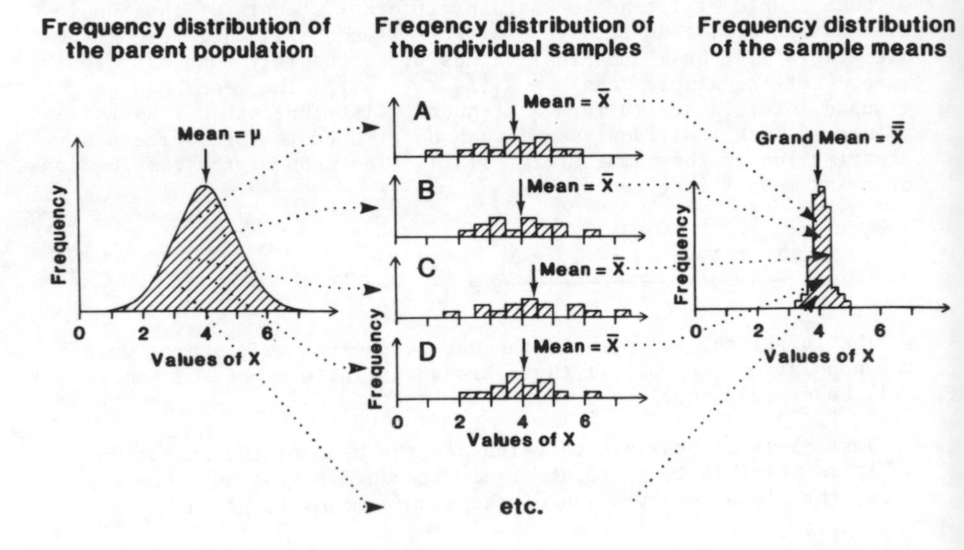

Fig. 11.1 - The frequency distribution of the means of k samples all drawn from the same population.

The two equations hold good whatever the form of the population distribution provided that σ and μ are finite numbers. The only other qualifications to note are that the values of the different members of the population must be independent of each other and the sampling must be random. The mathematical derivation of the equation will not be given here as it would involve too great a digression. Full details are provided by Kendall and Stuart (1977).

The frequency distribution of the means of an infinite number of

---

[1]The distinction between sampling with and without individual replacements applies only to the procedure in drawing a sample. It is assumed here that any sample whether drawn with or without replacements is restored to the population before the next sample is drawn.

samples is called the *sampling distribution of the mean*.[1] If the samples are drawn from a population that is normally distributed, the sampling distribution of the mean will also be normal. Even if the samples are drawn from a population that is not normal, the sampling distribution will display some tendency towards normality. The larger the size of the samples the more closely the distribution of means will approach normality.

Figure 11.2A is a frequency distribution of the sizes of the populations of all local authority areas in England and Wales excluding the two largest, Leeds (population: 748,000) and Birmingham (1,087,700). The distribution is strongly right-skewed.

One hundred random samples, each consisting of the sizes of the populations of 25 local authority areas, were drawn with replacements from the population represented by Fig. 11.2A. The resulting distribution of means (Fig. 11.2B) is not as badly skewed as the parent population, but it is still far removed from normality. A second set of fifty samples, this time of size 50, yielded the distribution of means shown in Fig. 11.2C. Although the distribution is more normal than that obtained with the smaller samples it is still somewhat asymmetrical. Apparently samples of size 100 or more would be needed to give a reasonably normal distribution of means.

If n is sufficiently large, the distribution of the means of samples drawn from any population will be approximately normal regardless of the form of the population distribution. This important result is known as the *Central Limit Theorem*. The more a population departs from normality the larger n needs to be before the normal distribution is a good approximation to the distribution of sample means. It is not possible to lay down any hard and fast rules, but if the population is unimodal and only moderately skewed, samples of size 10 or 20 may be sufficient to give a reasonably normal distribution. If the population is severely skewed or markedly bimodal, even samples containing as many as 50 members of the population may fail to give a distribution that is reasonably normal as Fig. 11.2 illustrates.

At first encounter the Central Limit Theorem may not seem very convincing. Why should a distribution that is patently non-normal yield a normal distribution of sample means? Unfortunately it is not easy to answer this question. Most modern proofs of the Central Limit Theorem rely on very advanced mathematics, and though concise and elegant are far beyond the scope of this book. An elementary

---

[1]The qualification "of the mean" is important. A sampling distribution can be constructed for any statistic calculated from sample data, for example the median or the standard deviation. The sampling distribution is merely the frequency distribution of an infinite number of values of the statistic, values derived by notionally drawing an infinite number of samples from the population. All sampling distributions, it is important to note, are mathematical abstractions. They are deduced from the known facts about the population and the method of sampling. By experimentally drawing a large number of samples from a population an approximate sampling distribution can be obtained, but not one that is exact. Unlike the true sampling distribution the empirical distribution is subject to sampling error, particularly in the tails where there are relatively few values.

proof exists (see Bradley, 1976), but it is rather long and for this reason will not be presented here. If the Central Limit Theorem seems unconvincing, an attempt should be made to verify it experimentally by drawing samples from a test population that lacks normality. In the process some intuitive understanding will be gained of how the theorem works. It will be found that calculating means tends to suppress the extreme values in the population and favour the more central values. To some extent very large and very small values cancel out; in addition, the extreme values tend to get swamped by the addition of the more numerous central values. It is for these reasons that the standard error $\sigma_{\bar{X}}$ is less than σ.

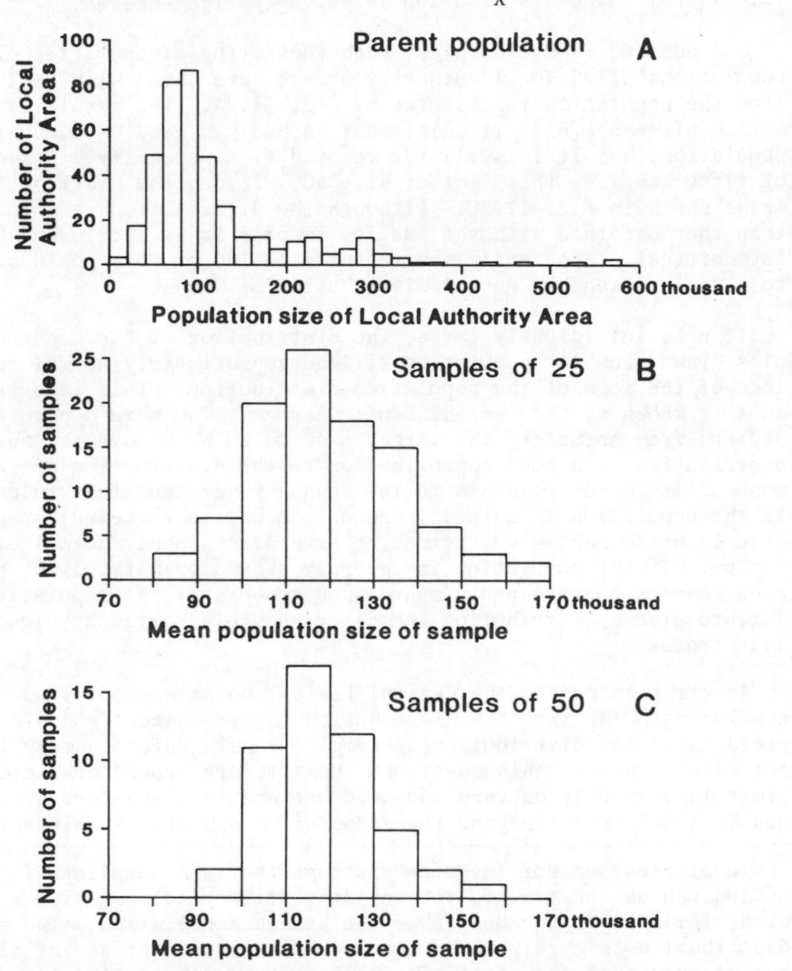

Fig. 11.2 - Repeated sampling from a skewed population.

A: Histogram of the population sizes of 402 Local Authority areas in England and Wales (mid-1973 estimates).

B: Histogram of the means of 100 samples each containing 25 Local Authority areas.

C: Histogram of the means of 100 samples each containing 50 Local Authority areas.

By increasing the size of the samples it is easy to verify that $\sigma_{\bar{X}}$ is inversely related to sample size. With large samples there is less chance than with small samples of obtaining means as extreme as the more extreme values in the population. The extremities of the population distribution are therefore not reflected in the distribution of means which approaches a unimodal symmetrical form as n is increased. In the limiting case, with samples of infinite size, it is easy to see that $\sigma_{\bar{X}}$ would be zero and all the sample means would have the same value μ.

The fact that the distribution of sample means approaches normality as sample size increases though the population from which the samples are drawn is not itself normal has far-reaching consequences for significance testing. Given a normal distribution of sample means, the probability that a particular mean will differ from the grand mean or population mean by Z standard errors or more can be determined by using Table B ("the Z-table"). Thus the probability that a particular mean will differ by 1 standard error or more is 0.3173, the probability that it will differ by 2 standard errors or more is 0.0455, while the probability that it will differ by 3 standard errors or more is 0.0027. Suppose, therefore, there exists a single sample of n = 100 items with a mean $\bar{X}$ = 16 units. It is desired to test the hypothesis that the sample comes from a population of infinite size with a mean μ of 15 units and standard deviation σ of 5 units. Assuming a normal distribution of sample means, the probability of obtaining a value of $\bar{X}$ differing from μ by 1 unit or more is the probability attached to the value of Z in Table B where

$$Z = \frac{|\bar{X} - \mu|}{\sigma_{\bar{X}}} \quad \text{and} \quad \sigma_{\bar{X}} = \frac{\sigma}{\sqrt{n}}$$

Substituting the values given above

$$\sigma_{\bar{X}} = \frac{5}{\sqrt{100}} = 0.5 \quad \text{and} \quad Z = \frac{16 - 15}{0.5} = 2$$

Referring to Table B, the probability P is found to be 0.0455. In other words, if repeated samples were to be drawn from the specified population, on average only about 4.5 in a hundred would differ in mean by as much as the sample in question or by a larger amount. Because P is low the hypothesis that the sample comes from the population is called into question. Using an 0.05 level of significance, the hypothesis can be rejected.

The arguments just outlined can be formalised as follows.

THE Z TEST[1]

TO TEST WHETHER A SAMPLE OF MEAN $\bar{X}$ DIFFERS SIGNIFICANTLY FROM A POPULATION OF MEAN μ AND STANDARD DEVIATION σ

THE DATA

A random sample of n items is drawn, either with individual replacements or without. Each of the n items is measured once in respect of

[1]The test discussed in the last chapter can be thought of as a special case of the Z test in which n = 1.

a variable X. The mean of the n measurements of X is $\bar{X}$.

It is desired to test the hypothesis that the mean of the population from which the sample is drawn is not $\bar{X}$ but μ, while the standard deviation of the population is σ. The population contains N items. It is normally distributed, or else the sample is large.

PROCEDURE

1. Set up the null hypothesis: the sample is drawn from a population whose mean is μ and whose standard deviation is σ. Sampling error accounts for the difference between the sample mean $\bar{X}$ and the population mean μ. The difference in other words is not statistically significant.

2. Calculate the standard error, $\sigma_{\bar{X}}$. If the sample is drawn with individual replacements, the standard error is

$$\sigma_{\bar{X}} = \frac{\sigma}{\sqrt{n}}$$

If the sample is drawn without individual replacements, the standard error is somewhat smaller

$$\sigma_{\bar{X}} = \frac{\sigma}{\sqrt{n}} \left(\sqrt{\frac{N - n}{N - 1}}\right)$$

3. Calculate Z, the difference between the sample and population means divided by the standard error

$$Z = \frac{|\bar{X} - \mu|}{\sigma_{\bar{X}}}$$

The modulus sign $||$ indicates that $\bar{X} - \mu$ is to be taken as a positive quantity.

The formula just given is valid only if X is a continuous variable. If X is limited to integer values (whole numbers), the correct formula is slightly more complicated

$$Z = \frac{|\bar{X} - \mu| - 1/(2n)}{\sigma_{\bar{X}}}$$

Somewhat confusingly, $1/(2n)$ is called a *correction for continuity* (or *continuity correction*), although it is, in fact, a correction for discontinuity. Obviously if n is large, the correction for continuity may be so small as to be not worth calculating.

4. Decide whether a one- or two-tailed test is appropriate.

5. Determine the value of P, the probability of obtaining the calculated value of Z, or one greater, as a result of sampling error. P is the probability of obtaining a sample with a mean differing from that of the population by $|\bar{X} - \mu|$ or a greater amount.

Values of P for two-tailed tests are listed in Table B (the "Z" table). Values of P for one-tailed tests are exactly half those listed in Table B.

6. Decide whether to retain or reject the null hypothesis. If P is low, serious doubt is cast on the null hypothesis. Reject the null if P is equal to or less than whatever level of significance is judged appropriate. e.g. 0.05 or 0.01. Conclude that the value of the sample mean is significantly different from that which would ordinarily occur if the mean of the parent population is μ and the standard deviation is σ. Assume that the population mean is not μ and/or the population standard deviation is not σ.

If P is greater than the level of significance, reserve judgement, i.e. regard the null hypothesis as possibly, but not necessarily, correct.

EXAMPLE (1)

In 1966 the Council of the London Borough of Camden asked the Centre for Urban Studies at University College, London, to produce a study of public (Council) and private housing in the Borough (Centre for Urban Studies, 1969; Glass, 1970). The study, which was completed in the following year, consisted of (a) a complete census of all 13,603 households living in housing provided by the Council and (b) a sample survey of 3019 households out of a total of about 67,089 living in privately owned housing. One of many variables studied was household size - the number of persons per household. Households living in Council-owned housing were found to contain 2.80 persons on average, the standard deviation being 1.64 persons. Households living in privately owned property contained 2.31 persons on average. Can it be said that the households living in public and privately owned housing differed in size?

There are two populations (in the statistical sense of this term) that need to be carefully distinguished. One is the population from which the sample is drawn, namely all households living in privately owned property. The other is the population of households living in property owned by the Council, the population examined by the census. The question to be answered is "Do households in these two populations differ in size?" The mean of the households living in privately owned property is 2.31, but this figure is tentative because it is liable to be affected by sampling error. The mean of the households living in Council-owned property (2.80) is free from sampling error because it is based on a complete census. It may, of course, be subject to other kinds of error, but there is no way in which these can be evaluated statistically.

The hypothesis to be tested (null hypothesis) may be stated as follows. The population from which the sample is drawn has a mean of 2.80. It is entirely a matter of chance that the mean of the sample is 2.31, not 2.80.

Now if an infinite number of samples, each of size 3019, were to be drawn without replacements from a population of size 67,089, mean 2.80, and standard deviation 1.64;[1] the means of the samples would

[1] The standard deviation of the population of households living in Council-owned property is 1.64 persons. Whether σ, the standard deviation of the population of households living in private property, is also 1.64 persons is questionable. The sample standard deviation is only 1.40 persons. Arguably it would be better to substitute s for the unknown value of σ rather than substitute the value 1.64. This alternative procedure is discussed later in the chapter.

vary with a standard error of

$$\sigma_{\overline{X}} = \frac{1.64}{\sqrt{3019}}\left(\sqrt{\frac{67089 - 3019}{67089 - 1}}\right) = \frac{1.64}{\sqrt{3019}}\left(\sqrt{\frac{64070}{67088}}\right) = 0.2851$$

The mean of the sample means would be 2.80. And because of the size of the samples, their means would follow an approximately normal distribution, even if the population was far from normal.

On the assumption that the distribution of sample means is exactly normal it is possible to calculate how many samples have means differing from that of the population by an amount equal to or greater than 2.80 - 2.31 = 0.49. To begin with

$$Z = \frac{2.80 - 2.31}{0.02851} = \frac{0.49}{0.02851} = 17.19 \text{ approximately.}$$

Strictly speaking a continuity correction ought to be employed because the number of persons per household is a discrete variable. However, the correction = 1/(2n) = 1/2(3019) = 0.000166 is so small that it can be omitted.

A two-tailed test is appropriate since there can be no certainty that households living in privately owned housing really are smaller than households living in Council-owned housing. The sample suggests that they are smaller, but the sample is subject to the vagaries of chance. There is no physical law that says that if the households differ in size those in privately owned housing must be smaller than those in Council housing.

The calculated value of Z exceeds the largest value shown in Table B (Z = 6.9). The probability P of obtaining a value of Z of 18.61 or more cannot therefore be higher than the lowest value of P shown in the table viz. 0.00000000000520. Fewer than 52 samples in ten million million drawn from a population with the same mean and standard deviation as the census would differ from it in mean by 0.49 or more. On the assumption that the sample in question is not one of these rare occurrences, the null hypothesis can be discarded. P is too minute for the hypothesis to be credible. Households living in Council-owned housing can be assumed to differ in size from households living in privately owned housing.

Observe carefully that the test does not enable one to say whether households living in Council and privately owned housing differ solely in mean, or solely in variance, or in both mean and variance. The test lacks precision; all one can say with confidence is that the households differ in size.

EXAMPLE (2)

The Camden Housing Study found that 51 per cent of all the households living in housing provided by the Council moved to their present address before 1960. The corresponding figure for the sample of households living in privately owned property was 41 per cent. Is the difference in percentages statistically significant?

The question can be put in a slightly different form: given a population proportion of 0.51, what is the probability of obtaining a sample proportion of 0.41, or a proportion less like the population proportion?

A proportion can be interpreted as a special case of the mean - a mean based on values of X that are either 0 or 1. If an item in a sample has a particular characteristic then its value is 1, and if it lacks the characteristic its value is 0. If there are n items in the sample, the total number with the characteristic is ΣX. The sample proportion $\hat{p}$ is simply the mean $\bar{X} = \Sigma X/n$.

Suppose that the items are drawn randomly from a population in which a proportion p of the items possesses the characteristic in question and a proportion $q = 1 - p$ lacks the characteristic. Each item in the sample thus has a probability p of possessing the characteristic and a probability $q = 1 - p$ of lacking it. Let ΣX be the number of items in the sample that possesses the characteristic. The probabilities associated with $\Sigma X = 0, 1, 2, 3, \ldots, n$ are the successive terms in the expansion of $(q + p)^n$, i.e.

$$q^n, \; npq^{n-1}, \; \frac{n(n - 1)}{2} p^2q^{n-2}, \; \frac{n(n - 1)(n - 2)}{6} p^3q^{n-3}, \; \ldots, \; p^n$$

These probabilities form a binomial distribution (see Chapter 8). The mean is np and the variance is $npq = np(1 - p)$.

Instead of recording ΣX, the number of items in the sample that have the characteristic, it is possible to record the proportion of items $\hat{p} = \Sigma X/n$. Since the distribution of ΣX is binomial, the distribution of the proportion $\hat{p}$ is also binomial. The mean proportion is $np/n = p$ and the variance is $npq/n^2 = pq/n$. It might perhaps be thought that the formula for the variance ought to be npq/n not npq/n^2, but if the values of a variable are altered by dividing by a constant, the variance must be altered by dividing by the square of the constant. For example, if the values of the variable are in miles they can be converted to kilometres by dividing by 0.6214. A variance in square miles can be converted into a variance in square kilometres only by dividing by $(0.6214)^2 = 0.3861$.

Now suppose that an infinite number of samples of size n are drawn from the population. The sampling distribution of the sample proportion $\hat{p}$ will have a mean p. If the sampling is with individual replacements the standard deviation (standard error) is simply

$$\sigma_{\hat{p}} = \sqrt{\frac{pq}{n}} = \sqrt{\frac{p(1 - p)}{n}}$$

If the sample is drawn without individual replacements,

$$\sigma_{\hat{p}} = \left[\sqrt{\frac{p(1 - p)}{n}}\right] \left[\sqrt{\frac{N - n}{N - 1}}\right]$$

where N is the size of the population.

Provided n is large, and p not too near 0 or 1, the sampling distribution will be approximately normal.

In the present example $\hat{p} = 0.41$ and $p = 0.51$. Because the sample is large ($n = 3019$) and the value of p is close to 0.5, the sampling distribution of $\hat{p}$ is almost exactly normal. The standard error associated with $\hat{p}$ is

$$\sigma_{\hat{p}} = \left[\sqrt{\frac{(0.51)(0.49)}{3019}}\right]\left[\sqrt{\frac{67089 - 3019}{67089 - 1}}\right]$$

$$= [0.00910]\ [0.97725]$$

$$= 0.00889$$

Computed with a continuity correction, the formula for Z is

$$Z = \frac{|\hat{p} - p| - \frac{1}{2n}}{\sigma_{\hat{p}}} = \frac{|0.41 - 0.51| - \frac{1}{2(3019)}}{0.00889}$$

$$= 11.23 \text{ approximately.}$$

The continuity correction, which in this instance is very small, allows for the fact that X is limited to the values 0 and 1.

A two-tailed test is appropriate since $\hat{p}$ could with equal probability have been 0.1 greater than p instead of smaller.

The calculated value of Z is greater than the largest value shown in Table B. The probability P associated with Z is negligibly small (less than 0.00000000000520). It is therefore extremely unlikely that as many as 51 per cent of households living in privately owned property moved to their present address before 1960. The evidence supplied by the sample suggests that households living in privately owned property moved more often than households living in property belonging to the council.

ASSUMPTIONS AND LIMITATIONS

1. Random sampling. The test assumes that the items (or individuals) in the sample are selected entirely at random.

2. Independence of the items in the sampled population. The values of the different items in the population are assumed to be independent of one another. The value of one item in no way determines the value of another.

Unfortunately a great many populations violate this assumption of independence; the values of the items are systematically distributed in time or space and are thus interdependent. Usually, the values of items that are close together are more alike than the values of items that are far apart; the populations exhibit what is called *positive autocorrelation*. It is also possible for the values of items that are close together to be less alike than the values of distant items, but this condition, called *negative autocorrelation*, is relatively rare.

Any variable that changes only slowly over time or space is likely to show positive autocorrelation. Population growth rates, for example, are usually relatively constant in the short term, but undergo long term trends in response to slowly changing economic and social forces. Rates in consecutive years tend to be broadly similar, whereas rates in years a decade or more apart may show major differences. It is also a fact that countries adjoining one another tend to have rates that are much the same, but distant states tend to have very different rates. Numerous other examples of positive autocorrelation could be cited, involving a wide variety of variables. Negative autocorrelation is more difficult to illustrate, but weekly grocery bills provide a case in point. Overspending by a household in one week is usually followed by underspending in the next, because there is an increased desire to economise and perhaps an excess of food to clear up. Bills in consecutive weeks tend to be very different, whereas bills in weeks that are far apart may be roughly comparable if there are no price increases.

Surprising though it may seem, the arithmetic mean of a random sample is an unbiased estimate of the population mean whether autocorrelation is present or not. Also, any proportion in the sample is an unbiased estimate of the corresponding proportion in the population. Autocorrelation is a nuisance in significance testing only because it alters the standard error of the estimates. The usual formulas understate the standard error when positive autocorrelation is present in the population and overstate it when negative autocorrelation is present.

It is easy to see why autocorrelation alters the standard error of a sample mean or a proportion. Consider first the case of a positively autocorrelated population. A random sample of n items does not provide the usual number of pieces of information about the population but a smaller number as rather more pieces than usual are overlapping or merely duplicatory. The formulas for the standard error of a sample mean and the standard error of a sample proportion have $\sqrt{n}$ as a divisor, and will understate the true standard error because the effective value of n will be less than the nominal value. The reverse argument applies if the population is negatively autocorrelated. The sample will contain more than the usual number of pieces of information since there will be less overlap or duplication. The effective value of n will be more than the nominal value and hence the formulas for the standard error will overstate the true standard error.

If one could only calculate the effective value of n, one could allow for autocorrelation when calculating the true standard error. Unfortunately, the effective value of n is usually not known with any accuracy. Statisticians almost always ignore autocorrelation when running Z tests because of the difficulty of making precise adjustments to the formulas for the standard error. However, the tests may give very inaccurate results if the autocorrelation is at all severe.

The author has used a computer to create experimental populations in which the first item is given an arbitrary value of 100 and subsequent items are given values that are positively autocorrelated according to the formula

$$X_t = X_{t-1} + \varepsilon$$

where X_t is the value of a given item, say, the tth item in the population, X_{t-1} is the value of the preceding or $(t-1)^{th}$ item, and ε is a number drawn randomly from a normal distribution with a mean of zero and a variance of 1. The computer created 120 populations, each containing 5000 autocorrelated items, and drew 1500 random samples with replacement from each population. It then calculated the standard deviation (standard error) of the 180 thousand sample means. The operation was repeated for varying sample sizes.

The mean standard error was 36.05 for samples containing 10 items. The mean of the variances of the 120 sampled populations was 398.74, and the mean of the means was 97.00. If the items in the populations had been independent, the standard error would have been $\sigma/\sqrt{n}$ = 398.74/$\sqrt{10}$ = 6.31. The standard error was increased 5.7 times as a result of the autocorrelation.

Little improvement was noted with samples of 15 items. The mean of the 120 population variances was 383.15 and the mean of the means was 100.79. The standard error was 23.25, as against a value of $\sqrt{383.15}/\sqrt{15}$ = 5.05 for independent items. Because of the autocorrelation the standard error was increased 4.6 times.

The experiment was repeated for samples of 50 items. The mean of the population variances was 372.71 and the mean of the means was 98.27. The standard error was 8.00, 2.9 times the value of $\sqrt{372.71}/\sqrt{50}$ = 2.73 for independent items.

It is clear from these results that positive autocorrelation can considerably increase the standard error for small samples. The increase in the standard error becomes less with increasing sample size. Theoretical calculations show that for a given sample size the increase in the standard error depends on the strength and spatial or temporal persistence of the autocorrelation. When the autocorrelation is confined to immediately adjacent items in the population, the increase in the standard error is less than when the same autocorrelation extends over a range of items.

Box *et al.* (1978) have calculated that if only immediately adjacent items are positively autocorrelated the standard error for a sample of n items is increased by a factor varying from 1 to $\sqrt{[(2n-1)/n]}$ depending on the strength of the autocorrelation. They have also calculated that if there is negative autocorrelation, and this only affects immediately adjacent items, the sampling error for a sample of size n is decreased by a factor varying from 1 to $1/\sqrt{n}$. For a sample of size 50, autocorrelation could therefore decrease the standard error by a factor as large as $1/\sqrt{50}$ = 0.14 or increase it by a factor as large as $\sqrt{[(100-1)/50]}$ = 1.4. These calculations do not conflict with the computer study, where a factor of 2.9 was recorded for samples of 50 items, because the autocorrelation was allowed to extend beyond immediately adjacent items.

3. Normality of sample means. As already discussed, the Z test assumes that if repeated samples the same size as the sample in

question were to be drawn from the population their means would plot as a normal distribution. The assumption is reasonably justified if the population is approximately normal because then the distribution of sample means is approximately normal no matter how small the samples.

Complications arise if the population is not approximately normal. The means of samples of any size (other than 1) necessarily have a more normal distribution than the population from which they are drawn. According to the Central Limit Theorem, the larger the samples the more closely their means will approximate to a normal distribution. Provided therefore that the sample in question is sufficiently large, the Z test can be applied with reasonable accuracy regardless of the precise form of the population. The snag lies in deciding what is sufficiently large.

Many writers suggest $n = 30$ or more as a rough rule of thumb. However, this is a dangerous generalisation. The more the population departs from normality, the larger the sample must be in order to achieve a given level of accuracy. If the population is far from normal, even samples of several hundred items may not be sufficient to produce a distribution of means that is reasonably normal.

Bradley (1963, 1971, 1973) has constructed 24 artificial populations of varying shape and, using a computer, has drawn large numbers of random samples of size $n = 2, 4, 8, 16, 32$, etc. from each population. He has calculated the value of $\bar{X}$ for all the samples, and plotted the frequency distribution of $\bar{X}$ for each value of n for each of the populations. Because of the large number of samples (at least 10,000 for each value of n) each frequency distribution can be presumed to be a very good approximation to the true sampling distribution of $\bar{X}$.

Bradley's experiments show that the sampling distribution of $\bar{X}$ approaches normality with increasing sample size the more slowly the more skewed is the population. The degree of asymmetry in the tails of the population is especially important. If one tail is much longer or thicker than the other, the distribution of $\bar{X}$ may be far from normal, the more so if n is very small.

In the case of a symmetrical population one must consider the degree of *kurtosis* (or peakedness). If the distribution is *platykurtic* (broader than the normal distribution in the central portion and higher in the tails) the sampling distribution of $\bar{X}$ may be noticeably non-normal, particularly if n is small.

Bradley points out that the farther one goes from $\bar{X}$ the less normal the sampling distribution of $\bar{X}$ tends to become. It is at the tails, in other words, that the distribution of $\bar{X}$ approaches normality most slowly. The larger the value of Z, the less reliable the value of P tends to be. A P value of 0.05, for example, is likely to be more in error than a P value of 0.5. This makes for difficulty since probabilities in the tails are crucial in significance testing. It matters little that the probabilities in the central part of the sampling distribution of $\bar{X}$ are likely to be fairly reliable. What one needs but cannot always obtain are fairly reliable values of P at the commonly used levels of significance.

Few of the populations studied by geographers are approximately normal; many are badly skewed or bimodal. Large samples are seldom available. Because it is often unclear whether the assumption of normality of sample means is sufficiently justified, the scope for applying the Z test is limited. Suppose, for instance, the normality assumption is ignored, the test is run with a small sample, and a P value of 0.04 obtained. The null hypothesis cannot be legitimately rejected, even using a level of significance of 0.05, since the normality assumption may be partly or wholly in error. The true distribution of sample means may be such that P ought to be 0.06 (or some other value greater than the chosen level of significance) thus enabling the null hypothesis to be retained. Unless the normality assumption can be shown to be reasonably justified, the test is robbed of practical usefulness.

If Z is sufficiently large, however, the foregoing remarks do not apply. It can be shown that no matter how greatly a distribution departs from normality, the area lying more than Z standard deviations from the mean cannot exceed $1/Z^2$ of the total area. Thus the area lying beyond 2 standard deviations cannot be more than $1/2^2 = 0.25$ of the total while the area beyond 3 standard deviations cannot be more than $1/3^2 = 0.11$ of the total. This relationship between Z and the maximum possible area in the tails of a distribution is known as *Tchebycheff's inequality* after the Russian mathematician who first demonstrated it.

The inequality can also be put in probability terms. If an individual or item is drawn at random from a population, the probability that it will have a value deviating from the population mean by Z or more standard deviations cannot be greater than $1/Z^2$, whatever the form of the distribution of the population. The probability that the mean of a sample will deviate from the mean of the population from which it is drawn by 4.47 standard errors or more is at the most $1/(4.47)^2 = 0.05$. Similarly, the probability that it will deviate by 10 standard errors or more is 0.01 or less. These values are useful upper limits in hypothesis testing. In general, if a sample mean falls further than $1/\sqrt{\alpha}$ standard errors from the population mean (where α is the chosen level of significance) the null hypothesis can be dismissed since P will be less than α whatever the form of the population. The normality assumption needs to be considered only if the sample mean lies within $1/\sqrt{\alpha}$ standard errors of the population mean.

4. Failure to take account of the sample standard deviation. The test fails to consider the value of s, the sample standard deviation. It considers σ, the population standard deviation, but only in relation to the difference between $\overline{X}$ and μ. The difference, if any, between s and σ is disregarded.

It may so happen that the sample differs greatly from the population in standard deviation but not in mean. In such circumstances, the Z test is liable to be misleading. Although the sample may be very unlikely to come from the population, the test may fail to make this clear. A way round the difficulty, provided the population is normal or near normal, is to use the chi-square test of variance described in Chapter 15. Alternatively, if the population and sample data are grouped into classes, the chi-square one-sample test can be used (again see Chapter 15), which has the advantage that it makes no assumption about normality. On the other hand it is much less powerful than the chi-square test of variance.

5. Combined effects of μ and σ. A low value of P suggests that the null hypothesis is wrong. It does not necessarily follow, however, that the sample is unlikely to come from a population of mean μ. The value of P reflects not only the value of μ but also the value of σ. Because interest usually focuses on the difference in means $|X - \mu|$, the role of σ in the calculations is easily overlooked. A low value of P indicates only that the difference in the means is unexpectedly large relative to σ. Although the sample is unlikely to represent a population that combines a mean of μ with a standard deviation as small as σ, there is no reason why it should not represent a population of mean μ and some larger standard deviation (Fig. 11.3). If σ is large the difference between $\bar{X}$ and μ is more likely to be due to sampling error than if σ is small.

The test is therefore somewhat inconclusive. A low value of P indicates that something is wrong with the null hypothesis, but it does not pinpoint the exact cause. The difference between the population and sample may arise through a difference in means or standard deviations or a combination of both.

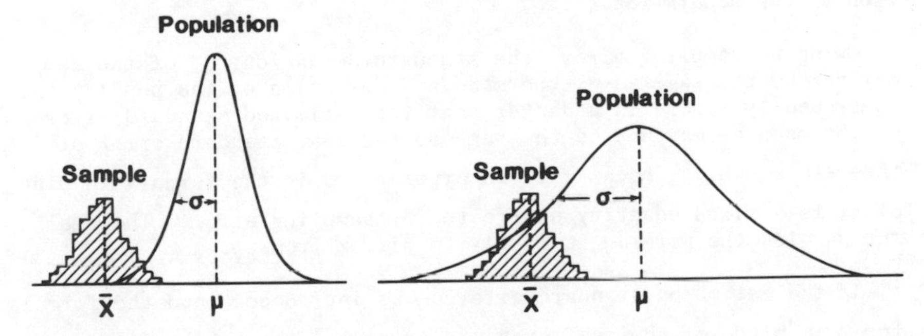

Fig. 11.3 - The difference between a sample mean $\bar{X}$ and a population mean μ.

Left: Small σ, $\bar{X}$ lies more than 1.96 standard deviations from μ, hence the difference in means is statistically significant.

Right: Large σ, $\bar{X}$ is nearer than 1.96 standard deviation to μ, hence the difference in means is not statistically significant.

6. Knowledge of the value of σ. In order to calculate the standard error the value of σ must be known. In practice information on σ is often lacking. For instance, the population may be theoretical in nature, and though the theory may give a value of μ it may not be precise enough to give a value of σ. Even if the population is a real one, there may be practical reasons why σ is unknown. Rather than list the raw data, census reports and similar compilations tend to provide summary tables where means are quoted but not standard

deviations. Unless the raw data can be obtained the standard deviations cannot be ascertained. Note also that in order to calculate μ it is not necessary to determine the values of individual members of the population, provided their total value is known. For instance, by comparing two maps of the coastline made at different dates it is possible to determine if a change in the position of the coastline has taken place and if so the mean rate of change over the interval between the dates of survey. There is, however, no way of determining the standard deviation.

If σ is missing, the Z test cannot be carried out. Provided the population is normal, however, another test can be substituted known as the t test.

THE t DISTRIBUTION

In the t test the standard error of the mean is estimated using s in place of σ. Provided the sample is drawn with individual replacements, or is negligibly small compared with the population, the estimated standard error $\hat{\sigma}_{\overline{X}}$ is given by

$$\hat{\sigma}_{\overline{X}} = \frac{s}{\sqrt{n}}$$

where n is the size of the sample.[1] The formula should not be used if the sample is drawn without replacements and forms a large fraction of the population.

Owing to sampling error, the standard deviation, s, of the sample may not be the same as σ, the standard deviation of the population. Consequently, there is a danger that the estimated standard error of the mean $\hat{\sigma}_{\overline{X}}$ may not be the same as the real standard error of the mean $\sigma_{\overline{X}}$ which, because it incorporates σ in the formula in place of s, is a fixed quantity unaffected by sampling error. The smaller the sample, the more $\hat{\sigma}_{\overline{X}}$ is likely to differ from $\sigma_{\overline{X}}$.

If the estimated standard error $\hat{\sigma}_{\overline{X}}$ is introduced into the formula for Z in place of the real standard error $\sigma_{\overline{X}}$, the result is a quantity known as t or Student's t which was discovered in 1908 by W.S. Gosset, who wrote under the pseudonym of "Student".

$$t = \frac{|\overline{X} - \mu|}{\hat{\sigma}_{\overline{X}}}$$

Note that all that has been done is to substitute $\hat{\sigma}_{\overline{X}}$ for $\sigma_{\overline{X}}$ in the formula for Z. Although this substitution may look trivial, it makes a considerable difference mathematically. Unlike the Z distribution, the t distribution varies in shape depending on the size of sample. For instance, drawing repeated samples of three items from a normal population would yield the t distribution labelled df = 2 in Fig. 11.4. Although unimodal and symmetrical about a mean of zero, the distribution is definitely not normal like the Z

[1]Many writers prefer to use the symbol $s_{\overline{X}}$ instead of $\hat{\sigma}_{\overline{X}}$.

distribution. The tails are noticeably larger than those of a normal curve, while the peak is both lower and narrower. The distribution is so shaped that 30% of samples have t values of 1.96 or more. It will be recalled that the corresponding figure for the Z distribution is only 5%.

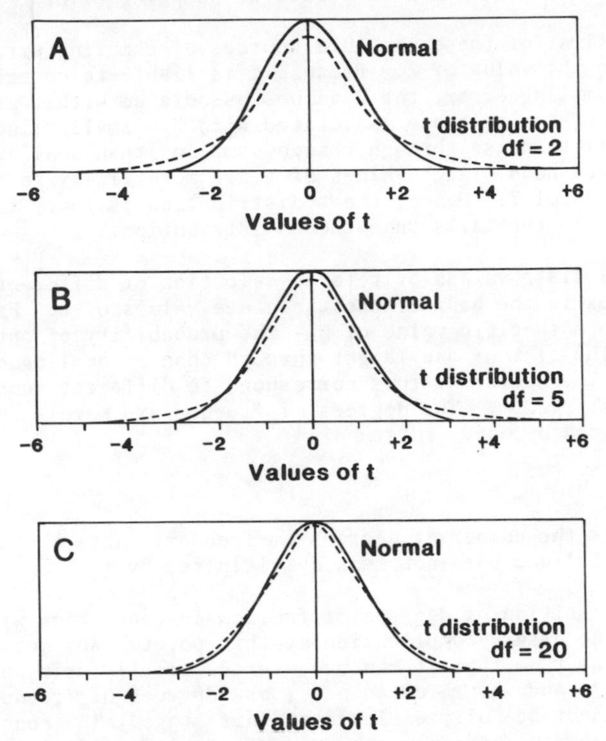

Fig. 11.4 – Comparison of the t distribution (broken line) with the normal distribution (solid line). A : samples of 3 items, B : samples of 6 items, C : samples of 21 items.

As Fig. 11.4 demonstrates, the shape of the t distribution becomes more normal with increasing size of sample. At the scale employed the t distribution for samples of 21 items can be barely distinguished from the normal distribution. The t distribution for samples of infinite size is precisely the same as the normal distribution.

P values associated with t reflect the area in the two tails of the t distribution in the same way as the P values associated with Z reflect the area in the two tails of the Z (normal) distribution. In the case of a sample consisting of just two items a t value of 1.96 or more is associated with a P value of 0.30, whereas a Z value of 1.96 is associated with a P of only 0.05. With increasing sample size the differences between the P values for t and those for Z become less and less. In the limiting case, with a sample of infinite size, the P values for t are identical with those for Z.

Given a sample of finite size why do the P values for t and Z differ? The answer is that the value of t, unlike Z, is affected by chance in two ways

1. $\overline{X}$ may differ from μ as a result of sampling error,

2. s may differ from σ as a result of sampling error.

Only the first of these possible sources of sampling error is reflected in the value of Z. Because t is liable to contain a double dose of sampling error, the P values associated with t are not in general the same as those associated with Z. Small values of t are less likely to arise through chance sampling than small values of Z; on the other hand, large values of t are more likely to occur than large values of Z. Hence, the t distribution is lower in the centre and higher in the tails than the Z distribution.

Table E lists values of t for a selection of different values of P. Figures in the body of the table are values of t. Each column relates to a specific value of P - the probability of obtaining the listed value of t or one larger through chance sampling of the population. The different rows correspond to different numbers of *degrees of freedom*. The degrees of freedom are merely the sample sizes reduced by one. In symbols

$$df = n - 1$$

where df is the number of degrees of freedom. Note that df is a single symbol and does not mean d multiplied by f.

The recognition of degrees of freedom in connection with t will seem an unnecessary complication at this point. Why not construct the table of t so that it can be entered directly using n rather than n - 1? And why refer to n - 1 as "degrees of freedom"? The reasons cannot be fully explored at this stage. The concept of degrees of freedom finds many uses in statistics, and although in the present context it serves only to complicate matters, later on situations will be encountered where the concept introduces useful simplification. The number of degrees of freedom can be thought of as the number of items or individuals in a sample that are determinable independently of t. In other words, if there are n items, then n - 1 are free to vary in value without affecting the value of t. Suppose, for example, there are 3 items, the value of t is 3.4641, and the population mean is 2 units. Of the 3 items, 3 - 1 = 2 can be assigned arbitrary (fictional) values without altering t. The value of the third item is in effect fixed by the value of the other two and by the value of t. Thus, if the first two items are assigned arbitrary values of 3 and 4 units, the value of the third item has to be 5 units if t is to equal 3.4641. The relevant calculations are as follows

Sample mean $\overline{X} = (3 + 4 + 5)/3 = 4$.

Sample variance $s^2 = [\Sigma X^2 - (\Sigma X)^2/n]/(n - 1) = [50 - 144/3]/2 = 1$.

Estimated standard error $\hat{\sigma}_{\overline{X}} = s/\sqrt{n} = 1/\sqrt{3} = 0.5774$

$t = |\overline{X} - \mu|/\hat{\sigma}_{\overline{X}} = |4 - 2|/0.5774 = 3.4641$.

Because only n - 1 items or individuals in a sample have the freedom to vary given the value of t, it is appropriate to refer to n - 1 as the number of degrees of freedom.

As will become clear later, the concept of degrees of freedom is employed in statistical work in connection with several other test statistics besides t. The number of degrees of freedom is always defined as the number of values that are free to vary given the value of the statistic in question. The number is not necessarily n - 1, but depends on the nature of the problem. Thus in certain applications of the t table, to be discussed later, the number is n - 2 not n - 1 (see Chapter 17).

THE ONE-SAMPLE t TEST

TO TEST WHETHER A SAMPLE OF MEAN $\bar{X}$ *AND STANDARD DEVIATION* s *DIFFERS SIGNIFICANTLY FROM A POPULATION OF MEAN* μ

THE DATA

A random sample of n items is drawn from a population. Each of the n items is measured once in respect of a continuous variable X. The mean of the n measurements is $\bar{X}$, and the standard deviation is s.

It is desired to test the hypothesis that the mean of the population is not $\bar{X}$, but μ. The population is normally distributed, or else n is large.

PROCEDURE

1. Set up the null hypothesis: the mean of the population is μ. Sampling error accounts for the difference between the sample mean $\bar{X}$ and the population mean μ. In other words the difference is not statistically significant.

2. Estimate the standard error, $\hat{\sigma}_{\bar{X}}$. If the sample is drawn with individual replacements (or, without replacements from an infinite population), the standard error is

$$\hat{\sigma}_{\bar{X}} = s/\sqrt{n}$$

3. Calculate t, the difference between the sample and population means divided by the estimated standard error[1]

$$t = \frac{|\bar{X} - \mu|}{\hat{\sigma}_{\bar{X}}}$$

4. Calculate n - 1, the number of degrees of freedom associated with t.

---

[1]The formula for t assumes that X is a continuous variable, but when X is discrete the value of t is usually so close to the true value that no continuity correction is necessary. For example, if X is restricted to integer values, empirical evidence suggests that t need be corrected only if n is less than 8 (Bradley, 1963). Discreteness has a much greater effect on Z than on t.

5. Decide whether a one- or two-tailed test is appropriate.

6. Determine P, the probability of obtaining the calculated value of t, or one greater, as a result of sampling error. P is the probability of obtaining a sample with a mean differing from that of the population by $|\bar{X} - \mu|$ or a greater amount

Selected values of P for two-tailed tests are given in Table E (the "t table"). Values of P for one-tailed tests are exactly half those listed in Table E.

7. Decide whether to retain or reject the null hypothesis. If P is low, serious doubt is cast on the null hypothesis. Reject the null if P is equal to or less than whatever level of significance is judged appropriate, e.g. 0.05 or 0.01. Conclude that the sample mean is significantly different from that which would ordinarily occur if the mean of the parent population is μ. Assume that the population mean is not μ but some other value closer to $\bar{X}$.

If P is greater than the level of significance, reserve judgement, i.e. regard the null hypothesis as possibly though not necessarily, correct.

EXAMPLE

A sample survey of 3019 households out of a total of about 67,089 living in privately owned housing in the London Borough of Camden found the mean number of persons per household to be 2.31 (Centre for Urban Studies, 1969; Glass, 1970). The standard deviation was found to be 1.40 persons. A complete census of the 13,603 households living in housing provided by Camden Council yielded a different mean, namely 2.80 persons per household. Is the difference in means statistically significant?

Null hypothesis: if a complete census of all households living in privately owned housing in Camden were to be carried out, it would yield a mean of 2.80 persons per household. The mean of the sample happens to be 2.31 rather than 2.80 simply because of sampling error. The difference in means is not statistically significant.

Assuming the sample to be drawn from a population that is normally distributed, the estimated standard error is given by

$$\hat{\sigma}_{\bar{X}} = \frac{s}{\sqrt{n}} = \frac{1.40}{\sqrt{3019}} = 0.02548$$

and the value of t by

$$t = \frac{|\bar{X} - \mu|}{\hat{\sigma}_{\bar{X}}} = \frac{|2.31 - 2.80|}{0.02548} = 19.62$$

with n - 1 = 3018 degrees of freedom.

Consulting Table E for t = 19.6 and df = 3018, P is found to be beyond the range of the table, i.e. less than 0.0000002. With repeated sampling of a normal population, samples yielding a t value of 19.6 or more would be obtained fewer than once in 5 million times. Since the chance of obtaining the observed mean, or one even more divergent from the hypothesised value of 2.80, is so remote, the null hypothesis can be disregarded as unreasonable. There is every reason to believe that households living in privately owned housing in Camden are smaller on average than households living in Council-owned housing.

In arriving at this conclusion, two technical matters have been glossed over. In the first place, the assumption has been made that the parent population is normally distributed. Although no mention has been made of the fact, the sample distribution has a reversed J-shape, and since the sample is large, this would suggest that the population also has a reversed J-shape. The t test can be safely employed, however, despite the probable lack of normality. The size of the sample is such that the Central Limit Theorem can be invoked. The distribution of the means of a series of samples of size n = 3019 can be assumed to be approximately normal even if the parent population has a reversed J-shape.

As has already been mentioned, the larger the sample is the more closely the distribution of t approaches that of Z. The sample standard deviation can be substituted for the population standard deviation σ without causing major loss of accuracy.

A second technical point is that the probabilities shown in Table E assume sampling with individual replacements, or equivalently sampling without replacements from an infinite population. The Camden sample was drawn without replacements from a population of finite size. However, the sample constitutes only a small fraction of the population (about 4.5%), and the test can therefore be presumed to be reasonably accurate, even though the exact value of P is doubtful.

ASSUMPTIONS AND LIMITATIONS OF THE ONE-SAMPLE t TEST

Like the Z test, the one-sample t test assumes that the sampling is random and that the values of the population items are independent of one another. If positive autocorrelation is present in the population, the estimated standard error will be too small and the t value will be too high. The test may give very inaccurate results if the autocorrelation is severe.[1]

As already mentioned, the one-sample t test assumes that the sampled population is normally distributed. Since no real population is exactly normal, it is useful to consider the errors that are likely to arise as a consequence of lack of normality. The *robustness* of the test (that is the extent to which the true value of P corresponds to the nominal value obtained from the t table) depends on the following factors:

[1] Lest there be any uncertainty, it should perhaps be pointed out that the autocorrelation only biases the standard error. The sample mean provides an unbiased estimate of the mean of the population (whether normal or non-normal). Also, the sample variance provides an unbiased estimate of the population variance.

1. The degree of non-normality of the sampled population. As a general rule the more the population deviates from normality, the less accurately the sampling distribution of t approximates to the theoretical t tabulated in Table E. The sampling distribution of t is symmetrical if, and only if, the population is symmetrical. If the population is skewed, the sampling distribution is also skewed, though to a lesser extent. The fit with the theoretical t distribution is likely to be worst if the population is severely skewed or U-shaped (see, for example, the empirical studies of Holzinger and Church, 1928; and Bradley, 1963).

2. The sample size, n. The robustness of the test increases slowly with sample size. Many statisticians have assumed that the sampling distribution of t is virtually the same as the theoretical t distribution whenever n is greater than 30. This is an over-simplification, however. If the population deviates greatly from normality, a sample of a hundred or more is required for the sampling distribution to approximate closely to the theoretical distribution.

3. The nominal value of P. The fit with the theoretical t distribution is usually closest in the central portion of the sampling distribution and tends to worsen as one proceeds outwards towards the tails. A nominal value of P of 0.05, for example, is likely to be less in error than a nominal value of 0.01, which in turn is likely to be less in error than one of 0.001.

4. The tail or tails selected for testing. If the sampled population is skewed, the left and right tails of the sampling distribution of t are not equal in size (they are only equal if the population is symmetrical). Empirical evidence suggests that in the case of most, if not all, right-skewed populations, the left tail of the sampling distribution of t is larger than the right tail (see Bradley, 1963). If μ exceeds $\bar{X}$, and a one-tailed test is run using the left tail, then one must expect the true value of P to be larger than the nominal value obtained using the t table. The test, in other words, is biased against the null hypothesis, which is a dangerous characteristic. If $\bar{X}$ exceeds μ, and a one-tailed test is run using the right tail, one must expect the true value of P to be smaller than the nominal value. The test is biased in favour of the null hypothesis. If a two-tailed test is run, the biases partly cancel out, and the test is more robust at the nominal value of P obtained from the t table than either of the one-tailed tests at half the value of P. Unfortunately, because the bias in the left-tail exceeds the bias in the right, a two-tailed test ends up biased against the null hypothesis. It would greatly assist significance testing if the bias were the other way.

The one-sample t test tends to be highly inaccurate when n is small and the sampled population is far from normal. Bradley (1963), for example, investigated an extremely skewed and bimodal population and found that for n = 32 the true value of P in a two-tailed test was about twice what the "t" table supposed it to be. When n = 16 the true value of P was around four times the nominal value. These errors are very considerable and show that the t test is not to be trusted if the sampled population deviates greatly from normality and the sample size is small.

The one-sample t test is distinctly less robust than the Z test for a given n. It is also less powerful when n is small, allowing one to reject the null hypothesis only if the difference in means is very large. As Bradley has written (1963, p.24) the experimenter who proposes to use a one-sample t test "is very nearly wasting his time by collecting only a small amount of data unless he is interested only in huge effects".

The one-sample t test takes the sample standard deviation into account, unlike the Z test. It fails, however, to consider σ, the population standard deviation. The difference, if any, between s and σ is ignored.

Although lacking both power and robustness, the t test allows more definite conclusions to be drawn than the Z test. A low value of P indicates that the sample is unlikely to come from a normal or near normal population with a mean of μ, even one with a standard deviation σ exactly equal to s. The Z test is more ambiguous. If P is low, this indicates that the sample is unlikely to come from a population with a mean of μ and a standard deviation of σ. But it might come from a population with a mean of μ and a standard deviation greater than σ. Or from a population with a standard deviation of σ and a mean closer to the sample mean than μ. Or from a population with a mean other than μ and a standard deviation other than σ. It is impossible to reach a definite conclusion.

12 Interval Estimation Based on Z and t

INTRODUCTION

The tests discussed in the last chapter enable one to decide whether a specific hypothesis about the value of a population mean is reasonable given the information contained in a sample. Having drawn a sample, however, one often has no specific hypothesis to test. One simply wishes to arrive at an estimate of the population mean on the basis of the information in the sample.

If the population is symmetrically distributed, there is no problem estimating μ. The best estimate (most probable value) of μ is the sample mean $\bar{X}$. $\bar{X}$ is an *unbiased* estimator in that it is no more likely to be larger than μ than smaller than μ. If an infinite number of random samples were to be taken from the population the mean value of $\bar{X}$ would exactly equal μ. $\bar{X}$ is also the most *efficient* estimator of μ that can be devised: no other estimator is available with a sampling distribution more closely concentrated around the value of μ. It is not possible, in other words, to devise a measure that will estimate μ with greater accuracy than $\bar{X}$.

If the population is skewed, difficulties immediately arise. Although $\bar{X}$ provides an unbiased estimate of μ, it is not the most efficient that can be devised. Provided the precise form of the population distribution is known, one can always arrive at a better estimate of μ than $\bar{X}$, although the mathematics may be complicated. The case of a log-normal population was considered in detail in Chapter 9.

CONFIDENCE LIMITS FOR A POPULATION MEAN

Often one wishes to estimate an interval or range of values within which the true value of μ is likely to lie. Such an interval provides a measure of the degree of certainty or uncertainty that attaches to the value of μ. The methods of estimating intervals are an extension of the hypothesis testing procedures discussed in the last chapter.

Before considering how best to estimate an interval for μ, it may prove helpful to consider how to estimate an interval for $\bar{X}$. Suppose that one is dealing with a continuous variable. There exists a normally distributed population with a mean of μ and a standard deviation of σ. A sample of n items is drawn randomly from this population, either with individual replacements or without. Within what interval or range of values is the value of the sample mean likely to lie?

As Table B shows, a value of Z of 1.96 corresponds to a probability P of 0.05. There are 5 chances in 100 of securing a sample mean deviating from the population mean by 1.96 standard errors or more. Conversely, there are 95 chances in 100 of securing a sample mean lying in the range $\mu - 1.96\sigma_{\bar{X}}$ to $\mu + 1.96\sigma_{\bar{X}}$.

If desired, different intervals or ranges of values can be calculated for the sample mean $\bar{X}$ that reflect higher or lower levels of probability P. Referring to Table B again, one finds that when Z is 2.57 the value of P is 0.01. There are thus 99 chances in 100 that $\bar{X}$ will fall within the range $\mu - 2.57\sigma_{\bar{X}}$ to $\mu + 2.57\sigma_{\bar{X}}$.

Calculations such as these can be readily reversed. Instead of starting with the parameter μ and proceeding to estimate a range of values for the statistic $\bar{X}$, it is possible to start with the statistic $\bar{X}$ and estimate a range of values for the parameter μ. For instance, the range

$$\bar{X} - 1.96\sigma_{\bar{X}} \text{ to } \bar{X} + 1.96\sigma_{\bar{X}}$$

may be called the *95% confidence interval* for μ. On the evidence supplied by the sample one can be 95% confident that the value of μ falls within the two limits $\bar{X} \pm 1.96\sigma_{\bar{X}}$. These limits, which bound the confidence interval, may be called the *95% confidence limits*. They are of course subject to sampling error like any other statistic. If repeated samples were to be drawn from the same population, and 95% confidence limits calculated for each sample, the limits would be found to vary merely because of chance sampling. In the long run one would expect 95 out of 100 confidence intervals to include μ, and 5 not to include it. The probability of μ being included in any one interval is therefore 0.95.

Using Table B, confidence intervals can be established corresponding to any selected percentage *level of confidence*. The 99% confidence interval, for instance, is

$$\bar{X} - 2.57\sigma_{\bar{X}} \text{ to } \bar{X} + 2.57\sigma_{\bar{X}}$$

In general, if α% is the chosen level of confidence, the formula for calculating the α% confidence interval is $\bar{X} - Z\sigma_{\bar{X}}$ to $\bar{X} + Z\sigma_{\bar{X}}$ where Z is the value of Z given in Table B corresponding to a value of P of $(100 - \alpha)/100$. If X is not a continuous variable, a correction for continuity must be applied. For instance, if X is limited to integer values, the α% confidence interval is $(\bar{X} - \frac{1}{2}n) - Z\sigma_{\bar{X}}$ to $(\bar{X} + \frac{1}{2}n) + Z\sigma_{\bar{X}}$.

In constructing confidence limits using the above formulas one encounters the same difficulties as when performing the Z test. The population must be normally distributed, or else the sample must be so large that the Central Limit Theorem can be invoked. The values of the items forming the population must be independent of each other, and the sampling must be random. In addition, it is necessary to know the value of σ, the population standard deviation, in order to calculate the value of the standard error $\sigma_{\bar{X}}$. In practice the value of σ is often unknown. And if σ is known then μ is almost always known, making the calculation of a confidence interval unnecessary.

In the absence of a value of σ it is possible to substitute the sample standard deviation s. The confidence interval must then be based on the t distribution. The 95% interval is

$$\bar{X} - t_{(0.05)}\, \hat{\sigma}_{\bar{X}} \text{ to } \bar{X} + t_{(0.05)}\, \hat{\sigma}_{\bar{X}}$$

where $t_{(0.05)}$ is the value of t in Table E that for n - 1 degrees of freedom just reaches significance at the 5% level, i.e. corresponds to a P value of 0.05.

In general, if α% is the chosen level of confidence, the interval is $\bar{X} - t_{(100-\alpha)/100}\hat{\sigma}_{\bar{X}}$ to $\bar{X} + t_{(100-\alpha)/100}\hat{\sigma}_{\bar{X}}$ where $t_{(100-\alpha)/100}$ is the value of t that for n - 1 degrees of freedom just reaches significance at the $(100-\alpha)$% level, i.e. corresponds to a P value of $(100-\alpha)/100$.

The above confidence limits based on t assume that the population is normally distributed or else that the sample is very large. The limits also assume sampling with individual replacements, or alternatively sampling without replacements for an infinite population.

EXAMPLE

Find the 95% confidence limits for the mean size of households living in privately owned accommodation in Camden (see the last chapter for details).

The mean size of households was found to be 2.31 persons, and the standard deviation 1.40. As 3019 households were included in the sample, the degrees of freedom were 3019 - 1 = 3018. Referring to Table E, the value of t corresponding to P = 0.05 is approximately 1.96. The 95% confidence interval is $2.31 - 1.96\hat{\sigma}_{\bar{X}}$ to $2.31 + 1.96\hat{\sigma}_{\bar{X}}$ persons per household where $\hat{\sigma}_{\bar{X}}$, the estimated standard error, $= s\sqrt{n} = 1.40/\sqrt{3019} = 0.02548$. The interval is therefore 2.31 - 1.96 (0.02548) to 2.31 + 1.96 (0.02548) persons per household = 2.31 - 0.0499 to 2.31 + 0.0499 persons per household = 2.26 to 2.36 persons per household. One can with 95% confidence assert that the true mean size of households lies between these limits.

Since the sample is large, it is permissible to use the Z formula to calculate the confidence interval. Applying a correction for continuity, the interval is

$$(\bar{X} - \tfrac{1}{2}n) - 1.96\,\frac{\sigma}{\sqrt{n}}\left(\sqrt{\frac{N-n}{N-1}}\right) \text{ to } (\bar{X} + \tfrac{1}{2}n) + 1.96\,\frac{\sigma}{\sqrt{n}}\left(\sqrt{\frac{N-n}{N-1}}\right)$$

$$= \left(2.31 - \frac{1}{6038}\right) - \frac{1.96(1.64)}{\sqrt{3019}}\left[\sqrt{\frac{64070}{67088}}\right] \text{ to } \left(2.31 + \frac{1}{6038}\right)$$

$$+ \frac{1.96(1.64)}{\sqrt{3019}}\left[\sqrt{\frac{64070}{67088}}\right] = (2.3098 - 0.0572) \text{ to } (2.3102 + 0.0572)$$

or 2.25 to 2.37 persons per household.

These limits are very slightly different from those calculated using t partly because of the correction for continuity and partly because allowance has been made for the finite size of the population.

CONFIDENCE LIMITS FOR A POPULATION PROPORTION

The Z distribution can be used to provide confidence limits for a population proportion p. As explained previously, provided the sample is large and p not too near 0 or 1, the sampling distribution of the sample proportion $\hat{p}$ is approximately normal with mean p and variance $p(1 - p)/n$. An unbiased estimate of p is the sample proportion $\hat{p}$. Assuming sampling with individual replacements (or equivalently sampling without replacements from an infinite population) an unbiased estimate of the variance of the sampling distribution of $\hat{p}$ is $\hat{p}(1 - \hat{p})/(n - 1)$. The 95% confidence interval is therefore approximately

$$\hat{p} - 1.96\sqrt{\frac{\hat{p}(1-\hat{p})}{n-1}} \text{ to } \hat{p} + 1.96\sqrt{\frac{\hat{p}(1-\hat{p})}{n-1}}$$

With a correction for continuity this becomes

$$(\hat{p} - \tfrac{1}{2}n) - 1.96\sqrt{\frac{\hat{p}(1-\hat{p})}{n-1}} \text{ to } (\hat{p} + \tfrac{1}{2}n) + 1.96\sqrt{\frac{\hat{p}(1-\hat{p})}{n-1}}$$

If the sampling is without individual replacements, and the population is finite, the confidence interval is slightly smaller

$$(\hat{p} - \tfrac{1}{2}n) - 1.96\sqrt{\left\{\left[\frac{\hat{p}(1-\hat{p})}{n-1}\right]\left[\frac{N-n}{N-1}\right]\right\}}$$

$$\text{to } (\hat{p} + \tfrac{1}{2}n) + 1.96\sqrt{\left\{\left[\frac{\hat{p}(1-\hat{p})}{n-1}\right]\left[\frac{N-n}{N-1}\right]\right\}}$$

If n is small and p is close to either 0 or 1, the sampling distribution of $\hat{p}$ is appreciably skewed. The normal distribution no longer serves as a useful approximation. Cochran (1953, p.41) suggests the working rules given in Table 12.1. If the value of n is too small for the normal approximation to be useful, binomial probabilities can be used to calculate confidence intervals. The interested reader should consult Yamane (1964, pp. 576-578).

TABLE 12.1 - LIMITATIONS CONCERNING THE NORMAL APPROXIMATION

If $\hat{p}$ equals:	*Use the normal approximation only if n is equal to or greater than:*
0.5	30
0.4 or 0.6	50
0.3 or 0.7	80
0.2 or 0.8	200
0.1 or 0.9	600
0.05 or 0.95	1400

EXAMPLE

Find the 95% confidence limits for the proportion of households in Camden living in privately owned accommodation (see the last chapter for details).

Because the sample is large (3019 households) and the sample proportion $\hat{p} = 0.41$ is nowhere near 0 or 1 the sampling distribution of $\hat{p}$ is approximately normal. The 95% limits are therefore

$$(\hat{p} - \tfrac{1}{2}n) - 1.96 \sqrt{\left\{ \left[\frac{\hat{p}(1-\hat{p})}{n-1}\right] \left[\frac{N-n}{N-1}\right] \right\}}$$

$$\text{to } (\hat{p} + \tfrac{1}{2}n) + 1.96 \sqrt{\left\{ \left[\frac{p(1-p)}{n-1}\right] \left[\frac{N-n}{N-1}\right] \right\}}$$

$$= \left(0.41 - \frac{1}{6038}\right) - 1.96 \sqrt{\frac{(0.41)(0.59)(67089-3019)}{(3018)(67088)}} \text{ to}$$

$$\left(0.41 + \frac{1}{6038}\right) + 1.96 \sqrt{\frac{(0.41)(0.59)(67089-3019)}{(3018)(67088)}}$$

$$= 0.40983 - 1.96\ (0.00875) \text{ to } 0.41017 + 1.96\ (0.000875)$$

or 0.39 to 0.43.

One can state with 95% confidence that the true proportion lies between these limits.

13 Paired-Sample Tests Based on t and W

Sometimes it makes sense to measure each of the items or individuals forming a sample not once, as has been assumed in previous chapters, but twice, first under one set of conditions, then under another. For example, the gradient of a beach might be measured at the same randomly selected locations both before and after a storm to see if there was any change. Or a random sample of commuters might be approached and asked how long they take to travel to work using (a) public transport, (b) their own cars - the objective being to determine which is the quicker method of travel. Or two different instruments for measuring soil moisture might be tried out side by side at a number of sites selected at random in order to see if there is any consistent difference in the readings.

Alternatively, the items or individuals in a sample might be divided into two parts, and each part measured separately. For instance, a number of mountains might be selected at random in a certain region and the altitude of the tree line determined on the north and south side of each. Or certain cities might be randomly selected and the density of roads calculated for the inner and outer zones. Or a number of marriages might be chosen at random from those entered on a parish marriage register. For each marriage partner the place of birth might be noted and a calculation made of its distance from the parish. The question might be asked, "Do husbands move greater distances than their wives?".

Such investigations have in common the collection of a pair of measurements for each item or individual included in the sample. The measurements relate to the same variable. Although they are divisible into two sets (before and after, husbands and wives, etc.) the measurements are not independent. For every measurement in one set there is a corresponding measurement in the other set.

The first step in the analysis of paired measurements is to calculate the difference between the measurements in each pair.[1] The null hypothesis is adopted that the differences are due to chance sampling. In other words, they constitute a sample drawn from a population of differences whose mean is zero. A test of significance is run to see whether the null hypothesis is tenable. If the answer is no, the differences between the two sets of measurements are interpreted as at least partly real.

---

[1]Alternatively, the relationship between the pairs of measurements may be investigated using regression and correlation analysis. See Chapters 17 - 20.

The simplest test of significance is the *paired-sample t test*. It is no different in principle from the one-sample t test.

THE PAIRED-SAMPLE t TEST

TO TEST WHETHER TWO PAIRED SAMPLES DIFFER SIGNIFICANTLY IN MEAN

THE DATA

A random sample of n items is drawn from a population. Each of the items is measured twice in respect of a continuous variable X. The first measurement relates to one set of conditions (or to part of the item only), the second measurement relates to another set of conditions (or to another part of the item). The purpose of making the two measurements is to compare the two sets of conditions (or parts).

PROCEDURE

1. Set up the null hypothesis: the measurements forming each pair differ purely because of chance sampling. The differences represent a sample drawn from a population of normally distributed differences whose mean is zero.

2. Calculate the difference between the first and second measurement in each pair. The difference d is equal to $(X_1 - X_2)$ where X_1 is the first measurement and X_2 the second.

3. Taking due account of sign, add up all the differences, so obtaining Σd. Next calculate the mean difference, $\bar{d} = \Sigma d/n$ where n is the number of differences, i.e. pairs of measurements.

4. Calculate s_d, the standard deviation of the differences

$$s_d = \sqrt{\frac{\left[\Sigma d^2 - \frac{(\Sigma d)^2}{n}\right]}{n - 1}}$$

5. Estimate $\hat{\sigma}_{\bar{d}}$, the standard error of the mean difference

$$\hat{\sigma}_{\bar{d}} = s_d/\sqrt{n}$$

6. Calculate t, the difference between the observed mean $(\bar{d})$ and the assumed population mean (zero), divided by the estimated standard error

$$t = |\bar{d} - 0|/\hat{\sigma}_{\bar{d}} = |\bar{d}|/\hat{\sigma}_{\bar{d}}$$

7. Calculate df, the number of degrees of freedom associated with t

$$df = n - 1.$$

8. Decide whether a one- or two-tailed test is appropriate.

9. Determine P, the probability of obtaining the calculated value of t, or one greater, as a result of sampling error. P is the prob-

ability of obtaining a sample of differences with a mean differing from zero by the observed amount or a greater amount.

Selected values of P appropriate for two-tailed tests are given in Table E. Values of P for one-tailed tests are exactly half those listed in Table E.

10. Decide whether to retain or reject the null hypothesis. If P is low, serious doubt is cast on the null hypothesis. Reject the null if P is equal to or less than whatever level of significance is judged appropriate, e.g. 0.05 or 0.01. Assume that the differences were drawn from a population whose mean was not zero, that is to say the two sets of conditions (or parts of the items) differ. If P is greater than the level of significance reserve judgement, i.e. regard the null hypothesis as possibly, but not necessarily, correct.

EXAMPLE

It is difficult to determine stream lengths accurately from topographic maps. Even large-scale maps cannot always be relied upon to show streams correctly. In a paper published in 1957, Marie Morisawa describes how she visited 11 small drainage basins in the Appalachian Plateau of northern Pennsylvania, and measured the total length of the streams in each basin using a tape. She also measured the stream lengths as marked in blue on the best-available topographic maps of the area (scale 1:24,000). Table 13.1 shows the two sets of measurements.

TABLE 13.1 - STREAM LENGTHS, APPALACHIAN PLATEAU

Drainage basin	*Length of streams in km as measured in the field* X_1	*Length in km as measured from published maps* X_2	$d=X_1-X_2$	d^2
1.	0.747	0.000	0.747	0.558009
2.	1.080	0.000	1.080	1.166400
3.	1.138	0.914	0.224	0.050176
4.	0.700	0.732	-0.032	0.001024
5.	1.223	1.097	0.126	0.015876
6.	1.924	0.853	1.071	1.147041
7.	1.540	0.793	0.747	0.558009
8.	1.432	0.914	0.518	0.268324
9.	2.251	1.706	0.545	0.297025
10.	1.125	1.220	-0.095	0.009025
11.	2.198	2.317	-0.119	0.014161
	15.358	10.546	4.812	4.085070

Source: Morisawa (1957). The lengths measured in the field are corrected to horizontal distances to facilitate comparison with the measurements taken from the maps. Streams in the first two basins are not marked on the maps.

According to the field survey the total length of all the streams was 15.358 km, but the total length as measured from the maps was only 10.546. This appears to be a major difference, but one has to ask whether it is not perhaps an accident of sampling. Since only 11 basins were visited can one be reasonably certain that a significant difference exists? If 11 different basins (of comparable size) were selected, might not the measurements taken from the maps exceed on average the measurements made in the field? A t test is needed to check that the apparent difference between the two methods of measurement is not solely due to chance.

1. Null hypothesis

As a starting point suppose that the total length of the streams in all the small drainage basins shown on the maps is the same whether measured in the field or from the maps. It is pure chance that the total length of the streams included in the sample is greater when measured in the field than when measured from the maps; the difference could with equal probability have been the opposite way round.

2. Let d be the difference between the length of the streams in each drainage basin as measured in the field and as measured from the maps. $\Sigma d = 4.812$.

3. The mean difference in the measurements, $\bar{d} = \Sigma d/n = 4.812/11 = 0.43745$ approximately.

4. The standard deviation of the differences, s_d, is given by the formula

$$s_d = \sqrt{\frac{\left[\Sigma d^2 - \frac{(\Sigma d)^2}{n}\right]}{n - 1}} = \sqrt{\frac{\left[4.08507 - \frac{(4.812)^2}{11}\right]}{10}}$$

$= 0.44498$ approximately.

5. The standard error of the mean is estimated by $\hat{\sigma}_{\bar{d}} = 0.44498/\sqrt{11} = 0.13417$ approximately.

6. The value of t is $|\bar{d}|/\hat{\sigma}_{\bar{d}} = 0.43745/0.13417 = 3.261$ approximately.

7. The degrees of freedom df are $11 - 1 = 10$.

8. A two-tailed test is appropriate because the total length of the streams in all the small drainage basins in the area may be (1) greater when measured in the field than when measured from maps, which is what the sample tends to suggest, or (2) greater when measured from maps than when measured in the field. The second possibility is not supported by the sample data, but cannot be ruled out as logically or physically impossible.

9. Referring to the t table (Table E) P is found to lie between 0.01 and 0.005. In other words, the observed value of t, or one greater, would turn up on average only about once or twice in every 200 samples, assuming the null hypothesis to be correct. The value of P is so low that the null hypothesis can be discounted. One can be practically certain that the apparent difference between the field and map measurements is not due to chance alone.

The field measurements are presumably more accurate than the map measurements though this is not something that can be deduced from the data. The maps are compiled from aerial photographs and contain many small errors, and some large ones. They show only streams that carried water at the times the photographs were taken. Streams that happened to be dry because of drought conditions are not included on the maps, nor are streams in thick forest that were invisible on the photographs. In certain instances the cartographers have inserted blue lines on the maps up valley of the places where the streams actually begin. Marie Morisawa concludes that streams in small drainage basins should not be measured using the lines printed in blue when accurate stream lengths or drainage densities are required.

ASSUMPTIONS AND LIMITATIONS

The paired-sample t test presupposes that the sample differences are drawn randomly from a population of differences. The differences in the population are assumed to have a normal distribution and to be independent of each other. There must be no positive or negative autocorrelation.

Exact normality is, of course, an impossibility. All real differences are finite in size, yet the tails of a normal distribution stretch to minus infinity and plus infinity. The paired t test can nevertheless be relied upon to give reasonably accurate results provided the differences do not depart too greatly from a normal distribution.

If the sample is small, it may not be clear whether the differences form part of a distribution that approaches the normal in form. The paired t test can be run only at the risk of serious error. If the sample is large it may be clear that the differences are part of a distribution that is far from normal. Before the test is used the sample differences must be transformed (e.g. by taking logarithms) so that they become approximately normal. The necessary calculations can often be tedious and time consuming.

The following test makes no assumption of normality, and in this respect is preferable to the paired t test. It is not practicable, however, if n is at all large.

*THE RANDOMISATION TEST FOR PAIRED MEASUREMENTS

TO TEST WHETHER TWO PAIRED SAMPLES DIFFER SIGNIFICANTLY IN MEAN

THE DATA

A random sample of items is drawn from a population. Each of the items is measured twice in respect of a continuous variable X. The first measurement relates to one set of conditions (or to part of the item only), the second measurement relates to another set of conditions (or to another part of the item). The purpose of making the two measurements is to compare the two sets of conditions (or parts).

PROCEDURE

1. Set up the null hypothesis: the measurements forming each pair differ purely because of chance sampling. The differences represent a sample drawn from a population of differences whose mean is zero.

2. For each pair calculate d, the difference in the measurements. $d = X_1 - X_2$ where X_1 is the first measurement and X_2 the second.

3. Add up all the d's so obtaining Σd.

4. Discarding any d that happens to be zero, count how many d's are left, say n.

5. Remove the signs (+ or -) from the non-zero d's, leaving just the numbers.

6. Decide whether a one- or two-tailed test is appropriate.

7. Determine how many of the 2^n separate ways of randomly assigning + or - signs to the non-zero d's yield totals that have the same sign as the observed total Σd and equal or exceed it in numerical value. Let K be the number of ways.

If the null hypothesis is true, all 2^n arrangements of signs are equally likely. The observed arrangement of signs in the sample of d's is purely accidental. Each plus sign could equally well have been a minus sign, and vice versa. Hence, for a one-tailed test, $P = K/2^n$ and, for a two-tailed test, $P = 2K/2^n$.

8. Decide whether to retain or reject the null hypothesis. If P is low serious doubt is cast on the null hypothesis. Reject the null if P is equal to or less than whatever level of significance is judged appropriate, e.g. 0.05 or 0.01. Assume that the differences were drawn from a population whose mean was not zero. If P is greater than the level of significance reserve judgement, i.e. regard the null as possibly, but not necessarily, correct.

EXAMPLE

To revert to the data on stream lengths: the hypothesis to be tested is that the total length of streams in all small drainage basins shown on the maps is the same whether measured in the field or from maps; it is only the lengths of individual streams that vary according to the method of measurement. Although the field measurements for the 11 sample basins mostly exceed the map measurements, the differences could equally easily have been the other way round. Thus the difference of +0.747 for the first basin could with equal probability have been -0.747. Similarly, the difference of +1.080 for the second basin could just as well have been -1.080.

There are 11 basins and $2^{11} = 2048$ possible arrangements of the signs of the differences between the two methods of measurement. Table 13.2 sets out the arrangements with the highest and lowest total scores.

TABLE 13.2 - ARRANGEMENTS OF THE SIGNS OF THE DIFFERENCES BETWEEN THE MEASUREMENTS

Rank	*Possible arrangements of the signs of the differences*											*Total*
1	+0.747	+1.080	+0.224	+0.032	+0.126	+1.071	+0.747	+0.518	+0.545	+0.095	+0.119	+5.304
2	+0.747	+1.080	+0.224	-0.032	+0.126	+1.071	+0.747	+0.518	+0.545	+0.095	+0.119	+5.240
3	+0.747	+1.080	+0.224	+0.032	+0.126	+1.071	+0.747	+0.518	+0.545	-0.095	+0.119	+5.114
4	+0.747	+1.080	+0.224	+0.032	+0.126	+1.071	+0.747	+0.518	+0.545	+0.095	-0.119	+5.066
5	+0.747	+1.080	+0.224	+0.032	-0.126	+1.071	+0.747	+0.518	+0.545	+0.095	+0.119	+5.052
6	+0.747	+1.080	+0.224	-0.032	+0.126	+1.071	+0.747	+0.518	+0.545	-0.095	+0.119	+5.050
7	+0.747	+1.080	+0.224	-0.032	+0.126	+1.071	+0.747	+0.518	+0.545	+0.095	-0.119	+5.002
8	+0.747	+1.080	+0.224	-0.032	-0.126	+1.071	+0.747	+0.518	+0.545	+0.095	+0.119	+4.988
9	+0.747	+1.080	+0.224	+0.032	+0.126	+1.071	+0.747	+0.518	+0.545	-0.095	-0.119	+4.876
10	+0.747	+1.080	+0.224	+0.032	-0.126	+1.071	+0.747	+0.518	+0.545	-0.095	+0.119	+4.862
11	+0.747	+1.080	-0.224	+0.032	+0.126	+1.071	+0.747	+0.518	+0.545	+0.095	+0.119	+4.856
12	+0.747	+1.080	+0.224	+0.032	-0.126	+1.071	+0.747	+0.518	+0.545	+0.095	-0.119	+4.814
13	+0.747	+1.080	+0.224	-0.032	+0.126	+1.071	+0.747	+0.518	+0.545	-0.095	-0.119	+4.812
etc												
2036	-0.747	-1.080	-0.224	+0.032	-0.126	-1.071	-0.747	-0.518	-0.545	+0.095	+0.119	-4.812
2037	-0.747	-1.080	-0.224	-0.032	+0.126	-1.071	-0.747	-0.518	-0.545	-0.095	+0.119	-4.814
2038	-0.747	-1.080	+0.224	-0.032	-0.126	-1.071	-0.747	-0.518	-0.545	-0.095	-0.119	-4.856
2039	-0.747	-1.080	-0.224	-0.032	+0.126	-1.071	-0.747	-0.518	-0.545	+0.095	-0.119	-4.862
2040	-0.747	-1.080	-0.224	-0.032	-0.126	-1.071	-0.747	-0.518	-0.545	+0.095	+0.119	-4.876
2041	-0.747	-1.080	-0.224	+0.032	+0.126	-1.071	-0.747	-0.518	-0.545	-0.095	-0.119	-4.988
2042	-0.747	-1.080	-0.224	+0.032	-0.126	-1.071	-0.747	-0.518	-0.545	-0.095	+0.119	-5.002
2043	-0.747	-1.080	-0.224	+0.032	-0.126	-1.071	-0.747	-0.518	-0.545	+0.095	-0.119	-5.050
2044	-0.747	-1.080	-0.224	-0.032	+0.126	-1.071	-0.747	-0.518	-0.545	-0.095	-0.119	-5.052
2045	-0.747	-1.080	-0.224	-0.032	-0.126	-1.071	-0.747	-0.518	-0.545	-0.095	+0.119	-5.066
2046	-0.747	-1.080	-0.224	-0.032	-0.126	-1.071	-0.747	-0.518	-0.545	+0.095	-0.119	-5.114
2047	-0.747	-1.080	-0.224	+0.032	-0.126	-1.071	-0.747	-0.518	-0.545	-0.095	-0.119	-5.240
2048	-0.747	-1.080	-0.224	-0.032	-0.126	-1.071	-0.747	-0.518	-0.545	-0.095	-0.119	-5.304

As regards signs, the lower part of the table is the exact reverse of the upper part. The sample under study ranks 13th in the sequence of arrangements. There are 26 arrangements that give totals deviating from zero by as much as, or more than, the sample under study (including the sample itself). The probability P of obtaining the sample or one more extreme is therefore 26/2048 = 0.0127 approximately. Because P is minute the hypothesis under test can be discarded. It is reasonable to assume that stream lengths measured from the maps will be shorter on average than measurements made in the field.

The same conclusion was drawn from the one-sample t test. Note, however, that the present test yields a slightly higher value of P, reflecting its rather different operating assumptions.

ASSUMPTIONS AND LIMITATIONS

1. Independence of the d's. Like the paired-sample t test the randomisation test assumes that the sample differences are drawn randomly from a population of differences that are independent of one another.

2. Symmetry of the d's. The randomisation test is frequently described as an "exact" test. It is exact in the sense that it utilizes all the information contained in the sample. However, it is also approximate in that it fails to consider all the samples that could be drawn from the population of differences assumed by the null hypothesis. The number of possible samples is of course infinite. Yet the randomisation test considers only the 2^n samples (arrangements) created by randomising the signs of the observed differences. In this sense it is radically different from conventional statistical tests such as the Z test or the t test.

The test will yield accurate values of P only if the 2^n samples obtained by randomisation are representative of the infinite number of samples that might be drawn from the population of differences. Obviously it is essential that the distribution of differences is symmetrical. If the distribution is asymmetric the test will not necessarily give correct values of P.[1] Unfortunately, there is generally no way of knowing for certain whether the distribution is symmetrical. The observed differences may suggest that the distribution is symmetrical but they cannot prove it absolutely.

Subject to this qualification, the randomisation test would appear to be a very powerful and useful test. It does not assume normality as does the paired-sample t test.

3. Exclusion of zero d's. The exclusion of zero d's is somewhat arbitrary and difficult to justify.

4. Computational complexity. The test is impractical if the number of pairs is at all large. For instance, if n is 20, the total number of possible arrangements of signs is 1,048,576. Thus, if one is working to a 5% level of significance, it may be necessary to sort through up to 27,000 arrangements in order to establish whether or

[1]Not all statisticians accept this. See, for example, Bradley (1968).

not the sample is significant (54,000 in the case of a one-tailed test). Unless the sample represents a very extreme arrangement (i.e. one where all or nearly all the signs are the same), the amount of work will be so great as to require a computer. If a computer is not available, there are two main alternatives. The first is to use the paired-sample t test. The second alternative is to substitute ranks for absolute numbers. The resulting test is called the "Wilcoxon signed-ranks matched-pairs test", or more briefly, the Wilcoxon signed-ranks test. Despite its separate name, it is in principle the same as the randomisation test.

*THE WILCOXON SIGNED-RANKS TEST

TO TEST WHETHER TWO PAIRED SAMPLES DIFFER SIGNIFICANTLY IN MEAN

THE DATA

A random sample of items is drawn from a population. Each of the items is measured twice in respect of a continuous variable X. The first measurement relates to one set of conditions (or to part of the item only), the second measurement relates to another set of conditions (or to another part of the item). The purpose of making the two measurements is to compare the two sets of conditions (or parts).

PROCEDURE

1. Set up the null hypothesis: the measurements forming each pair differ purely because of chance sampling. The differences represent a sample drawn from a population of normally distributed differences whose mean is zero.

2. For each pair of measurements calculate the difference, d. $d = X_1 - X_2$ where X_1 is the first measurement and X_2 the second.

3. Rank the d's in order of size. Ignore any d that is zero. Assign the rank of 1 to the d that is smallest numerically (regardless of sign), the rank of 2 to the next smallest, and so on. If two or more d's are the same numerically, assign to each the average of the ranks that would have been assigned had they been distinguishable. For example, if a d ranks 4 and the next three d's tie, the three ought each to be given a rank of $(5 + 6 + 7)/3 = 6$.

4. Attach the sign of the difference to each rank. Attach a plus sign to each rank that represents a positive d and a minus sign to each rank that represents a negative d.

5. Calculate W, the sum of the positive ranks or the sum of the negative ranks whichever is the smaller. Discard the sign of W if it is negative.

6. Decide whether a one- or two-tailed test is appropriate.

7. Determine the probability, P, of getting the calculated value of W or one smaller, through chance sampling. Let n be the total number of d's bearing a sign (i.e. the total number of pairs of measurements minus any for which d = 0). If n is 20 or less consult

Table F which lists values of P and W for one-tailed tests. Note that the value of P for a two-tailed test is twice that for a one-tailed test, i.e. twice the value shown in Table F.

If n is larger than 20, W is approximately normally distributed with mean $\mu_W = n(n + 1)/4$ and standard deviation $\sigma_W = \sqrt{[n(n + 1)(2n + 1)/24]}$. Estimate the value of P associated with W by calculating

$$Z = \frac{|W - \mu_W| - \frac{1}{2}}{\sigma_W} = \frac{\left|W - \frac{n(n + 1)}{4}\right| - \frac{1}{2}}{\sqrt{\frac{n(n + 1)(2n + 1)}{24}}}$$

and enter Table B (the Z table) to determine the approximate value of P. The 1/2 in the numerator of Z is a continuity correction to allow for the fact that W is limited to integer values.

EXAMPLE

The test will be applied to the same data as the randomisation and one-sample t tests. The unsigned ranks of the differences between the lengths of the streams as measured in the field and from maps are calculated in Table 13.3. The signed ranks are simply the unsigned ranks with the signs of the differences attached. The sum

TABLE 13.3 - RANKS OF THE DIFFERENCES BETWEEN THE MEASUREMENTS

Drainage basin	*d*	*Unsigned rank*	*signed rank*	*Drainage basin*	*d*	*Unsigned rank*	*Signed rank*
1	0.747	8.5	+8.5	7	0.747	8.5	+8.5
2	1.080	11	+11	8	0.518	6	+6
3	0.224	5	+5	9	0.545	7	+7
4	-0.032	1	-1	10	-0.095	2	-2
5	0.126	4	+4	11	-0.119	3	-3
6	1.071	10	+10				

of the negative ranks is -6. The sum of the positive ranks is 60. Hence W = 6. Referring to Table F, the values of P for a two-tailed test and n = 11, W = 6 is found to be 2(0.0068) = 0.0136. In other words, there is less than one chance in fifty of obtaining the observed value of W, or an even smaller value, through chance sampling. The evidence suggests that stream lengths measured from the maps will be shorter on average than measurements made in the field.

Values of P obtained from Table F are correct to 4 decimal places for a one-tailed test, but only two decimal places for a two-tailed test because the table entries have to be doubled. In the present instance, although P is nominally 0.0136, only the first two decimal places are significant, the last two may or may not be in error.

If desired, the precise value of P can be found by randomising the signs of the ranks as demonstrated in the accompanying table (Table 13.4). There are 28 arrangements of the signs of the ranks that yield values of W of 6 or less (14 at the top of the table and 14, not shown, at the bottom). The total number of possible arrangements is $2^{11} = 2048$. Hence P is exactly 28/2048 = 0.0137 to four significant decimal places.

It is worth noting that the Z formula gives a fairly accurate estimate of P even though n is only 11

$$Z = \frac{\left|W - \mu_W\right| - \frac{1}{2}}{\sigma_W} = \frac{\left|6 - \frac{11(12)}{4}\right| - \frac{1}{2}}{\sqrt{\frac{11(12)(23)}{24}}} = \frac{26.5}{\sqrt{\frac{11(23)}{2}}}$$

$$= 2.3561 \text{ approximately.}$$

Referring to Table B, P is found to be 0.0188 which for most purposes would be an acceptable approximation to the true value of 0.0127 yielded by the randomisation test.

TABLE 13.4 - ARRANGEMENTS OF THE SIGNS OF THE RANKED DIFFERENCES IN THE MEASUREMENTS

Permutations of signs of ranks											*Sum of positive ranks or negative ranks, whichever is smaller* W	*Sum of positive and negative ranks*	*Probability of obtaining this sum or one greater (sign ignored)* P
+1	+2	+3	+4	+5	+6	+7	+8.5	+8.5	+10	+11	0	66	2/2048
-1	+2	+3	+4	+5	+6	+7	+8.5	+8.5	+10	+11	1	64	4/2048
+1	-2	+3	+4	+5	+6	+7	+8.5	+8.5	+10	+11	2	62	6/2048
-1	-2	+3	+4	+5	+6	+7	+8.5	+8.5	+10	+11	3	60	10/2048
+1	+2	-3	+4	+5	+6	+7	+8.5	+8.5	+10	+11	3	60	10/2048
+1	+2	+3	-4	+5	+6	+7	+8.5	+8.5	+10	+11	4	58	14/2048
-1	+2	-3	+4	+5	+6	+7	+8.5	+8.5	+10	+11	4	58	14/2048
+1	+2	+3	+4	-5	+6	+7	+8.5	+8.5	+10	+11	5	56	20/2048
+1	-2	-3	+4	+5	+6	+7	+8.5	+8.5	+10	+11	5	56	20/2048
-1	+2	+3	-4	+5	+6	+7	+8.5	+8.5	+10	+11	5	56	20/2048
+1	+2	+3	+4	+5	-6	+7	+8.5	+8.5	+10	+11	6	54	28/2048
-1	-2	-3	+4	+5	+6	+7	+8.5	+8.5	+10	+11	6	54	28/2048
-1	+2	+3	+4	-5	+6	+7	+8.5	+8.5	+10	+11	6	54	28/2048
+1	-2	+3	-4	+5	+6	+7	+8.5	+8.5	+10	+11	6	54	28/2048
+1	+2	+3	+4	+5	+6	-7	+8.5	+8.5	+10	+11	7	52	38/2048
-1	+2	+3	+4	+5	-6	+7	+8.5	+8.5	+10	+11	7	52	38/2048
+1	-2	+3	+4	-5	+6	+7	+8.5	+8.5	+10	+11	7	52	38/2048
+1	+2	-3	-4	+5	+6	+7	+8.5	+8.5	+10	+11	7	52	38/2048
-1	-2	+3	-4	+5	+6	+7	+8.5	+8.5	+10	+11	7	52	38/2048

etc.

ASSUMPTIONS AND LIMITATIONS

The Wilcoxon signed-ranks test makes the same assumptions and suffers from the same limitations as the randomisation test. In addition, by substituting ranks for actual differences, it introduces an element of approximation that is absent from the randomisation test. Its great merit is that it is relatively simple to carry out.

The test assumes that X is a continuous variable. In theory, there ought not to be any d's that are tied or exactly zero. In practice, however, ties and zeros do occur even if X is a continuous variable, simply because of rounding errors. In the worked example, two drainage basins tie (Numbers 1 and 7) although it is unlikely that they are exactly equal. If there is a difference, however, it is so slight that it does not show up with 3 places of decimals.

The distribution of W under the null hypothesis is of course altered if ties or zeros occur. Unless a large proportion of d's are tied or are zero, however, the effect on the value of P is small and can be ignored. If none of the d's are zero, the value of P will only need to be corrected if there are ties that involve the more extreme ranks, so affecting the tails of the distribution of W.

Table F does not allow for ties and zeros because this would make it too complicated. If the number of d's is small, the exact value of P can be found by listing all the relevant permutations, as demonstrated in Table 13.4.[1]

If the number of d's is large, listing the permutations is impractical. Instead, one can modify the formula for Z

$$Z = \frac{\left| W - \dfrac{n(n+1) - d_0(d_0+1)}{4} \right| - \frac{1}{2}}{\sqrt{\left\{ \dfrac{n(n+1)(2n+1) - d_0(d_0+1)(2d_0+1)}{24} - \dfrac{\Sigma[d_i(d_i-1)(d_i+1)]}{48} \right\}}}$$

where n is the total number of differences (including zero differences), d_0 is the number of zero differences, and d_i is the number of non-zero differences, forming each tie. For example, suppose the differences are -7, -6, -3, -3, 0, 0, 2, 2, 2, 5, 8 and 10. There are 12 differences, hence n = 12. Two differences are zero, hence $d_0 = 2$. There are two ties amongst the non-zero differences, one composed of 3 differences (2, 2 and 2) and one composed of two differences (-3 and -3). Hence $\Sigma d_i(d_i - 1)(d_i + 1) = 3(3 - 1)(3 + 1) + 2(2 - 1)(2 + 1) = 24 + 6 = 30$. The formula for Z in this instance reduces to

$$Z = \frac{\left| W - \dfrac{12(13) - 2(3)}{4} \right| - \frac{1}{2}}{\sqrt{\dfrac{12(13)(25) - 2(3)(5)}{24} - \dfrac{30}{48}}} = \frac{|W - 38|}{\sqrt{\dfrac{3855}{24}}} = 1.105$$

[1]Tied differences are treated as if they are fractionally different. Thus +0.5, +7.2, +7.2, +10.3 and +12.4 and +0.5, +7.2, +7.2, +10.3 and +12.4 count as two separate permutations since each 7.2 represents a different d.

It is worth perhaps noting that there are alternative procedures for handling ties and zeros. Pratt (1959) and Conover (1973) give extended discussions. The procedure adopted here follows Lehmann (1975).

14 Two-Sample Tests Based on F, t and *U*

Many research studies seek to determine whether two populations differ in mean, standard deviation or other numerical properties. Do people in Wales and Scotland differ in mean income? Do people living in urban areas devote more time on average to recreational activities than people living in rural areas? Are slope angles on chalk more variable than those on granite? Usually, the populations in question are so big that they can only be studied by using samples. This introduces uncertainty. If the samples are found to differ in mean, standard deviation etc. this may be because the populations from which they are drawn differ or it may be just a matter of chance (sampling error). Even if two samples are drawn from the same population, they are liable to differ simply because they include different members of the population.

The smaller the difference between two samples, the more likely it is that the difference is due to chance. In other words, if the difference is small enough, the two samples can be plausibly regarded as belonging to populations that are identical as regards the variable under study. On the other hand, if the difference is large, then the samples are more likely to represent populations that are distinct. Figure 14.1A for example, represents two normally distributed samples that differ slightly in variance, though not in mean. Obviously, they are more likely to be drawn from identical populations than the two samples shown in Fig. 14.1B which differ markedly in variance.

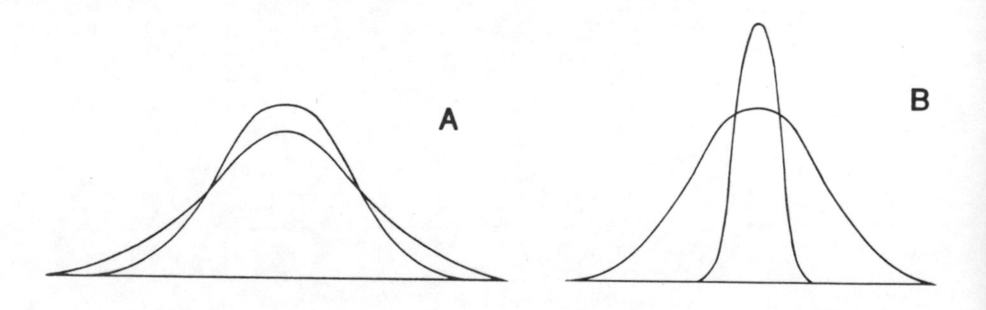

Fig. 14.1 - Comparison of two normally distributed samples. (A) two samples that differ slightly in variance, (B) two samples that differ markedly in variance.

Statisticians have invented a great many tests to decide whether two samples are likely to represent populations that are identical or different. Only a few of the more important tests are described in this chapter. The *F test*, the first to be considered, is used to compare the variances of two samples drawn from normal or near normal populations. The second test to be described, the *two-sample t test*, is used to compare the means of two samples drawn from normal or near normal populations. It is one of the most widely used of all statistical tests.

If the two samples under investigation are far from normal, they can often be transformed into approximately normal ones by various devices such as taking logarithms. However, the same transformation has to be applied to both samples. If one sample is skewed whereas the other is normal or skewed the other way, transformation will not help. One of the best tests in the circumstances is the *Mann-Whitney U test*, described at the end of the chapter. The test is a very general one in that it considers differences in mean, variance, and also shape.

THE F RATIO

Imagine two normal populations, A and B, of infinite size. Both populations have the same mean μ and variance σ^2. A series of random samples, each consisting of n_a items, are drawn from Population A and the variance s_a^2 of each sample is calculated. A second series of random samples, each containing n_b items, are drawn from Population B. The sample variances s_b^2 are then calculated.

Now consider a pair of samples, one sample representing Population A and the other Population B. Let F be the ratio of the sample variances, s_a^2/s_b^2. Suppose one calculates the F ratio for all possible pairs of samples representing the two populations. The frequency distribution (sampling distribution) of the F ratio is called the F distribution.

There are as many F distributions as there are sample sizes. Some representative examples are shown in Fig. 14.2. The mean of every F distribution is exactly 1, because the two populations share a common variance σ^2 and in the absence of sampling error s_a^2/s_b^2 is 1. The distributions are asymmetrical with a tail to the right since F values of less than zero cannot occur, but large positive values are possible.

The F ratio can be used for testing whether the variances of two samples differ significantly. The nearer the ratio is to unity the more reasonable it is to assume that the samples are drawn from populations with the same variance. If the ratio departs greatly from unity because the sample variances are very dissimilar, the assumption can be made that the samples are drawn from populations with different variances.

When running a test of significance it is customary to designate the larger of the variances s_a^2 and the smaller s_b^2. This means that the value of F cannot be less than 1. Always placing the larger variance over the smaller one suppresses the lower (left) tail of the F distribution. Abnormally small F values are converted to abnormally large ones.

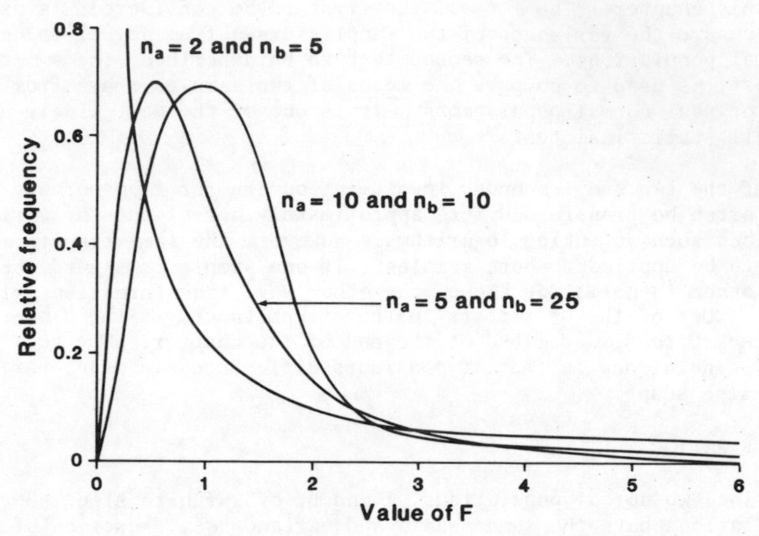

Fig. 14.2 - Some F distributions for different sample sizes.

THE F TEST

TO TEST WHETHER TWO SAMPLES FROM NORMALLY DISTRIBUTED POPULATIONS DIFFER SIGNIFICANTLY IN VARIANCE

THE DATA

Two random samples A and B are drawn independently of one another from different populations. Each of the items making up the two samples is measured once in respect of a continuous variable X. Both populations possess normal distributions of X. It is desired to test the hypothesis that the populations differ in variance.

PROCEDURE

1. Set up the null hypothesis: the two samples are drawn from normal populations with the same variance. The variances of the samples differ merely because of chance. The difference is not statistically significant.

2. Calculate the variance of each sample in the usual way.

3. Calculate the upper (= right) tail value of F

$$F = \frac{\text{larger variance}}{\text{smaller variance}}$$

4. For both samples calculate the degrees of freedom

$$df = n - 1$$

where n is the size of the sample.

5. Decide whether a one- or two-tailed test is appropriate.

6. Using Table G find P, the probability of obtaining the calculated value of F, or one greater, if the population variances are equal. Table G lists only upper-tail values of F since the calculated value of F cannot fall in the lower tail. When consulting the table use the column corresponding to the degrees of freedom in the sample with the larger variance (the numerator in the formula for F) and the row corresponding to the degrees of freedom in the sample with the smaller variance (the denominator in the F formula). The entries in the body of the table are values of F. The corresponding values of P for two-tailed tests are listed on the left and right sides of the table. Values of P for one-tailed tests are exactly half those shown in the table.

If the calculated value of F falls between two values shown in the table, the value of P can be estimated by interpolation.

7. Decide whether to retain or reject the null hypothesis. If P is low, serious doubt is cast on the null hypothesis. Reject the null if P is equal to or less than whatever level of significance is judged appropriate, e.g. 0.05 or 0.01. Conclude that the two samples differ in variance by more than can be reasonably attributed to chance. Assume that the samples were derived from populations that differ in variance.

If P is greater than the level of significance (because F is near 1), reserve judgement, i.e. regard the null hypothesis as possibly, but not necessarily, correct.

EXAMPLE

The Verdugo Hills near Burbank, California, are cut by numerous valleys whose sides tend to be steep and straight in vertical profile. The maximum angle of the valley sides was measured by Strahler (1950) at 171 separate locations where streams were actively cutting a channel at their base. The maximum angles were found to comprise an approximately normal distribution, mean 44.8 degrees and standard deviation 3.27 degrees (Fig. 14.3). Measurements were also made at 33 other locations where the valley sides had escaped basal cutting for a greater or lesser period of time and had deposits of alluvium, slope wash, etc. protecting their base. The maximum angles were again found to comprise an approximately normal distribution, but their mean was somewhat less at 38.2 degrees, as was the standard deviation at 2.70 degrees.

The question arises whether the maximum angles of slopes experiencing basal cutting are really different from the maximum angles of the slopes where basal cutting has ceased. Could the difference in variance, for example, be the result of chance sampling?

1. Null hypothesis: the variance of Sample A (slopes currently experiencing basal cutting) differs from that of Sample B (slopes currently escaping basal cutting) purely because of chance. The two samples belong to populations that have the same variance.

2. The variance of each sample is merely the square of the standard

deviation. The variance of A = $(3.27)^2$ = 10.6929 degrees squared. The variance of B = $(2.70)^2$ = 7.2900 degrees squared.

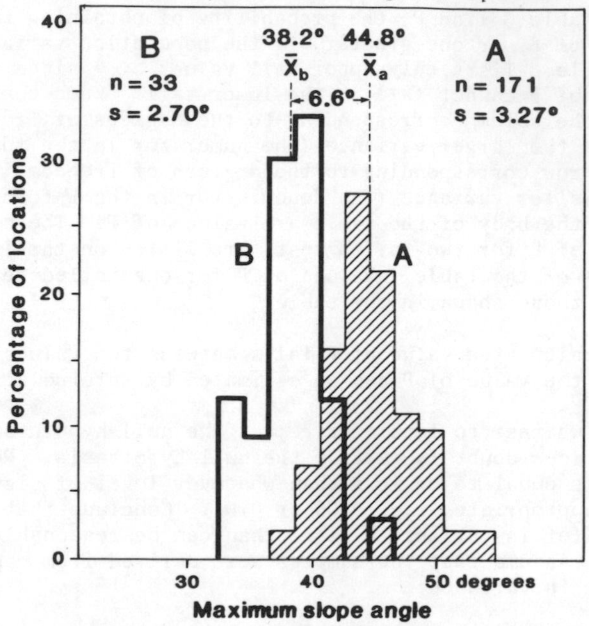

Fig. 14.3 - Histograms of maximum slope angles, Verdugo Hills, California. A (shaded): valley sides currently experiencing basal cutting; B (outline): valley sides that have escaped basal cutting for some time past.

3. F = 10.6929/7.2900 = 1.467.

4. The degrees of freedom associated with Sample A, whose variance is the numerator of F, are df = n - 1 = 170. The degrees of freedom associated with Sample B, whose variance forms the denominator of F, are df = n - 1 = 32.

5. A two-tailed test is appropriate. There was no reason to suppose before the samples were taken that the variance of A would exceed that of B rather than vice versa.

6. The value of P corresponding to F = 1.467 with df = 170 and 32 is approximately 0.10 (Table G). There is about a 1 in 10 chance of obtaining the observed difference between the sample variances, or a greater difference, if the null hypothesis is true.

7. Because the value of P is relatively large the null hypothesis cannot be rejected without further investigation. The observed difference in the variances can plausibly be explained by chance sampling.

The difference in the sample means will be evaluated later using the two-sample t test.

ASSUMPTIONS AND LIMITATIONS

1. The items forming each sample must be randomly selected; they

must not be paired in any way, but must be selected completely independently of one another.

The sampling must be with replacements, or else the populations must be infinite in size. (Sampling without replacements will not cause serious errors unless the samples are very small.)

2. The values of the items in each sampled population are assumed to be independent of one another.

The effect of autocorrelation on the accuracy of the F test has not been investigated in any detail. Some authors suggest that even slight autocorrelation may cause serious errors, particularly if the samples are small. However, further experimental and theoretical work is needed to check this suggestion.

3. Each sample is assumed to be drawn randomly from a population that is normally distributed.

The accuracy of the F test is known to be seriously affected by quite minor departures from normality. Even fairly slight skewness, for example, may render the test inaccurate. The standard deviation (standard error) of the variances of samples drawn from a skewed population tends to be much greater than the standard deviation (standard error) of the variances of samples drawn from a normal population. Very large and very small variances tend to occur with far greater frequency if the sampled population is skewed.

The accuracy of the test does not tend to improve much with increasing sample size, in sharp contrast to the Z and t tests. Large samples, in other words, do not compensate much for non-normality. As Bradley (1968, p.32) has said, "there is no 'Central Limit effect' for sample variances". A further depressing fact is that the accuracy of the F test tends to worsen towards the tails of the sampling distribution. A P value of, say, 0.05 is less trustworthy than a P value of 0.50. Unfortunately, it is the tails that matter in significance testing, not the central portion of the sampling distribution.

Mood (1954), and Siegel and Tukey (1960), have devised alternative tests, based on ranks, that do not require the populations to be normal, although they do require the populations to possess identical means. A small difference in the means will cause the tests to give results that are quite seriously in error. The tests are not very practical because if the sampled populations possess unequal variances it is likely that they will also differ in means. The great virtue of the F test is that it is unaffected by differences in means.

THE TWO-SAMPLE t TEST

*THEORY

Suppose there exist two populations, A and B, each consisting of an infinite number of values of a continuous variable X. The mean of the X values of Population A is μ_a and the variance is σ_a^2, whereas the mean of Population B is μ_b and the variance is σ_b^2. A random sample consisting of n_a items is drawn from Population A and a second random sample consisting of n_b items is drawn from Population B. The difference in the means $(\bar{X}_a - \bar{X}_b)$ is calculated, and the samples then discarded. Two more random samples are drawn, containing the same numbers of items as before, and $(\bar{X}_a - \bar{X}_b)$ is calculated anew. The process is repeated again and again, the idea being to obtain an infinite number of values of $(\bar{X}_a - \bar{X}_b)$ representing all possible pairs of samples from the populations A and B. The values of $(\bar{X}_a - \bar{X}_b)$ can be grouped so as to form a frequency distribution, or *sampling distribution of the difference in means.*

It is possible to prove with a little algebra what is intuitively obvious: that the mean of the sampling distribution is the difference between the population means $(\mu_a - \mu_b)$. The standard error of the difference between the sample means (i.e. the standard deviation of the sampling distribution) is given by the formula

$$\text{Standard error of the difference} = \sqrt{\left(\sigma_{\bar{X}_a}^{\ 2} + \sigma_{\bar{X}_b}^{\ 2}\right)}$$

where $\sigma_{\bar{X}_a}$ is the standard error of the mean of samples drawn from Population A and $\sigma_{\bar{X}_b}$ is the standard error of the mean of samples drawn from Population B. Since $\sigma_{\bar{X}_a}^{\ 2} = \sigma_a^{\ 2}/n_a$ and $\sigma_{\bar{X}_b}^{\ 2} = \sigma_b^{\ 2}/n_b$ an alternative formula for the standard error of the difference between the means is

$$\text{Standard error of the difference} = \sqrt{\left(\frac{\sigma_a^{\ 2}}{n_a} + \frac{\sigma_b^{\ 2}}{n_b}\right)}$$

These formulas for the mean and standard error of the sampling distribution hold good regardless of the form of the population distributions. There is no requirement that the populations be normally distributed. One does not need to know anything about the distributions of the populations in order to use the formulas.

A knowledge of the distributions is useful, however, in that it enables one to predict the form of the sampling distribution of the difference in the means. The form of the sampling distribution varies according to the form of the population distributions - if the samples are of finite size. It is only the mean and the standard error of the sampling distribution that are unvarying.

If both sampled populations are normally distributed, the sampling distribution of the difference in the means is also normal. Moreover, the Central Limit Theorem (see Chapter 11) operates, just as it does for the sampling distributions of the individual means, $\bar{X}_a$ and $\bar{X}_b$. If n_a and n_b are made infinitely large, the sampling distribution of the difference between the means $(\bar{X}_a - \bar{X}_b)$ becomes exactly normal, regardless of the distributions of the sampled populations. If one has two infinitely large samples drawn randomly from different populations, the quantity

$$Z = \frac{|(\bar{X}_a - \bar{X}_b) - (\mu_a - \mu_b)|}{\text{standard error of the difference}} = \frac{|(\bar{X}_a - \bar{X}_b) - (\mu_a - \mu_b)|}{\sqrt{\left(\dfrac{\sigma_a^2}{n_a} + \dfrac{\sigma_b^2}{n_b}\right)}}$$

is distributed normally, and is identical with the Z whose probabilities are listed in Table B (the Z table). And, if the samples are reasonably large, but not infinite, the Z in the above formula will still approximate closely to the Z in Table B. One has here the basis of a significance test. Suppose it is important to know whether or not the two population means μ_a and μ_b can be considered identical. Provided σ_a^2 and σ_b^2 are known, the standard error of the difference in the means can be calculated. The null hypothesis that $\mu_a = \mu_b$ (or equivalently that $\mu_a - \mu_b = 0$) can be tested by calculating Z

$$Z = \frac{|\bar{X}_a - \bar{X}_b|}{\sqrt{\left(\dfrac{\sigma_a^2}{n_a} + \dfrac{\sigma_b^2}{n_b}\right)}}$$

and looking up the value of P in Table B. If P is small, the null hypothesis can be abandoned and μ_a can be assumed to differ from μ_b. If P is large, the null hypothesis is best retained, even though μ_a may not be exactly equal to μ_b.

In most sampling situations σ_a^2 and σ_b^2 are unknown. Unbiased estimates, however, are provided by the sample variances s_a^2 and s_b^2. Provided the samples are reasonably large

$$\hat{Z} = \frac{|\bar{X}_a - \bar{X}_b|}{\sqrt{\left(\dfrac{s_a^2}{n_a} + \dfrac{s_b^2}{n_b}\right)}}$$

will approximate closely to the Z of Table B. The approximate probability P of obtaining a given value of $\hat{Z}$, or one greater, as a result of chance sampling can be found from Table B by simply substituting $\hat{Z}$ for Z.

If the two samples are small, Table B will not yield an accurate enough value of P. However, provided the populations are normally distributed, recourse may be had to the t distribution. Regardless of the size of the samples, the quantity

$$t = \frac{|\bar{X}_a - \bar{X}_b|}{\sqrt{\left(\dfrac{s_a^2}{n_a} + \dfrac{s_b^2}{n_b}\right)}}$$

will be approximately distributed as Student's t with

$$\left(\frac{s_a^2}{n_a} + \frac{s_b^2}{n_b}\right) \Bigg/ \left(\frac{s_a^4}{n_a^3} + \frac{s_b^4}{n_b^3}\right)$$

degrees of freedom. The degrees of freedom will not necessarily be an integer and the value of P may need to be found by interpolation in the t table.

If the parent populations have the same variance σ^2, the variances of the two samples can be combined to obtain a "pooled" estimate of σ^2 given by

$$\hat{\sigma}^2 = \frac{s_a^2(n_a - 1) + s_b^2(n_b - 1)}{(n_a + n_b - 2)}$$

where $\hat{\sigma}^2$, the estimate of σ^2, is pronounced "sigma hat squared". The value of t for testing the null hypothesis that the two populations have the same mean is

$$t = \frac{|\bar{X}_a - \bar{X}_b|}{\sqrt{\left(\dfrac{\hat{\sigma}^2}{n_a} + \dfrac{\hat{\sigma}^2}{n_b}\right)}} = \frac{|\bar{X}_a - \bar{X}_b|}{\hat{\sigma}\sqrt{\left(\dfrac{n_a + n_b}{n_a n_b}\right)}}$$

with $n_a - n_b - 2$ degrees of freedom.

PROCEDURE

1. Set up the null hypothesis: the two samples are drawn from normal populations with the same mean and variance. The means of the samples differ merely because of chance.

2. Pool the variances of the two samples to obtain an estimate of the variance of the hypothesised parent populations. Let the estimated variance be $\hat{\sigma}^2$ where

$$\hat{\sigma}^2 = \frac{s_a^2(n_a - 1) + s_b^2(n_b - 1)}{(n_a + n_b - 2)}$$

If n_a happens to equal n_b, the formula simplifies to $(s_a^2 + s_b^2)/2$, the mean of the sample variances.

If the sample variances have not already been calculated, there is no need to calculate them separately from the means. The following relation may be used to calculate $\hat{\sigma}^2$

$$\hat{\sigma}^2 = \frac{[\Sigma X_a^{\ 2} - \bar{X}_a(\Sigma X_a)] + [\Sigma X_b^{\ 2} - \bar{X}_b(\Sigma X_b)]}{(n_a + n_b - 2)}$$

where $\Sigma X_a^{\ 2}$ and $\Sigma X_b^{\ 2}$ are the sums of the squares of the X values making up each sample, and ΣX_a and ΣX_b are the sums of the X values.

3. Estimate the standard error of the difference between the means

$$\text{Estimated standard error} = \sqrt{\left(\frac{\hat{\sigma}^2}{n_a} + \frac{\hat{\sigma}^2}{n_b}\right)} = \hat{\sigma}\sqrt{\left(\frac{n_a + n_b}{n_a n_b}\right)}$$

4. Calculate t

$$t = \frac{|\bar{X}_a - \bar{X}_b|}{\text{estimated standard error}}$$

5. Calculate the degrees of freedom (df) associated with t

$$df = n_a + n_b - 2$$

6. Decide whether a one- or two-tailed test is appropriate.

7. Determine P, the probability of obtaining the calculated value of t, or one greater, as a result of sampling error. P is the probability given identical populations of obtaining two samples with means differing by the observed amount $|\bar{X}_a - \bar{X}_b|$ or a greater amount.

Selected values of P appropriate for two-tailed tests are given in Table E (the "t table"). Values of P for one-tailed tests are exactly half those listed in Table E.

8. Decide whether to retain or reject the null hypothesis. If P is low, serious doubt is cast on the null hypothesis. Reject the null if P is equal to or less than whatever level of significance is judged appropriate, e.g. 0.05 or 0.01. Conclude that the two samples differ in mean by more than can be reasonably attributed to chance. Assume the samples were derived from populations that differ in mean.

If P is greater than the level of significance reserve judgement, i.e. regard the null hypothesis as possibly, but not necessarily, correct.

EXAMPLE

The data discussed under the F test will be examined further. Recall that the mean angle of valley sides experiencing basal cutting was found to be 44.8 degrees with a standard deviation of 3.27 degrees. Valley sides protected from active stream erosion were found to have a mean angle of 38.2 degrees and a standard deviation of

2.70 degrees. Both samples were distributed approximately normally, and the F test suggested that the variances were not significantly different. In order to determine whether the difference in the means is statistically significant a two-sample t test can be run.

1. Null hypothesis: the difference between the two means results from chance sampling. The difference is not statistically significant.

2. $s_a^2 = (3.27)^2 = 10.6929$ degrees squared. $n_a = 171$.

$s_b^2 = (2.70)^2 = 7.2900$ degrees squared. $n_b = 33$.

The pooled variance $\hat{\sigma}^2$ is calculated as follows

$$\hat{\sigma}^2 = \frac{10.6929(170) + 7.2900(32)}{(171 + 33 - 2)} = \frac{2051.0730}{202}$$

$$= 10.1538 \text{ degrees squared.}$$

Hence $\hat{\sigma}$, the standard deviation = 3.1865 degrees.

3. The estimated standard error of the difference between the means

$$= 3.1865 \sqrt{\left(\frac{171 + 33}{(171)(33)}\right)} = 0.60586 \text{ degrees.}$$

4. $t = \dfrac{44.8 - 38.2}{0.60586} = 10.894$ approximately.

5. The degrees of freedom are df = 170 + 32 = 202.

6. The problem is unusual in being one-sided. If the streams had any effect at all, it was to steepen the slopes, not to make them more gentle. Had the slopes experiencing basal cutting proved to have been gentler than the protected slopes the results would have been instantly dismissed as a freak of sampling.

7. Because the problem is one-sided, a one-tail test is required. According to Table E, the probability P of getting a value of t as large as the calculated value or larger is substantially less than 0.0000001 (half the lowest value of P shown in the table which is 0.0000002).

8. Since the chance is less than one in ten million of getting this value of t, or one greater, the null hypothesis has to be abandoned as unreasonable. The difference between the means would appear not to be the result of chance sampling.

ASSUMPTIONS AND LIMITATIONS

1. The items or individuals forming each sample must be randomly selected; the items or individuals must not be paired in any way, but must be selected completely independently of one another. If the populations are finite, the sampling must be with replacements; sampling without replacements is permissible only if the populations are infinite.

2. The values of the items in each population are assumed to be independent of one another. There must be no positive or negative autocorrelation.

3. Each sample is assumed to come from a population that is normally distributed.

4. The variances of the populations must be identical. This identity of variance is often referred to (perhaps somewhat confusingly) as *homogeneity of variances*.

Effects of violations of assumptions 1 and 2

There is little point carrying out a t test if the sampling is not random or approximately random. The results are almost certain to be worthless. Although sampling without replacements in theory is allowable only if the populations are infinite, no great error is likely to result unless the populations are very small.

The formula for the standard error of the difference in the means tends to produce an under-estimate when positive autocorrelation is present in the populations and an over-estimate when negative autocorrelation is present. Cliff and Ord (1975) have experimented with positively autocorrelated populations, drawing repeated pairs of samples and calculating the value of t. They show that the accuracy of the t test is greatly affected by positive autocorrelation. The real value of P may be much higher than the tabled value. For instance, if the correlation between successive members of each sampled population is + 0.5 (for an explanation of how correlation is measured see Chapter 17), the true 5% value of t for a two-tailed test with 96 degrees of freedom is 3.63 when the tabled value is 1.98. If the correlation is + 0.9, the true 5% value of t is 12.88.

Effects of violations of assumptions 3 and 4

No real population is exactly normal, and few are even approximately normal. The two-sample t test is nevertheless a fairly *robust* test in that it often gives results that are reasonably accurate even when the assumption of normality is not fully met. The reader is referred to papers by Welch (1937), Boneau (1960) and Bradley (1963) for an extended discussion of the robustness of the test.

The following is a brief summary of the views that have been expressed.

1. Samples of equal size.

If the sampled populations are approximately normal and have only slightly different variances, the accuracy of the two-sample t test is very high. The accuracy is not much less if the variances are markedly unequal. Welch has shown, for example, that if repeated pairs of samples of size 10 are taken from two normally distributed populations that possess unequal variance, they will give t values that, according to the t table, are associated with P values of 5%, or less, no more than 6.5% of the time. In other words, if the nominal value of P is 0.05, the true probability cannot be more than 0.065, however much the variances differ.

Provided the samples are equal in size, the accuracy of the t test is little affected even if the sampled populations are only very approximately normal. The accuracy increases steadily with increasing sample size, becoming perfect when both sample sizes become infinite. Boneau suggests that samples of size 30 are sufficient to cope with very considerable departures from normality, but this figure may well be an under-estimate. A lot depends upon what level of significance is chosen. As Bradley has pointed out, the accuracy of the t test tends to worsen rapidly as one moves towards the tails. The smaller the nominal value of P, the more the true value of P tends to differ from it, and the less robust the test tends to be. While values of P around the commonly used levels of significance of 0.05, or even 0.01, may be fairly accurate if n is 30, values of P of 0.005 or 0.001 may be much less accurate.

A two-tailed test tends to be less robust than a one-tailed test yielding the same value of P, if the sampled populations are symmetrical. If the sampled populations are asymmetrical, the distribution of the t statistic will be asymmetrical, unlike the distribution of the theoretical t shown in Table F. The difference between the true value of P and the nominal value may be much greater in one tail than in the other. Thus the true value may be quite close to the nominal value for a left-tail test, but far removed from the nominal value for a right-tail test. Or the reverse condition may prevail; a left-tail test may be much less accurate than a right-tail test. Because the true and nominal values deviate in different directions in the two tails the robustness of a two-tailed test yielding a nominal probability P may be very much better than that for either of the one-tailed tests yielding half the value of P.

2. Samples of unequal size.

If the samples are unequal in size, the t test is not nearly so robust in the face of non-normality or unequal population variances. When the variances are different the process of pooling is no longer a valid means of estimating the population variance. Welch has shown that if samples of size 5 and 15 are drawn from two normal populations that differ in variance, the percentage of t's exceeding the nominal 5% value varies between 0 and 31% depending upon the amount of difference between the population variances. Given the combination of unequal population variances and unequal sample sizes, the t test is liable to produce results that are seriously in error. This is true whether the populations are normal or non-normal.

The robustness of the t test tends to worsen towards the tails of the sampling distribution just as it does when the sample sizes are equal. A value of P of 0.05 is usually more trustworthy than one of 0.01, for example. The robustness of the test also depends on whether one or two tails are used for testing. As stated above, a two-tailed test tends to be less robust than a one-tailed test yielding the same value of P if both populations are symmetrical. If the sampled populations are asymmetrical, a test using the left tail may be much less accurate than a test using the right tail, or vice versa. A two-tailed test may be more robust than either of the one-tailed tests at half the nominal probability of the two-tailed test.

Letting the sample sizes become infinite does not make the t test perfectly accurate. When the samples are infinite, but unequal (in mathematics this is not a contradiction), the t test is still sensitive to unequal population variances. Bradley (1976) has proved that if the populations are normally distributed the true value of P can be obtained from the Z table (Table B) by looking up the probability associated with

$$Z = \frac{t}{\sqrt{\frac{\gamma + R}{1 + \gamma R}}}$$

where t is the value of t yielded by the two sample t test, γ is the ratio of the sample sizes n_a/n_b, and R is the ratio of the population variances $\sigma_b{}^2/\sigma_a{}^2$. As an example, suppose that Sample A is twice as large as Sample B, and is drawn from a normally distributed population whose variance is twice that of the normally distributed population from which Sample B is drawn

$$\gamma = \frac{n_a}{n_b} = 2, \; R = \frac{\sigma_b}{\sigma_a} = \frac{1}{2} \text{ and } \sqrt{\frac{\gamma + R}{1 + \gamma R}} = \sqrt{\frac{2.5}{2}} = 1.1180 \text{ approx.}$$

Suppose also that the value of t is 1.960. The degrees of freedom are infinite since both samples are infinite. The nominal value of P obtained from Table E (the t table) is 0.05. The true value can be obtained by first calculating Z = 1.960/1.1180 = 1.7531 and then consulting Table B (the Z table). The true value turns out to be 0.0796. The ratio between the nominal and true values is 0.05/0.0796 = 0.6281.

Bradley provides a graph of the ratio between the nominal and true values of P for selected values of γ and R. If there is a considerable difference in sample size, and the population variances are markedly unequal, the nominal value of P may bear no resemblance to the true value, even though both samples are infinite.

SOME ALTERNATIVES TO THE TWO-SAMPLE t TEST

Over the years several substitutes for the t test have been proposed, but none has proved entirely satisfactory.

A test that does not require the sampled populations to be normally or near normally distributed (and is valid even if the population variances are unequal) was devised by Fisher in 1935. Based on randomisation principles it is too time consuming for general use. A more convenient test is the Mann-Whitney *U* test, but even this has major limitations. Both the Fisher and Mann-Whitney tests will be described later in this chapter.

If the sampled populations are approximately normal but differ in variance, there are three further tests that can be applied. Seemingly the most sophisticated is the Fisher-Behrens test (for details, see Fisher and Yates, 1957), but its validity has been questioned (Bartlett, 1936). An alternative test due to Welch (1947), Aspin (1949) and Trickett *et al.* (1956) is preferred by some statisticians. Both tests require special tables. The following test uses the ordinary t table. It makes no claim to precision, but provides a rough and ready test of means in place of the standard t test. The greater the difference in variances, or sample sizes, the more approximate the test becomes.

PROCEDURE FOR A MODIFIED t TEST (PARENT POPULATIONS WITH DIFFERENT VARIANCES BUT BOTH NORMALLY DISTRIBUTED)

1. Calculate the value of t where

$$t = \frac{|\bar{X}_a - \bar{X}_b|}{\sqrt{\left(\frac{s_a^2}{n_a} + \frac{s_b^2}{n_b}\right)}}$$

2. Calculate the degrees of freedom

$$df = \left(\frac{s_a^2}{n_a} + \frac{s_b^2}{n_b}\right)^2 \bigg/ \left(\frac{s_a^4}{n_a^3} + \frac{s_b^4}{n_b^3}\right)$$

3. Decide whether a one- or two-tailed test is appropriate.

4. Using Table E determine P, the probability of obtaining the calculated value of t, or one greater, as a result of sampling error.

5. Decide whether to retain or reject the null hypothesis. Reject the null if P is equal to or less than whatever level of significance is judged appropriate, e.g. 0.05 or 0.01. Assume the samples were derived from populations that differ in mean. If P is greater than the level of significance reserve judgement, i.e. regard the null hypothesis as possibly, but not necessarily, correct.

EXAMPLE

Consider again the two samples of slope angles. Although the F test demonstrated that the difference in variance could well be due to chance it did not prove conclusively that the samples were drawn from two populations with the same variance. Had the samples been larger the difference in variance might have been statistically significant. In order to guard against the possibility of unequal population variances the two-sample t test can be modified as follows

$$t = \frac{|\bar{X}_a - \bar{X}_b|}{\sqrt{\left(\frac{s_a^2}{n_a} + \frac{s_b^2}{n_b}\right)}} = \frac{6.6}{\sqrt{\left[\frac{(3.27)^2}{171} + \frac{(2.70)^2}{33}\right]}} = \frac{6.6}{\sqrt{0.28344}} = 12.397$$

$$df = \left(\frac{s_a^2}{n_a} + \frac{s_b^2}{n_b}\right)^2 \bigg/ \left(\frac{s_a^4}{n_a^3} + \frac{s_b^4}{n_b^3}\right)$$

$$= \left[\frac{(3.27)^2}{171} + \frac{(2.70)^2}{33}\right]^2 \Bigg/ \left[\frac{(3.27)^4}{171^3} + \frac{(2.70)^4}{33^3}\right]$$

$$= (0.28344)^2/(0.00150) = 53.5 \text{ approximately.}$$

For a one-tailed test the value of P is again less than 0.0000001 (half the lowest value shown in Table E). The null hypothesis can be confidently rejected. The observed difference in the sample means cannot be plausibly explained on a basis of chance sampling.

CONFIDENCE LIMITS FOR THE DIFFERENCE IN TWO MEANS

Provided the populations are normally distributed and the variances are equal, the α per cent confidence interval for the difference in the population means $\mu_a - \mu_b$ is given by

$$(\bar{X}_a - \bar{X}_b) - t_{(100-\alpha)} \text{ (estimated standard error)}$$

to $$(\bar{X}_a - \bar{X}_b) + t_{(100-\alpha)} \text{ (estimated standard error)}$$

where $t_{(100-\alpha)}$ is the value of t that for $(n_a + n_b - 2)$ degrees of freedom just reaches significance at the $(100 - \alpha)\%$ level. As in the t test the estimated standard error is

$$\sqrt{\left(\frac{\hat{\sigma}^2}{n_a} + \frac{\hat{\sigma}^2}{n_b}\right)} = \hat{\sigma}\sqrt{\left(\frac{n_a + n_b}{n_a n_b}\right)}$$

One can be 95% confident that the actual difference in the population mean $\mu_a - \mu_b$ lies within the calculated interval.

If the population variances are unequal, an approximate α per cent confidence interval is given by

$$(\bar{X}_a - \bar{X}_b) - t_{(100-\alpha)}\sqrt{\left(\frac{s_a^2}{n_a} + \frac{s_b^2}{n_b}\right)}$$

to $$(\bar{X}_a - \bar{X}_b) + t_{(100-\alpha)}\sqrt{\left(\frac{s_a^2}{n_a} + \frac{s_b^2}{n_b}\right)}$$

where $t_{(100-\alpha)}$ is the value of t that for

$$df = \left(\frac{s_a^2}{n_a} + \frac{s_b^2}{n_b}\right)^2 \Bigg/ \left(\frac{s_a^4}{n_a^3} + \frac{s_b^4}{n_b^3}\right)$$

degrees of freedom just reaches significance at the $(100 - \alpha)\%$ level.

Referring to the two samples of slope angles, the estimated standard error = 0.60586 degrees, $\bar{X}_a - \bar{X}_b$ = 6.6 degrees and $n_a + n_b - 2$ = 170 + 32 = 202.

The value of t that for 202 degrees of freedom just reaches significance at the 5% level, i.e. corresponds to a value of P of 0.05 can be estimated from Table E to be 1.652. Hence, using the first formula the 95% confidence interval for the difference in the population means is 6.6 - 1.652 (0.60586) to 6.6 + 1.652 (0.60586) degrees or 5.6 to 7.6 degrees.

If the second formula is used, the interval becomes 6.6 - 1.674 (0.53239) to 6.6 + 1.674 (0.53239) degrees or 5.7 to 7.5 degrees.

The difference in the sample variances is too slight to affect the confidence limits to any great extent.

The rest of this chapter may be omitted on first reading.

*THE RANDOMISATION TEST FOR DIFFERENCES IN MEANS

A major weakness of the two sample t test is its assumption of normality. As Geary (1947, p.241) has remarked, "normality is a myth; there never was, and never will be, a normal distribution". No real population is ever exactly normal, and only a minority are even approximately normal.

Many statisticians have tried to devise an alternative to the t test capable of giving accurate results even if the populations are far from normal. They have not had much success to date. In 1935, R.A. Fisher proposed a clever but generally impractical test based on randomisation. The basic principle of the test is as follows. Suppose that there exist two populations, A and B, both consisting of items measured once in respect of a continuous variable X. The measurements in each population are independent of one another. A random sample of n_a items is drawn from Population A and a second random sample of n_b items is drawn independently of the first from Population B. It is suspected that the populations possess identical means, variances and distributional shapes, but this is not certain.

If the populations really are identical, the samples can be legitimately combined to form a single sample, size $n_a + n_b$. The two populations can be looked upon as forming a single, homogeneous population (homogeneous, that is, as regards X, the variable under study though not necessarily homogeneous in any other respect). The pair of samples under study can be assumed to have had the same chance of being drawn as any other pair of samples of size n_a and n_b containing the observed values of X but in different combinations. It can be readily proved that there are $(n_a + n_b)!/(n_a!n_b!)$ ways of dividing the $(n_a + n_b)$ values of X into two samples of sizes n_a and n_b containing different combinations of the same $(n_a + n_b)$ values.

Each of these $(n_a + n_b)!/(n_a!n_b!)$ pairs of samples will yield a difference in means, $\bar{X}_a - \bar{X}_b$. The differences for some pairs of samples may be small, the differences for others may be large. One can use the $(n_a + n_b)!/(n_a!n_b!)$ differences to construct a frequency distribution or "sampling distribution".[1]

Let it be assumed provisionally that the sampled populations are identical with respect to X (null hypothesis). The next step is to calculate P, the proportion of the $(n_a + n_b)!/(n_a!n_b!)$ pairs of samples that yield a difference in means $(\bar{X}_a - \bar{X}_b)$ equal to or greater than the difference actually observed. P is the probability under the null hypothesis of obtaining the observed difference or one that is greater. If P is small, this suggests that the null hypothesis is faulty and that the populations really are different. On the other hand, if P is large, the null hypothesis may be correct At any rate there are no good grounds for abandoning the null hypothesis.

Fisher's test is perhaps best understood by considering an example. E.W. Anderson (1975) has measured the rate of soil creep on a number of different slopes in a small drainage basin in the Pennine Uplands. Some of his results are given in Table 14.1. Do rates of soil creep vary according to the type of vegetation?

TABLE 14.1 - RATES OF SOIL CREEP ON SLOPES AROUND ROOKHOPE, UPPER WEARDALE

Sample	*Annual rate of creep (mm) on each slope*	*Mean*	*Variance*
A, grassland dominated by *Nardus stricta*	1.12,0.95,1.46,1.29, 0.85,1.47,1.06	1.1714	0.0590
B, stands of bracken *(Pteridium aquilinum)*	0.33,0.86,1.11,1.04 0.41,0.55	0.7167	0.1102

Difference in means = 0.4548

If the rates of soil creep are unrelated to the type of vegetation, the samples can be combined to form a single, homogeneous sample. Each of the 13 measurements listed in Table 14.1 had a 7/13 chance of being included amongst the 7 measurements in Sample A and a 6/13 chance of being included amongst the 6 measurements in Sample B. There are (7 + 6)!/(7!6!) = 1716 different ways of allocating the (7 + 6) = 13 measurements of soil creep into two

---

[1] It is worth noting that the $(n_a + n_b)!/(n_a!n_b!)$ differences do not constitute a complete sampling distribution, as there are only a finite number, and not all possible X values from the common population are considered, only those X values that happen to be included in the two samples under study.

samples of sizes 7 and 5. Each of these 1716 pairs of samples yields a difference in means. What one needs to know is whether the difference for the pair of samples under study (= +0.4548) is exceptional compared with the differences for the other pairs of samples obtained by randomisation.

One needs to consider differences in means in both the upper and lower tails of the "sampling distribution": that is to say, differences equal to or greater than +0.4548, and equal to or less than -0.4548. A two-tailed test is appropriate since prior to sampling there was no way of knowing what effect, if any, the type of vegetation would have on the rate of soil creep. To judge from the sample data, the rate of creep is greater on grassy slopes than under bracken, which is perhaps what one would expect, but there is no rule that says that the rate has to be the same or greater and cannot be lower. If the data had tended to suggest a lower rate of creep on grassy slopes, a test of significance would still have been necessary. The data could not have been instantly dismissed as an accident of sampling.

Table 14.2 gives some examples of pairs of samples yielding differences in means lying in the upper and lower tails of the sampling distribution. There is not room, unfortunately, to show all pairs of samples.

TABLE 14.2 - SOME PAIRS OF SAMPLES YIELDING DIFFERENCES IN MEANS EQUAL TO OR GREATER THAN +0.4548, AND EQUAL TO OR LESS THAN -0.4548

		Difference in Means
A	1.12, 1.11, 1.46, 1.29, 1.04, 1.47, 1.06	+0.5631
B	0.33, 0.86, 0.85, 0.95, 0.41, 0.55	
A	1.12, 0.95, 1.46, 1.29, 1.11, 1.47, 1.06	+0.5352
B	0.33, 0.86, 0.85, 1.04, 0.41, 0.55	
A	1.12, 0.95, 1.46, 1.29, 1.04, 1.47, 1.06	+0.5136
B	0.33, 0.86, 1.11, 0.85, 0.41, 0.55	
A	1.12, 1.11, 1.46, 1.29, 0.85, 1.47, 1.06	+0.5043
B	0.33, 0.86, 0.95, 1.04, 0.41, 0.55	
A	1.12, 1.04, 1.46, 1.29, 0.85, 1.47, 1.06	+0.4826
B	0.33, 0.86, 1.11, 0.95, 0.41, 0.55	
A	1.12, 0.95, 1.46, 1.29, 0.85, 1.47, 1.11	+0.4702
B	0.33, 0.86, 1.06, 1.04, 0.41, 0.55	
A	1.04, 0.95, 0.86, 0.55, 0.41, 0.33, 0.85	-0.5388
B	1.46, 1.47, 1.29, 1.12, 1.11, 1.06	

With patience one can find 59 pairs of samples (including the pair under study) that yield differences in means equal to or greater than +0.4548 and equal to or less than -0.4548. The value of P is simply this number of pairs divided by the total number of pairs or 59/1716 = 0.034 approximately.

Since P is low one can safely assume that rates of soil creep really do vary according to the type of vegetation covering the slope. Despite the small sizes of the samples it is unreasonable to suppose that there is no connection between the type of vegetation and the rate of creep. Some of the variation in the rate of creep between different slopes would seem to be related either directly or indirectly to the type of vegetation cover.

Fisher's test has some serious drawbacks that need to be examined. It is especially sensitive to differences in means, but differences in distributions and variances may also affect the value of P, despite the fact that the test statistic is the difference in means ($\bar{X}_a - \bar{X}_b$). The sampled populations may have identical means and yet the value of P may be very low if the distributions or variances differ. For example, suppose one of the populations consists of the integers from 1 to 100 together with the one integer 20,200, while the other population consists of the integers from 200 to 300. Both populations have means of 250, but they are otherwise very dissimilar. Pairs of samples drawn from the populations will tend to differ substantially in means, and Fisher's test will tend to reject the null hypothesis that the samples are drawn from identical populations, even though the means are identical.

Only if there is good reason to believe that two sampled populations have identical (or near identical) distributions and variances, can one confidently assume that a low value of P signifies a real difference in means. In the cast of the worked example, the measurements of soil creep are so few, one cannot draw any definite inferences about the population shapes and variances. The value of P, though low, is not especially enlightening. The populations can be assumed to differ, but whether they differ in means, variances and/or distributional shape is unclear.

Because of its nature and its limitations Fisher's test does not find many uses. It can be employed as an "omnibus" or "non-specific" test to decide whether two sampled populations are likely to differ, but not how they are likely to differ. It can also be employed as a substitute for the two-sample t test provided there is good evidence that the sampled populations have essentially the same distributions and variances. As explained earlier in the chapter, the t test is subject to almost identical restrictions. The variances must be more or less equal, and the populations must be distributed more or less normally. Fisher's test is slightly less restrictive in that it makes no assumptions about normality, although it does require that the populations have more or less identical distributions.

When applying Fisher's test to the measurements of soil creep the difficulty of working out the different permutations was glossed over. Unfortunately the permutations take some time to write out, and mistakes are easily made. Fisher's test can be performed quickly only if the samples are small and the observed difference in means lies at the extreme edge of the "sampling distribution". With samples of moderate or large size the amount of work needed to calculate P tends to be prohibitively large unless a computer is to hand.

The computational difficulties associated with Fisher's test disappear if ranks are substituted for the actual measurements of X. Tables can be prepared that list the value of P for different sample sizes.[1] When using ranks, therefore, one has no need to write down the different permutations - the values of P can be found directly from the tables. There is less chance of making mistakes, and the test can be performed very quickly.

The first person to devise a substitute test based on ranks was Frank Wilcoxon (1945), but he wrote so briefly that the importance of his work was not grasped at first. Wilcoxon's test (the *Wilcoxon rank-sum test*) utilises as a test statistic the sum of the ranks of whichever sample yields the smaller sum. H.B. Mann and D.R. Whitney (1947) introduced some minor modifications of Wilcoxon's test, including the use of a slightly different test statistic. It is their modified test that will now be described.

THE MANN-WHITNEY TEST

TO TEST WHETHER TWO SAMPLES DIFFER IN ANY SIGNIFICANT RESPECT

THE DATA

Two random samples are drawn independently from different populations. Each of the items making up the two samples is measured once in respect of a continuous variable X. Let n_a be the size of the sample yielding Σr (see below) and n_b the size of the other sample.

PROCEDURE

1. Set up the null hypothesis: the two populations are identical as regards X, the variable under study. They have identical distributions, and share the same mean and variance. Any differences between the samples are due to sampling error.

2. Combine the two samples and rank the values of X in order of increasing size, i.e. assign the rank 1 to the smallest value in the combined sample, the rank 2 to the next smallest value, and so on.

If two or more values of X are the same, assign to each the average of the ranks that would have been assigned had they been distinguishable. For example, if one X value ranks 6, and the next three values tie, the three ought each to be given a rank of $(7 + 8 + 9)/3 = 8$.

3. Calculate Σr, the sum of the ranks (not the sum of the measurements) of Sample A or Sample B, whichever sum is the smaller. If the samples are large and it is not immediately obvious which sum of ranks is the smaller proceed as follows

(a) Calculate the sum of the ranks of the first sample.
(b) Calculate the total sum of the ranks of the combined samples using the formula $(n_a + n_b)(n_a + n_b + 1)/2$.

[1] No tables can be devised for Fisher's test since each pair of samples involves a unique assortment of X values.

(c) Find the sum of the ranks of the second sample by subtracting the sum of the first sample from the total sum calculated in step (b).

4. Calculate the Mann-Whitney test statistic U

$$U = \Sigma r - \frac{n_a(n_a + 1)}{2}$$

where n_a is the size of the sample yielding Σr.

5. Decide whether a one- or two-tailed test is appropriate.

6. Determine the probability P of obtaining the calculated value of U, or one greater, through chance sampling.

If neither sample contains more than 10 items, consult Table H which lists all possible values of U and the associated probabilities P. The probabilities are one-tailed; for a two-tailed test they should be doubled.

If one sample contains more than 10 items, but neither sample contains more than 20, consult Table I. Entries in the body of the table are the smallest values of U yielding one-tailed values of P equal to or greater than the values listed in the left-hand margin, which are the commonly used levels of significance (0.05 etc.). For example, if the value of U entered in the table is 6 and the value of P shown in the margin is 0.05, this means that when $U = 6$ the value of P is at least 0.05. Again if the value of U is shown as 3, and the value of P as 0.01, the true value of P when U is 3 is at least 0.01. When consulting Table I, use the row corresponding to n_a, the size of the sample yielding Σr, and the column corresponding to n_b, the size of the other sample. If the calculated value of U is less than the value entered in the table for a given P, the calculated value is associated with a one-tailed probability less than P. For a two-tailed test, double the values of P shown in the left-hand margin.

If both samples contain more than 20 items, consult Table B (the Z table). The sampling distribution of U becomes approximately normal as n increases with mean μ_U and standard deviation σ_U where

$$\mu_U = \frac{n_a n_b}{2}$$

and $$\sigma_U = \sqrt{\frac{n_a n_b(n_a + n_b + 1)}{12}}$$

Hence the approximate value of P can be found by calculating

$$Z = \frac{|U - \mu_U| - \frac{1}{2}}{\sigma_U} = \frac{\left|U - \frac{n_a n_b}{2}\right| - \frac{1}{2}}{\sqrt{\frac{n_a n_b(n_a + n_b + 1)}{12}}}$$

and referring to Table B. The 1/2 in the numerator for Z is a continuity correction to allow for the fact that U is confined to whole numbers. Note carefully that Table B yields two-tailed probabilities; these must be halved for a one-tail test.

7. Decide whether to reject the null hypothesis. If P is greater than 0.05 (or whatever level of significance is judged appropriate) reserve judgement, i.e. regard the null hypothesis as possibly, though not necessarily, correct. If P is less than the level of significance reject the null. Conclude that the sampled populations differ as regards X.

If the sampled populations possess identical distributions and variances, the alternative to the null hypothesis is that they differ in means. Hence if P is less than the level of significance, and there is good evidence that the populations possess essentially the same distributions and variances, assume that the populations differ in means.

EXAMPLE

The measurements of soil creep used to illustrate Fisher's test will be re-examined using the Mann-Whitney test.

1. Null hypothesis: the rate of soil creep does not vary with the type of vegetation. It is purely fortuitous that the rates of movement on the 7 grass-covered slopes are rather different from the rates of movement on the 6 slopes under bracken.

2. In Table 14.3 the measurements of creep are replaced by ranks.

TABLE 14.3 - RATES OF SOIL CREEP ON SLOPES AROUND ROOKHOPE, MEASUREMENTS REPLACED BY RANKS

Sample	*Type of vegetation*	*Annual rate of creep expressed as a rank*
A	Grassland dominated by *Nardus stricta*	10, 6, 12, 11, 4, 13, 8,
B	Stands of bracken *(Pteridium aquilinum)*	1, 5, 9, 7, 2, 3,

3. The sum of the 7 ranks of the larger sample is 64, the sum of the 6 ranks of the smaller sample is 27. Hence $\Sigma r = 27$ and $n_a = 6$.

4. The Mann-Whitney test statistic U is given by

$$U = \Sigma r - \frac{n_a(n_a + 1)}{2} = 27 - \frac{6(7)}{2} = 6$$

According to Table H the one-tailed probability associated with $U = 6$, $n_a = 6$ and $n_b = 7$ is 0.0175. The two-tailed probability is $2(0.0175) = 0.035$. Since there are only 3.5 chances in 100 of obtaining the calculated value of U or one still smaller the null hypothesis can be rejected. The evidence suggests strongly that rates of soil creep vary according to the type of vegetation covering the slopes.

Table I could have been used to obtain the approximate probability associated with U. Entering the table using the row marked 6 ($=n_a$) and the column marked 7 ($=n_b$) one finds that the tabled value of U for a two-tailed probability of 0.02 is 5 increasing to 7 for a two-tailed probability of 0.05. The probability associated with the calculated value of U is therefore less than 0.05 but greater than 0.02. Table I does not allow the probability to be determined more exactly.

Although the low value of P of 0.035 suggests that the sampled populations are likely to differ, it does not define the way in which they are likely to differ. Because the samples are so small one cannot be at all sure whether the populations have similar distributions and variances. The low value of P is not really evidence that the means differ. The most one can say is that the test is especially sensitive to differences in means and hence P is more likely to be low because the population means differ than because the variances or distributions differ. But to arrive at a more definite conclusion one would need larger samples.

THE TREATMENT OF TIES

The method of assigning ranks to tied values has already been explained, but what has not been disclosed is that the tables and the Z formula are invalid if ties exist. The tables are constructed by first writing down all possible pairs of samples representing different permutations of the ranks, then calculating U for each pair of samples. The procedure is the same as that for Fisher's test, except that the permutations involve ranks not X values, and U is calculated rather than the difference in means.

Values of P given by the tables will not be greatly in error if only a few of the values of X are tied. One can use the tables and safely ignore the fact that the ties exist. If the proportion is large, and especially if the ties occur in the tails of the sampling distribution, the tables may give values of P that are seriously in error. The value of P can be found by writing down the various pairs of samples that yield a value of U equal to or less than the value in question. The proportion of such permutations is the value of P for a one-tailed test.

If n_a and n_b are fairly large, the approximate value of P for a two-tailed test (or double the value for a one-tailed test) can be obtained from the Z formula

$$Z = \frac{\left| U - \frac{n_a n_b}{2} \right| - \frac{1}{2}}{\sqrt{\left[\frac{n_a n_b (n_a + n_b + 1)}{12} - \frac{n_a n_b \Sigma (d_i^3 - d_i)}{12 (n_a + n_b)(n_a + n_b - 1)} \right]}}$$

where d_i is the number of items tied at a given rank. As an example, suppose the ranks are: 1, 3, 3, 3, 5.5, 5.5, 7, 9, 9 and 9. d_i for the first rank is 1, for the third rank 3, for the 5.5th rank 2, for the seventh rank 1, and for the ninth rank 3. Hence the sum $\Sigma(d_i^3 - d_i) = (1^3 - 1) + (3^3 - 3) + (2^3 - 2) + (1^3 - 1) + (3^3 - 3)$ $= 0 + 24 + 6 + 0 + 24 = 54$.

ASSUMPTIONS AND LIMITATIONS

The Mann-Whitney test assumes (as does Fisher's test) that

1. The samples are drawn randomly from the parent populations.

2. Sampling is with individual replacements, or else the populations are infinite (sampling without replacements is acceptable provided the populations are not too small).

3. The variable under investigation is continuous, not discrete. (If many X values are tied, some adjustment must be made as explained in the previous section.)

4. The values of the items in each population are independent of one another. There is no positive or negative autocorrelation.

5. The "sampling" distribution obtained by randomisation is representative of the much larger sampling distribution that would be obtained if an infinite number of pairs of samples were to be drawn from the parent populations. Some statisticians (e.g. Bradley, 1968, pp. 84-85) are convinced that there is no problem here: the sampling distribution based on the observed samples is automatically representative of the sampling distribution based on an infinite number of pairs of samples. Many statisticians remain unconvinced, however.

The Mann-Whitney test is almost as powerful as the t test if the sampled populations are normally distributed and the sample sizes are large. When the sample sizes are infinite, U yields a significant value of P for 995 of every 1000 pairs of samples for which t yields a significant value of P. The *asymptotic relative efficiency* (A.R.E.) of the Mann-Whitney test relative to the t test is said to be 0.955 (Noether, 1967a). If the populations are identical but not normal, the A.R.E. can never be less than 0.864 (Hodges and Lehmann, 1956). As Conover (1971, p.224) has remarked, "the A.R.E. of the Mann-Whitney test is never too bad when compared with the two-sample t test ... And yet the contrary is not true; the A.R.E. of the t test compared to the Mann-Whitney test may be as small as zero, or infinitely bad! So the Mann-Whitney test is a safer test to use".

It must be emphasised that this superiority applies only if the sampled populations have identical distributions. If the distributions differ, the t test may in certain circumstances be the safer test.

15 Chi-Square Tests

INTRODUCTION

The tests in this chapter (with one exception) are concerned with data in the form of frequencies, that is to say with counts of the numbers of items belonging to various groups or classes. The tests may not be used on data in the form of ranks or measurements. The tests therefore differ from those described in previous chapters, all of which were concerned with analysing ranks or measurements.

It may be helpful to begin by considering an example of the sort of problem involving frequency data that the tests are designed to solve. The Census of England and Wales in 1961 required one in ten persons to give details of their occupation or employment. Data for the male working population of Crawley New Town (Sussex) are shown in Table 15.1. Note that the data are in the form of counts. Economically active males are grouped into classes according to their occupation and are not separately specified as ranks or measurements.[1]

TABLE 15.1 - OCCUPATIONS OF A 10% SAMPLE OF MALES, CRAWLEY NEW TOWN, 1961

Sample survey 1961	*Occupations of economically active males*					*Total*
	Professional workers, employers and managers	*Other non-manual*	*Skilled manual*	*Semi-skilled manual*	*Unskilled manual*	
Numbers of males	240	359	708	214	87	1608

Some five years after the sample survey, a 100% survey of the male working population of Crawley New Town was organised by the Ministry of Labour. The results of this second survey are summarised in Table 15.2 (top two rows of figures).

[1]Because the figures represent frequencies it is tempting to describe Table 15.1 as a frequency distribution. However, the classes are not defined numerically in relation to some underlying variable as in the case of the frequency distributions discussed in previous chapters. In view of these differences Table 15.1 is best described as a *contingency table* rather than a frequency distribution.

TABLE 15.2 - OCCUPATIONS OF ALL WORKING MALES, CRAWLEY NEW TOWN, 1966

Complete survey 1966	*Occupations of economically active males*					*Total*
	Professional workers, employers and managers	*Other non-manual*	*Skilled manual*	*Semi-skilled manual*	*Unskilled manual*	
Numbers of males in the complete survey	3492	4190	7886	2876	976	19420
Proportions of total	0.180	0.216	0.406	0.148	0.050	1.000
Expected numbers in a sample of size 1608	289.142	346.937	652.970	238.136	80.814	1608

What one needs to decide is whether the proportion of males in each occupational category changed during the five years. By multiplying each proportion shown in Table 15.2 by 1608, it is possible to determine how many working males ought on average to belong to each class in a random sample of 1608 working males (bottom row of table).

These "expected frequencies" differ somewhat from the observed frequencies listed in Table 15.1. There would seem to be proportionately fewer working males engaged in manual occupations after the lapse of 5 years and proportionately more in non-manual occupations. However, the possibility that the differences between the observed and expected frequencies are due merely to sampling error must be investigated. Assuming that the proportions of males in the different occupations remained unchanged for five years, what is the probability of obtaining the sample or one less likely? The exact probability is given by the *multinomial distribution*.

THE MULTINOMIAL DISTRIBUTION

The number of ways of grouping n items into j mutually exclusive classes to give n_1 items in the first class, n_2 in the second class, n_3 in the third, etc. can be shown with a little algebra to be

$$\frac{n!}{n_1!n_2!n_3!\ldots n_j!}$$

where $n_1 + n_2 + n_3 + \ldots + n_j = n$.

Suppose there exists a population of items grouped into j mutually exclusive classes such that p_1 is the proportion of items in the first class, p_2 is the proportion in the second class, p_3 the proportion in the third, and so on. These proportions can be looked upon as probabilities. If a single item is drawn at random, p_1 is the probability that it will belong to the first class, p_2 is the probability that it will belong to the second class, p_3 the probability that it will belong to the third class, and so on. If a random sample of n items is drawn from the population, the probability that n_2 items will belong to the first class, n_2 to the second, etc. is given by the formula

$$\frac{n!}{n_1!n_2!n_3!\dots n_j!}(p_1)^{n_1}(p_2)^{n_2}(p_3)^{n_3}\dots(p_j)^{n_j}$$

there being $n!/(n_1!n_2!n_3!\dots n_j!)$ separate ways of obtaining the sample, each having a probability of $(p_1)^{n_1}(p_2)^{n_2}(p_3)^{n_3}\dots(p_j)^{n_j}$.

The formula provides a specification of what is called the multinomial distribution. The binomial distribution presented in Chapter 8 is merely a special case of the multinomial distribution. Instead of an indefinite number of classes (j), the binomial distribution is concerned with just 2. Observe that if $j = 2$, the formula for the multinomial distribution reduces to

$$\frac{n!}{n_1!n_2!}(p_1)^{n_1}(p_2)^{n_2}$$

which is the binomial formula given previously, but with different symbols. K in the binomial formula is here replaced by n, p_1 replaces p, p_2 replaces q, n_1 replaces X and n_2 replaces (K - X).

The formula for the multinomial distribution generally leads to awkward arithmetic. For instance, according to the formula, the probability of obtaining the sample of working males in Crawley is

$$\frac{1608!}{240!359!708!214!87!}(0.180)^{240}(0.216)^{359}(0.406)^{708}(0.148)^{214}(0.050)^{87}$$

Working this out even with a pocket calculator takes a lot of time. The powers are relatively easy to handle (0.180^{240}, for example, is the antilog of 240 times log 0.180 = 1.42995×10^{-179} approximately), but the factorials lead to heavy arithmetic. With patience, however, it is possible to calculate that the probability is approximately 0.22×10^{-9}, a very small value indeed.

When carrying out a significance test it is necessary to calculate the probability of obtaining not only the set of frequencies that make up the particular sample in question but also all other sets of frequencies less probable than the observed set. Unless j and n happen to be small the amount of arithmetic is liable to be prodigous. With the Crawley data, for example, it would be necessary to calculate several tens of millions of probabilities besides the one just calculated. The arithmetic would be prohibitive unless a large and powerful computer could be used.

The computational difficulties with the multinomial distribution could be avoided if only the value of P could be looked up in tables like the value of P for the normal distribution. However, tables are not a practical proposition because of all the different combinations of values that are possible for j, n, n_1, n_2, p_1, p_2, etc. Although tables could be compiled using a computer, they would be very expensive to produce and far too bulky to be of much use.

THE CHI-SQUARED DISTRIBUTION

The awkward truth about the multinomial distribution is that it is virtually useless for significance testing unless a large computer is available. Fortunately, statisticians have managed to find an alternative called the *chi-squared distribution*, which provides a close approximation to the multinomial distribution for large samples, and is far less complicated computationally.

Although the chi-squared distribution is used primarily to analyse frequency data, it is defined in terms of measurements, not frequencies. Imagine a population consisting of an infinite number of items. Also imagine that each item is measured once in respect of a continuous variable X. The measurements are independent of each other and taken together form a normal distribution with a mean μ and standard deviation σ. A random sample of n items is drawn from the population. Let the n measurements be $X_1, X_2, X_3, \ldots, X_n$. Converted to Z values the measurements become

$$Z_1 = \frac{|X_1 - \mu|}{\sigma},\ Z_2 = \frac{|X_2 - \mu|}{\sigma},\ Z_3 = \frac{|X_3 - \mu|}{\sigma}, \ldots,\ Z_n = \frac{|X_n - \mu|}{\sigma}$$

Chi-square[1] is defined as the sum of the squares of the Z values. It is given the symbol χ^2, χ being the Greek lower-case letter *chi*. By definition

$$\chi^2 = (Z_1)^2 + (Z_2)^2 + (Z_3)^3 + \ldots + (Z_n)^2$$

It is clear from the formula that chi-square can take any value from zero to plus infinity. Unlike Z it can never be negative since it is made up of squares. And only if all the X values happen to equal μ will chi-square be zero. Otherwise chi-square will be positive. The more the X values deviate from the mean the greater will be the value of chi-square. The value of chi-square also depends on the number of X values, i.e. on the size of sample. If there are many X values, the value of chi-square is likely to be larger than if there are only a few. The number of X values is termed the number of *degrees of freedom* in chi-square.[2]

Figure 15.1A shows the distribution of chi-square for 1, 3, 5 and 10 degrees of freedom. For df = 1 and 2, the distribution has a reversed J-shape with a maximum ordinate at the origin. For df $\geq$ 3 the distribution assumes a hump-backed shape with a single mode of χ^2 = df - 2. The distribution is markedly right-skewed at first, but slowly becomes more symmetrical with increasing df. For degrees of freedom over about 40 the distribution is approximately normal.

Although the chi-squared distribution varies greatly in shape, the mean is always equal to the number of degrees of freedom and the variance to twice the number of degrees of freedom. A straightforward proof is given in Lancaster (1969).

The proportion of the area lying under the curve to the right of any selected value of χ^2 is the probability (P) of obtaining that value of χ^2, or one even larger, as a result of chance sampling.

---

[1]Not "chi-squared", although the distribution of chi-square is correctly called the "chi-squared distribution".

[2]The number of degrees of freedom associated with chi-square is not always equal to the sample size as will be explained later.

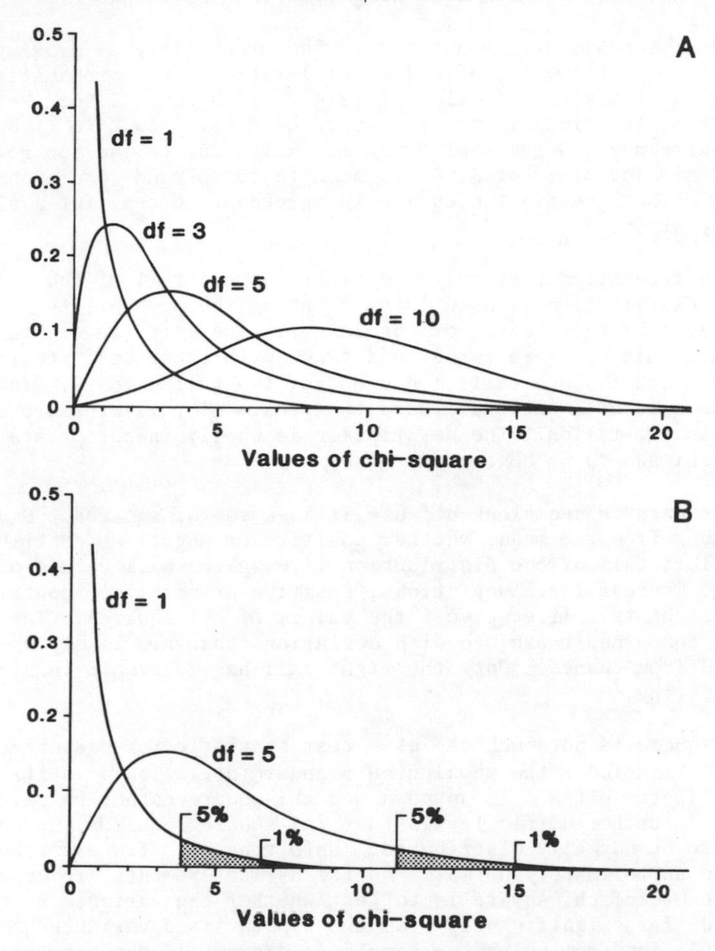

Fig. 15.1 - A: Distribution of chi-square for different degrees of freedom.

B: Selected areas in the right tail of the distribution of chi-square with 1 and 5 degrees of freedom.

As Fig. 15.1B demonstrates, a χ^2 of 3.841 cuts off 5% of the area under the curve for df = 1, while a χ^2 of 6.635 cuts off 1% of the area. Hence, for 1 df, the value of P corresponding to a χ^2 of 3.841 or more is 0.05, while that corresponding to a χ^2 of 6.635 or more is 0.01. In the case of the more symmetrical curve for 5 df, values of χ^2 of 11.070 or more correspond to a value of P of 0.05, and values of χ^2 of 15.086 or more to a value of P of 0.01. These values of P, along with others, are set out in Table J. The lay-out of the table resembles that of Table E (the "t-test"). Column headings refer to selected values of P and row headings to numbers of degrees of freedom. Figures in the body of the table are values of chi-square.

The probabilities for chi-square with one degree of freedom (Table J, top row) are the same as the probabilities for Z (Table B). For example, consider a normally distributed population with a mean of

235 and a standard deviation of 20. The probability of drawing a single item with a value of 274.2, or larger, is the probability of drawing an item with a Z value of $|274.2 - 235|/20 = 1.96$. Referring to Table B, the probability is seen to be 0.05. Now $Z^2 = (1.96)^2 = 3.8416$ with n = 1 degrees of freedom. Referring to the top row of Table J a value of χ^2 of 3.841 is seen to correspond to a probability of 0.05. Thus the two tables are in agreement, except for a slight rounding error.

For a two-tailed test only the right (upper) tail of the chi-squared distribution is used in calculating the probability P of a given value of chi-square, or one larger. The left (lower) tail is ignored. This may seem rather odd since with other test statistics, such as Z and t, both tails are used for two-tailed tests. Indeed, it is because both tails are used that tests of this kind are described as two-tailed. The description is wholly inappropriate as far as chi-square is concerned.

Chi-square is anomalous because it is a sum of squares. Small deviations from the mean, whether positive or negative, contribute to the left tail of the distribution (i.e. give small values of chi-square), whereas large deviations, positive or negative, contribute to the right tail (i.e. give large values of chi-square). The left tail is concerned therefore with deviations that are likely to have resulted from chance. Only the right tail has relevance in significance testing.[1]

Chi-square is not much use as a test statistic for measurements. Like Z, it includes the population standard deviation σ in its formula. All too often σ is unknown and chi-square cannot be calculated. A further difficulty is that chi-square requires the population to be normally distributed. Unfortunately, few distributions are even approximately normal. As far as measurements are concerned the main use of chi-square is to test whether the variance s^2 of a sample differs significantly from the hypothesised variance σ^2 of the population from which the sample is drawn. If the population is normally distributed, with variance σ^2, it can be shown mathematically that $\chi^2 = (n - 1)s^2/\sigma^2$, the number of degrees of freedom associated with χ^2 being $(n - 1)$. A complete proof of this result cannot be given here, but it may be helpful to note that by definition

$$\chi^2 = \Sigma\left(\frac{X - \mu}{\sigma}\right)^2 = \frac{\Sigma(X - \mu)^2}{\sigma^2}$$

[1]There is one exception, however, which should be noted carefully. If the X values deviate from μ by an amount that is substantially less than that which would ordinarily occur through chance sampling (or if the X values are the same as μ), this may be as remarkable a circumstance as if the X values deviate by a large amount. It may indicate, for instance, that the X values have been fraudulently invented to "support" the hypothesis under test. Given deviations that are very small, one may wish to use the left tail of the chi-square distribution to test the hypothesis that the deviations are too good to be true. With this one exception, however, it is the right tail of the chi-square distribution that is used in significance testing.

Also, by definition

$$s^2 = \frac{\Sigma(X - \overline{X})^2}{n - 1}$$

Rearranging gives $\Sigma(X - \overline{X})^2 = (n - 1)s^2$. Since $\overline{X}$ provides an unbiased estimate of μ it can be substituted for μ in the formula for χ^2 which therefore becomes

$$\chi^2 = \frac{\Sigma(X - \overline{X})^2}{\sigma^2} = \frac{(n - 1)s^2}{\sigma^2}$$

One degree of freedom is lost in substituting $\overline{X}$ for μ, which leaves $(n - 1)$ degrees of freedom associated with χ^2.

As an example suppose that according to some null hypothesis the variance of a population is 10 square units. A random sample of 6 items drawn from the population is found to have a variance of 8 square units. The value of $\chi^2 = (6 - 1)8/10 = 4$ with $(6 - 1) = 5$ degrees of freedom. Referring to Table J the value of P is found to exceed 0.50. The sample provides no grounds for doubting the null hypothesis. There are over 5 chances in 10 of obtaining the observed difference in variance or one greater through sampling error.

The chi-square test of variances is of theoretical interest but not of great practical importance since few hypotheses give a value of σ.

PEARSON'S CHI-SQUARE

It is in relation to frequency data that chi-square is invaluable as a test statistic. For large samples, the chi-squared distribution provides a good approximation to the multinomial distribution even though it is continuous, not discrete like the multinomial. A great advantage of the chi-squared distribution is that P values can be looked up in tables, and need not be calculated as in the case of the multinomial.

The way in which the chi-squared distribution approximates to the multinomial is unexpectedly simple having regard to the very different nature of the two distributions. Given a sample of n items grouped into j classes, let f represent the observed frequency in a class and F the expected frequency under the null hypothesis. Provided the sample is large the quantity

$$\mathbf{X}^2 = \Sigma\left[\frac{(f - F)^2}{F}\right]$$

is distributed approximately as chi-square with $(j - 1)$ degrees of freedom.[1] This important result will not be proved here as the mathematics are somewhat difficult. For an assortment of different proofs see Lancaster's monograph on chi-square (1969, Chapter V).

[1] $(j - 1)$ is the maximum number of degrees of freedom. In some applications the degrees of freedom are fewer as explained later.

Many statisticians refer to $\mathbf{X}^2$ as "chi-square" even though it is not quite the same as chi-square. It is discrete, whereas chi-square is continuous. To avoid confusion, $\mathbf{X}^2$ will be referred to in this book as *Pearson's chi-square*. The symbol χ^2 will be reserved for the continuous chi-square whose associated probabilities are shown in Table J.

The distinction between the two chi-squares is of no practical importance except in the case of small samples. With increasing sample size the distribution of Pearson's chi-square approximates more and more closely to that of the continuous chi-square. In the case of a sample of infinite size the two distributions are identical.

It is usually a simple matter to calculate Pearson's chi-square. Consider, for instance, the sample of economically active males in Crawley New Town

$$\mathbf{X}^2 = \frac{(240 - 289.142)^2}{289.142} + \frac{(359 - 346.937)^2}{346.937} + \frac{(708 - 652.971)^2}{652.971} + \frac{(214 - 238.136)^2}{238.136} + \frac{(87 - 80.814)^2}{80.814}$$

$$= 8.352 + 0.419 + 4.638 + 2.446 + 4.735$$

$$= 16.329 \text{ approximately.}$$

The degrees of freedom are (5 - 1) = 4. Because the sample is large, the distinction between Pearson's chi-square and the continuous chi-square can be safely ignored. Referring to Table J, the probability of obtaining a value of 16.329, or more, through chance sampling is found to be between 0.01 and 0.001. That is to say, the observed sample, or one less probable, would occur less than once in a hundred times if repeated samples were drawn from the statistical population represented by the 1966 survey.

The formula for Pearson's chi-square can be rewritten in a slightly different form which is sometimes simpler for computational purposes

$$\mathbf{X}^2 = \Sigma\left[\frac{f^2}{F}\right] - n$$

where n is the sample size or total frequency ($n = \Sigma f$ or ΣF). That this formula is algebraically equivalent to the standard formula can be shown as follows

$$\mathbf{X}^2 = \Sigma\left[\frac{(f - F)^2}{F}\right] = \Sigma\left[\frac{f^2 - 2fF + F^2}{F}\right] = \Sigma\left[\frac{f^2}{F}\right] - 2\Sigma f + \Sigma F$$

Since ΣF always equals Σf (or n) the expression on the right-hand side becomes

$$\Sigma\left[\frac{f^2}{F}\right] - 2n + n = \Sigma\left[\frac{f^2}{F}\right] - n$$

Pearson's chi-square finds a number of different applications as a test criterion:

A: to test whether the relative frequencies or proportions in a sample differ significantly from those in a hypothesised population,

B: to test whether two paired samples have significantly different relative frequencies or proportions,

C: to test whether two classification schemes produce significantly different relative frequencies or proportions OR (and this is only a slight variation) to test whether two or more independent samples differ significantly in relative frequencies or proportions.

The first of these applications is illustrated by the sample of males in Crawley New Town described above. A formal presentation of the test procedure and an additional example now follow.

A: *TO TEST WHETHER THE RELATIVE FREQUENCIES OR PROPORTIONS IN A SAMPLE DIFFER SIGNIFICANTLY FROM THOSE IN A HYPOTHESISED POPULATION*

THE DATA

A random sample is drawn consisting of n items. The items are then grouped into j mutually exclusive classes.

Let p denote the hypothesised proportion of items that belong to a given class in the population from which the sample is drawn. Then $F = pn$ is the expected number of items ("expected frequency") that belong to the class in the sample. Let f denote the actual number of items ("observed frequency") that belong to the class in the sample.

The expected frequencies for the different classes must add up to the same total (n) as the observed frequencies.

PROCEDURE

1. Set up the null hypothesis: chance alone is responsible for the differences between the observed and expected frequencies. The differences, in other words, are not statistically significant.

2. Calculate Pearson's chi-square

$$\chi^2 = \Sigma\left[\frac{(f - F)^2}{F}\right] = \Sigma\left[\frac{f^2}{F}\right] - n$$

3. Calculate the degrees of freedom associated with Pearson's chi-square. The degrees of freedom are the number of classes minus the number of constraints placed on the expected frequencies. In most applications of the test the sole constraint is that the expected

frequencies must add up to n, the sum of the observed frequencies. Hence, with j classes, there are (j - 1) degrees of freedom.

In so-called *goodness of fit* applications (see later) additional constraints are placed on the expected frequencies, and consequently the degrees of freedom are less than (j - 1).

4. Decide whether a one- or two-tailed test is appropriate.

5. Using Table J, estimate P, the probability of obtaining the calculated value of Pearson's chi-square, or one larger, through chance sampling (sampling error). The probabilities shown in Table J are for two-tailed tests; for a one-tailed test they should be halved.

6. Decide whether to retain or reject the null hypothesis. If P is low, serious doubt is cast on the null hypothesis. Reject the null if P is equal to or less than whatever level of significance is judged appropriate, e.g. 0.05 or 0.01. Conclude that the observed and expected frequencies differ by more than would ordinarily occur as a result of chance sampling.

If P exceeds the level of significance, reserve judgement, i.e. regard the null hypothesis as possibly, but not necessarily, correct.

EXAMPLE

The accompanying table lists the main types of soil in Essex and the area occupied by each soil type. It also shows the number of Roman settlements that are known to have existed on each soil type. Can one confidently assume that the Romans exercised an element of choice and did not just settle at random?

TABLE 15.3 - SOIL TYPES AND ROMAN SETTLEMENTS IN ESSEX

Soil type	*Area (sq km)*	*Percentage of total area*	*Number of settlements*	*Percentage number*
Alluvial soils	139	9.7%	17	6.5%
Light loams	118	8.2%	15	5.7%
Heavy loams and clay soils	682	47.3%	135	51.9%
Sandy or gravelly soils	400	27.7%	89	34.2%
Chalk soils	103	7.1%	4	1.5%
Total	1442	100.0%	260	100.0%

At first glance one might not think that there is any need to apply a significance test to the figures. After all, no sampling has been undertaken. The areas of the five soil types form a statistical population, not a sample. They may be subject to measurement error, but they are free from sampling error. The numbers of settlements on the five soil types also form a statistical population - the population of known Roman settlements. As there is no possibility of sampling error there might seem to be no justification for applying a significance test.

It is important to realise, however, that the population of settlements can be treated as if it were a sample. One can assume as a null hypothesis that each of the 260 settlements is randomly located with respect to soil type and on this hypothesis the population of 260 settlements becomes equivalent to a sample of 260 randomly distributed points. If one were to draw an infinite number of samples of such points within the area of Essex, one would find that the percentage number of points on each soil type was the same as the percentage area occupied by each soil type. Thus one would find that the percentage number of points on alluvial soils was 9.7, on light loams 8.2, and so on. The total number of points on alluvial soils would average $(139/1442)(260) \simeq 25.062$ per sample, and the total number on light loams would average $(118/1442)(260) \simeq 21.276$. These averages and the averages for the other types of soil are listed in Table 15.4. They are the most probable numbers of points in a single sample; in other words, they are the expected frequencies.

If the null hypothesis is correct, the differences between the actual numbers of the settlements and the averages (or expected frequencies) shown in Table 15.4 require no special explanation: they are due simply to chance sampling. The null hypothesis can be tested by calculating the value of chi-square which comes to just over 21. Table J shows that with $j - 1 = 4$ degrees of freedom the value of χ^2 is significant beyond the 0.001 level, the smallest probability shown in the table. This result suggests that the null hypothesis ought to be abandoned. It is most unlikely that the known settlements are located entirely at random.

TABLE 15.4 - CALCULATION OF CHI-SQUARE FOR THE DATA SHOWN IN TABLE 15.3

Soil type	*Observed frequency* f	*Expected frequency* F	*Difference* $(f - F)$	*Difference squared* $(f - F)^2$	*Contribution to chi-square* $(f - F)^2/F$
Alluvial soils	17	25.062	- 8.062	65.003	2.594
Light loams	15	21.276	- 6.276	39.388	1.851
Heavy loams and clay soils	135	122.968	+12.032	144.767	1.177
Sandy or gravelly soils	89	72.122	+16.878	284.865	3.950
Chalk soils	4	18.571	-14.571	212.327	11.433
Totals	260	260.000			21.005 = χ^2

Three weaknesses of the analysis need to be pointed out. In the first place, it is possible that only some of the Roman settlements in Essex have yet been discovered, and if all the Roman settlements could be analysed they might appear to be randomly distributed. Only if the known settlements are a random sample of all the settlements is it safe to conclude that the Romans exercised an element of choice in where they settled. There is a danger that the known settlements are very far from being a random sample, being better preserved or easier to find on certain types of soil.

A second weakness of the analysis is that it establishes only that the known settlements are unlikely to be located entirely at random. There is a possibility that some settlements are randomly located but others are the result of deliberate choice. One should also bear in mind that the Romans may have exercised limited choice over the location of each settlement, while allowing random forces some control.

The third weakness of the analysis is that it does not show whether the Romans took any account of soil conditions when choosing places to settle. The Romans may have settled in areas where there was existing settlement and may have completely ignored soil conditions. It may have been the existing settlers who took the soil into account and not the Romans. The analysis deals only with the Romans' actions and not with their motives or knowledge of soils.

It is interesting to note that the settlements on chalk soils provide by far the biggest contribution to the value of chi-square. Why have relatively few settlements been discovered on chalk soils in Essex? As a rule, evidence of early settlement is better preserved on chalk than on other soils. One thinks of Celtic field systems and Roman farms and villages that are so much a feature of the Chalk of southern England. It is not immediately obvious why there is relatively little evidence of Roman settlement on the Chalk in Essex.

B: *TO TEST WHETHER TWO PAIRED SAMPLES HAVE SIGNIFICANTLY DIFFERENT RELATIVE FREQUENCIES OR PROPORTIONS*

THEORY

The test procedure just described can be readily modified to compare paired samples. Suppose n items are selected at random from a population. Each item is classified on two separate occasions, A and B (e.g. before and after some treatment). The same j mutually exclusive classes are used on each occasion. Alternatively, each item is divided into two parts, A and B (e.g. upper and lower, or inner and outer), and each part is allocated to one of j mutually exclusive classes. The null hypothesis assumes that the relative frequencies or proportions of items in the j classes in the parent population are the same for A as for B. Any differences in the relative frequencies in the paired samples is due to sampling error.

Only items that are classified differently on A and B need be considered when testing the null hypothesis. The hypothesis assumes that all the possible differences in classification are equally likely. Any item in the population is as likely to change into any one class for B as change out of that class for A. The probable numbers of items (or expected frequencies) corresponding to each difference in classification can be compared with the actual numbers (or observed frequencies) using Pearson's chi-square. An example will help to make the procedure clear.

EXAMPLE

A group of children attending the primary school in a small village near Cambridge were shown a map and an aerial photograph of the village and asked to compare them (Dale, 1971). The children were then told that they would be given a test in which they would be required to identify well-known places in the village. They were asked to choose either the map or the aerial photograph for the test - whichever they thought would be easier to interpret. Twenty-seven children preferred the aerial photograph and only eleven the map. After the completion of the tests the children were again asked to choose either the map or the aerial photograph on the assumption that there were to be more tests. Seventeen children out of the 27 stuck to their original preference for the aerial photograph, whilst 10 changed their mind and thought that the map would be easier; 6 out of the 11 who originally chose the map stuck to their choice whereas 5 changed their minds and thought the photograph was preferable (Table 15.5).

TABLE 15.5 - PREFERENCES OF SCHOOL CHILDREN FOR MAPS AND AIR PHOTOGRAPHS

	No change of mind		*Change of mind*	
A. Initial choice	Photograph	Map	Photograph	Map
B. Final choice	Photograph	Map	Map	Photograph
Number of children	17	6	10	5

The changes in preference can be analysed using a paired-sample chi-square test.

1. Null hypothesis: the children in the school who changed their mind did so at random. It is pure chance that the majority of children who changed their mind ended up preferring the map.

2. The 15 children who changed their mind did so in 2 different ways (from the photograph to the map and from the map to the photograph).

3. Under the null hypothesis one would have expected 15/2 = 7.5 children to change from the photograph to the map and 15/2 = 7.5 to change from the map to the photograph.

4. $\chi^2 = \Sigma \left[\frac{(f - F)^2}{F} \right] = \frac{(10 - 7.5)^2}{7.5} + \frac{(5 - 7.5)^2}{7.5}$

$= 1.667$ approximately.

5. The degrees of freedom are the number of differences in classification minus one because there is one constraint: the expected frequencies are required to add up to the same total as the observed frequencies

df = 2 - 1 = 1

6. The value of P for a two-tailed test lies between 0.10 and 0.20 so it is unsafe to reject the null hypothesis. The children who changed their minds may have done so at random.

To supplement the paired-sample test, a one sample χ^2 test can be run to check whether more children in the school prefer the photograph to the map when they have time to compare the two. If the children in the school were equally divided one would expect that, out of a sample of 38, 19 on average would prefer the photograph and 19 would prefer the map. In the actual sample, however, 22 preferred the photograph and only 16 preferred the map. The null hypothesis that this difference is due to chance sampling can be tested by calculating χ^2

$$\chi^2 = \Sigma\left[\frac{(f - F)^2}{F}\right] = \frac{(22 - 19)^2}{19} + \frac{(16 - 19)^2}{19}$$

$$= 0.947 \text{ approximately.}$$

This value of χ^2 is so low that the null hypothesis cannot be discounted. The majority of children in the school may not prefer the photograph to the map when they have compared the two.

It is fairly safe to assume that the children in the school prefer the photograph to the map when they are first shown them. The possibility that the children are equally divided from the outset can be tested by running a second one-sample test

$$\chi^2 = \Sigma\frac{(f - F)^2}{F} = \frac{(27 - 19)^2}{19} + \frac{(11 - 19)^2}{19} = 6.737$$

with one degree of freedom this is significant at the 0.01 level. The data provide convincing evidence, therefore, that the majority of children in the school initially prefer the photograph to the map.

C: *TO TEST WHETHER TWO CLASSIFICATION SCHEMES DIFFER SIGNIFICANTLY OR TO TEST WHETHER TWO OR MORE INDEPENDENT SAMPLES DIFFER SIGNIFICANTLY IN RELATIVE FREQUENCIES OR PROPORTIONS*

THEORY

Consider in the first place a population of items classified according to two criteria, A and B. The classification can be presented as a *contingency table* with the classes of A and B arranged at right angles to one another. Table 15.6 is an example of what is called a *2 × 2 table*, there being just two classes of A and two classes of B. This is the simplest case: often contingency tables are more elaborate with many classes of A and B.

TABLE 15.6 - TWO-WAY CLASSIFICATION OF A POPULATION OF N = 48 ITEMS

		Classes of B		Totals
		b_1	b_2	
Classes of A	a_1	12	4	16
	a_2	8	24	32
Totals		20	28	N = 48

Given a two-way classification such as that shown in Table 15.6 an obvious question arises. Is there an *association* between A and B, that is to say, are the A and B classifications interdependent, or are they independent? In Table 15.6 A and B are clearly associated. If an item belongs to class a_1 it is three times more likely to belong to class b_1 than to class b_2. On the other hand, if it belongs to class a_2, it is three times more likely to belong to class b_2 as class b_1. Again, if an item belongs to class b_1, it is half as likely again to belong to class a_1 as to class a_2. If it belongs to class b_2, it is six times as likely to belong to class a_2 as a_1.

Table 15.7 illustrates total independence. The A class that an item belongs to has no bearing on the B class that it belongs to. In other words, there is no association between A and B. Three quarters of the items in class a_1 fall in class b_1 and one quarter fall in class b_2. Exactly three quarters of the items in class a_2 fall in class b_1 and one quarter in class b_2. Similarly one third of the items in class b_1 fall in class a_1 and two thirds in class b_2. This is repeated in class b_2, one third of the items falling in class a_1 and two thirds in class a_2.

TABLE 15.7 - CONTINGENCY TABLE ILLUSTRATING INDEPENDENCE OF A AND B

		Classes of B		Totals
		b_1	b_2	
Classes of A	a_1	12	4	16
	a_2	24	8	32
Totals		36	12	N = 48

It is an easy matter to decide whether two classifications are independent or not if the items under investigation form a population. With sample data problems arise, however. The numbers of items in the various classes may vary because of sampling error.

Three distinct sampling situations can be identified:

Case 1: the row and column totals of the contingency table are allowed to vary randomly (subject to the restriction that they must add up to the same grand total n).

This situation arises if n items are selected at random from a population and each item is placed in one class according to a criterion A and another class according to a second criterion B.

Case 2: the column totals are allowed to vary randomly, but the row totals are fixed beforehand by the person doing the sampling.

This situation arises if two or more populations are distinguished according to some criterion A. A separate random sample is drawn from each population, the size of the sample being set by the person doing the sampling. Each item in each sample is classified according to a second criterion B. There are n items in the combined samples.

The situation also arises in other ways. For instance, in the case of a single population a stratified random sample of size n can be drawn, the strata representing criterion A. Each item in each strata is then classified according to criterion B.

Case 3: both the column and row totals are fixed.

This situation is very unusual. N items are selected at random either from a single population or from several populations. Neither the row nor the column totals are allowed to vary randomly.

An example may help to make the foregoing distinctions clearer. A Professor of Geography claims to be able to judge entirely by eye whether one irregularly shaped area on a map is larger than another. A suspicious student asks the Professor to demonstrate his powers in regard to 20 pairs of areas closely matched in size. One area forming each pair is labelled X and the other Y. The Professor examines each pair of areas and states whether X is larger than Y, or vice versa. The areas are then measured to check whether he is right or wrong. Table 15.8 shows the results. The Professor is right 9 + 4 = 13 times out of 20.

TABLE 15.8 - HYPOTHETICAL DATA TO ILLUSTRATE TWO SAMPLING SITUATIONS

Result of measurements	*Identification made by the Professor*		*Totals*
	X > Y	*Y > X*	
X > Y	9	3	12
Y > X	4	4	8
Totals	13	7	n = 20

The sampling situation is readily identified as Case 1. Neither the row nor the column totals can be regarded as fixed. The Professor thought that area X was larger than Y 13 times out of 20, but he could have chosen any number between 0 and 20. And although in fact area X was larger than area Y 12 times out of 20 it could have been any number between 0 and 20. The marginal totals were free to vary randomly within the constraint that they had to add up to 20.

Now consider a slightly different test of the Professor's skill. The student decides in advance to make X larger than Y 12 times out of 20 and Y larger than X 8 times out of 20. In other words, he decides to fix the row totals. The experiment ceases to be Case 1 and becomes Case 2. The data perhaps remain the same but the underlying sampling situation is slightly different.

Case 3 is obtained if the Professor is told in advance that area X is greater than Y 12 times out of 20, and Y is greater than X 8 times out of 20. Being rational he takes care to match his guesses to these ratios, as illustrated in Table 15.9. In effect, both the row and column totals are fixed, the first by the student and the second by the Professor.

TABLE 15.9 - HYPOTHETICAL DATA TO ILLUSTRATE A THIRD SAMPLING SITUATION

Result of measurements	*Identification made by the Professor* $X > Y$	$Y > X$	*Totals*
X > Y	9	3	12
Y > X	3	5	8
Totals	12	8	n = 20

Sample data arranged in 2 × 2 (and larger) contingency tables can by analysed using Pearson's chi-square. The probabilities associated with Pearson's chi-square are quite different in the three sampling situations just discussed if n is small (how small will be discussed presently). If n is large, the probabilities become more or less the same, and are closely approximated by the continuous chi-square distribution. The test procedure discussed below assumes that n is large, and that the three sampling situations can be treated similarly.

THE DATA

n items are classified according to two criteria, A and B. Each item is placed in one of r classes representing A, and at the same time placed in one of c classes representing B.

In Table 15.10 the data are arranged to form what is called an *r × c contingency table*. There are r rows and c columns forming the body of the Table and defining *r × c cells*. The numbers of items in the cells are designated f_{11}, f_{12}, etc. The subscripts denote the class combinations.

TABLE 15.10 - CLASSIFICATION OF n ITEMS ACCORDING TO TWO CRITERIA, A AND B

A	*B* *Class 1*	*Class 2*	*etc.*	*Class C*	*Totals*
Class 1	f_{11}	f_{12}	etc.	f_{1c}	r_1
Class 2	f_{12}	f_{22}	etc.	f_{2c}	r_2
etc.	etc.	etc.	etc.	etc.	etc.
Class r	f_{r1}	f_{r2}	etc.	f_{rc}	r_r
Totals	c_1	c_2	etc.	c_c	n

PROCEDURE

1. Set up the null hypothesis: there is no relationship between A and B. The class that an item is referred to under A is independent of the class that it is referred to under B. Any seeming relationship between A and B is due entirely to chance sampling.

The exact wording of the null hypothesis can be adjusted to fit the circumstances of the individual problem (see examples).

2. Calculate the frequencies expected under the null hypothesis. If A and B are independent of each other, one would expect that on average the n items would be arranged as in Table 15.11. Under this arrangement the A classification is independent of the B classification, and vice versa. The classes of A appear with same relative frequencies (i.e. in the same proportions) in each of the columns (classes of B). Similarly, each of the classes of B appears with the same relative frequency in each of the rows (classes of A).

TABLE 15.11 - EXPECTED FREQUENCIES CORRESPONDING TO THE OBSERVED FREQUENCIES

A	*B*				*Totals*
	Class 1	*Class 2*	*etc.*	*Class C*	
Class 1	c_1r_1/n	c_2r_1/n	etc.	c_Cr_1/n	r_1
Class 2	c_1r_2/n	c_2r_2/n	etc.	c_Cr_2/n	r_2
etc.	etc.	etc.	etc.	etc.	etc.
Class r	c_1r_r/n	c_2r_r/n	etc.	c_Cr_r/n	r_r
Totals	c_1	c_2	etc.	c_C	n

It should be noted that the expected frequencies add up to the same marginal totals as the observed frequencies. The expected frequency in each cell is the product of the totals of the row and column forming the cell divided by n.

Under the null hypothesis the observed frequencies differ from the expected frequencies purely because of chance sampling.

3. Calculate Pearson's chi-square

$$\chi^2 = \Sigma\left[\frac{(f - F)^2}{F}\right] = \Sigma\left[\frac{f^2}{F}\right] - n$$

where f is the observed frequency in a cell and F the corresponding expected frequency. To be more specific

$$\chi^2 = \frac{(f_{11})^2}{c_1r_1/n} + \frac{(f_{12})^2}{c_2r_1/n} + \frac{(f_{21})^2}{c_1r_2/n} + \frac{(f_{22})^2}{c_2r_2/n} + \ldots + \frac{(f_{rc})^2}{c_cr_r/n}$$

4. Calculate the degrees of freedom. The degrees of freedom are the number of cells in the table minus the number of constraints placed on the expected frequencies. The number of cells is r times c, or rc. The number of constraints is (r + c - 1), as shown below. Hence the number of degrees of freedom is rc - (r + c - 1) = (r - 1)(c - 1).

At first glance the number of constraints would appear to be (r + c), not (r + c - 1). The expected frequencies are constrained to agree with the observed frequencies in respect of each of the r row totals and each of the c column totals. But once agreement has been secured in respect of all but one of these totals, agreement is automatic concerning the remaining total. For example, suppose the expected frequencies agree with the observed frequencies regarding all r row totals and all but one of the c column totals. The remaining column total is not free to vary randomly, but is fixed in value, being the sum of the row totals minus the sum of the other column totals.

Alternatively, the degrees of freedom may be thought of as the number of cells that can be assigned arbitrary (i.e. invented) expected frequencies and still yield the observed row and column totals. For example, if there are three rows and two columns, only two cells can be arbitrarily assigned a frequency - the frequencies in the remaining cells can then be calculated by subtraction from the marginal totals. In general, if there are r rows and c columns (r - 1)(c - 1) cells can be assigned arbitrary frequencies.

5. Decide whether a one- or two-tailed test is appropriate.

6. Using Table J estimate the probability of obtaining the calculated value of Pearson's chi-square, or one larger, through chance sampling. Note that the tabled values of P relate to a two-tailed test. For a one-tailed test they should be halved.

7. Decide whether to retain or reject the null hypothesis. If P is low, serious doubt is cast on the null hypothesis. Reject the null if P is equal to or less than whatever level of significance is judged appropriate, e.g. 0.05 or 0.01. Conclude that the two criteria A and B are not independent of each other, but are related.

If P exceeds the level of significance, reserve judgement, i.e. regard the null hypothesis as possibly, but not necessarily, correct.

EXAMPLE (CASE 1: RANDOM MARGINAL TOTALS)

1. In 1975-76 a sample of households in Great Britain who had moved during the preceding 12 months were asked their main reason for moving. The accompanying table summarises the replies of 1044 households who had the same head of household as 12 months previously. The households are divided into two groups according to the nature of the employment of the head of household.

TABLE 15.12 - MAIN REASON FOR MOVE OF 1044 HOUSEHOLDS WITH CONTINUING HEAD OF HOUSEHOLD. DATA FOR 1975 AND 1976 COMBINED

Nature of employment of head of household	*Main reason for moving*					*Total*
	Housing	*Environment*	*Job*	*Personal*	*Other*	
Manual (skilled and unskilled)	221	83	99	86	118	607
Non-manual (professional, managerial, etc.)	133	66	78	49	111	437
Total	354	149	177	135	229	1044

Source: General Household Survey (1976).[1]

2. Null hypothesis: a survey of all households in Great Britain of the type studied in the sample would find no relationship between the main reason for moving and the nature of the employment of the heads of household. The two schemes of classification are wholly independent; the apparent interdependence shown in Table 15.12 is the result of chance sampling.

3. Under the null hypothesis, 354 households moving for housing reasons ought on average to comprise (607/1044)(354) = 205.822 households with heads in manual employment and (437/1044)(354) = 148.178 households with heads in non-manual employment. Similarly, 149 households moving for environmental reasons ought on average to comprise (607/1044)(149) = 86.631 households with heads in manual employment and (437/1044)(149) = 62.369 households with heads in non-manual employment. Again, 177 households moving for job reasons ought on average to comprise (607/1044)(177) = 102.911 households with heads in manual employment and (437/1044)(177) = 74.089 households with heads in non-manual employment. These expected frequencies are listed in Table 15.13 together with the expected frequencies for households moving for personal and other reasons.

---

[1]Housing reasons include accommodation that is too large, too small or too expensive. Environmental reasons include poor neighbourhood and poor shopping facilities. Job reasons include change of job and desire to live near place of work. Personal and other reasons include retirement, ill-health and eviction.

Percentage frequencies given in the General Household Survey have been changed to actual frequencies. Slight rounding errors may therefore have occurred.

The G.H.S. uses systematic random sampling, but for purposes of illustrating Pearson's chi-square test the data shown above are assumed to constitute an ordinary random sample.

TABLE 15.13 - EXPECTED FREQUENCIES CORRESPONDING TO THE OBSERVED FREQUENCIES LISTED IN TABLE 15.12

Nature of employment of head of household	*Main reason for moving*					*Total*
	Housing	*Environment*	*Job*	*Personal*	*Other*	
Manual (skilled and unskilled	205.82184	86.63123	102.91092	78.49138	133.14464	607
Non-manual (professional, managerial, etc.)	148.17816	62.36877	74.08908	56.50862	92.85536	437
Total	354	149	177	135	229	1044

4. $$\chi^2 = \Sigma\left[\frac{(f - F)^2}{F}\right]$$

$$= \frac{(221 - 205.822)^2}{205.822} + \frac{(133 - 148.178)^2}{148.178} + \frac{(83 - 86.631)^2}{86.631}$$

$$+ \frac{(66 - 62.369)^2}{62.369} + \frac{(99 - 102.911)^2}{102.911} + \frac{(78 - 74.089)^2}{74.089}$$

$$+ \frac{(86 - 78.491)^2}{78.491} + \frac{(49 - 56.509)^2}{56.509} + \frac{(118 - 133.145)^2}{133.145}$$

$$+ \frac{(111 - 95.855)^2}{95.855}$$

$$= 1.119 + 0.152 + 0.149 + 0.718 + 1.723 + 1.555 + 0.211$$

$$+ 0.206 + 0.998 + 2.393 = 9.224 \text{ approximately.}$$

5. df = (2 - 1)(5 - 1) = 4.

6. With df = 4 and $\chi^2 = 9.224$, P is slightly greater than 0.05 (around 0.06).

7. This result casts much doubt on the null hypothesis but not enough to enable one to discount it. Households with heads in manual employment may have rather different reasons for moving from households with heads in non-manual employment. However, the evidence against the null hypothesis is not wholly convincing. Because the value of P exceeds the 0.05 level of significance, one must accept that the null hypothesis may just possibly be correct. The test is thus somewhat inconclusive.

EXAMPLE (CASE 2: FIXED ROW TOTALS, RANDOM COLUMN TOTALS)

1. Separate samples of householders living in "slum" and "twilight" districts of the industrial town of Batley, Yorkshire, were asked "Do you wish to move from this district?" Their replies are shown in the accompanying table.

TABLE 15.14 - ATTITUDES TO MOVING FROM DISTRICT: HOUSEHOLDS IN SLUM AND TWILIGHT DISTRICTS, BATLEY

Location of householder	*Attitude to moving from district* *In favour*	*Against*	*Don't know*	*Total*
Slum district (Sample 1)	59	72	7	138
Twilight district (Sample 2)	139	487	7	633
Total	198	559	14	771

Source: Wilkinson and Sigsworth (1972, p.206).

The table appears to show that the proportion of householders in favour of moving is much greater in slum districts than in twilight districts. Also the proportion of householders who are against moving is apparently greater in twilight districts than in slum districts. Although the numbers are very small, the proportion of "don't knows" is seemingly greater in slum districts.

To what extent are the differences recorded in the table purely the result of chance sampling? A test of significance is needed to decide whether there is a real difference in the attitudes of slum and twilight householders.

2. Null hypothesis: it is chance alone (apart from sample size) that distinguishes the two samples. There is no difference in attitudes between slum and twilight districts, that is to say the proportions of householders in favour of moving, against moving, and undecided are actually the same. The two samples are, in effect, drawn from a single population.

3. Assuming the null hypothesis to be true, the 198 householders in favour of moving could be expected to comprise 198 (138/771) = 35.440 slum householders and 198 (633/771) = 162.560 twilight householders. Similarly, the 559 householders against moving could be expected to comprise 559 (138/771) = 100.054 slum householders and 559 (633/771) = 458.946 twilight householders. Also, the 14 undecided householders could be expected to comprise 14 (138/771) = 2.506 slum householders and 14 (633/771) = 11.494 twilight householders. These expected frequencies are set out in Table 15.15.

TABLE 15.15 - EXPECTED ATTITUDES TO MOVING, BATLEY

Location of householder	*Attitude to moving from district* *In favour*	*Against*	*Don't know*	*Total*
Slum district	35.440	100.054	2.506	138.000
Twilight district	162.560	458.946	11.494	633.000
Total	198.000	559.000	14.000	771.000

4. $$\chi^2 = \frac{(59)^2}{35.440} + \frac{(139)^2}{162.560} + \frac{(72)^2}{100.054} + \frac{(487)^2}{458.946} + \frac{(7)^2}{2.506}$$

$$+ \frac{(7)^2}{11.494} - 771 = 809.476 - 771 = 38.476 \text{ approximately.}$$

5. The degrees of freedom are (2 - 1)(3 - 1) = 2.

6. With 2 degrees of freedom and χ^2 = 38.476, P is less than 0.001.

7. Because the value of P is so low, the null hypothesis is highly suspect and can be discarded. Householders living in slum districts of Batley almost certainly have different attitudes from householders living in twilight districts.

It should perhaps be mentioned that the foregoing test does not establish that the differences recorded in Table 15.14 are entirely real. The test shows only that the differences are unlikely to be due entirely to chance sampling. The differences may be partly real and partly due to chance sampling.

ASSUMPTIONS AND LIMITATIONS

Tests based on Pearson's chi-square are used extensively in the physical and social sciences. They are simple computationally and lend themselves to the solution of a varied range of problems. However, they are valid only if certain conditions are met:

1. Randomness. The data must ordinarily be obtained by simple random sampling. However, stratified random sampling is permissible with Procedure C, each stratum corresponding to a separate row of the contingency table.

2. Independence. The items in the sampled population must be classified independently of each other. One item's class or cell must not influence another item's class or cell.

3. Choice of classes. The classes or cells must be mutually exclusive and exhaustive. In other words, an item must qualify for one and only one class or cell.

By combining (or dividing) classes or cells it is possible to change the value of Pearson's chi-square as well as the number of degrees of freedom. The value of P can thus be manipulated. Consider, for instance, the hypothetical samples shown in Table 15.16. Each household is asked what would be its main reason for wishing to move.

If each of the 14 reasons is considered separately, the value of Pearson's chi-square is 18.076, which with 13 degrees of freedom is not significant at the 5% level (P lies between 0.10 and 0.25). However, if only the five generalised classes shown in Table 15.16 ("housing factors" etc.) are considered, and the frequencies within each of the subsidiary classes are consolidated, the value of Pearson's chi-square becomes 10.726. With 4 degrees of freedom this is significant at the 5% level (P lies between 0.025 and 0.050).

TABLE 15.16 - MAIN REASON FOR WISHING TO MOVE: HOUSEHOLDS IN TWO TOWNS

		Main reason for wishing to move	*Town A*	*Town B*
A		Housing factors		
	i	Accommodation too large, too expensive	13	8
	ii	Accommodation too small, lacking amenities	25	20
	iii	Accommodation in bad state of repair	10	4
	iv	Lack of privacy, garden, etc.	12	15
B		Environmental factors		
	i	Atmospheric pollution	7	8
	ii	High traffic density, noise	6	5
	iii	High population density, poor neighbourhood	5	7
	iv	Close proximity to industry	8	13
	v	Poor shopping facilities, schools, etc.	23	29
C		Employment factors		
	i	Change of job, lack of employment	10	11
	ii	Excessive distance to work	2	11
D		Personal factors		
	i	Retirement	10	16
	ii	Ill-health	10	15
E		Other factors	9	3
Totals (sample sizes)			150	165

To what extent is it fair to tamper with the classes once the data have been inspected? At one extreme, some statisticians argue that combining classes is perfectly fair provided the combinations are logical. Indeed the researcher has a duty to combine classes to obtain the lowest possible value of P, i.e. the highest probability that the differences are real and not due to sampling error. If there are too many classes yielding non-significant differences, they may completely mask a few classes yielding significant differences. By combining all the classes that yield non-significant differences, the researcher is able to focus better on the remaining classes.

At the other extreme there are statisticians who argue with equal force that once the data have been collected the classes ought not to be changed. Juggling with the data after they have been scrutinised undermines the randomness of the sample (or samples) and prejudices the outcome of the test. Given enough classes to play with, one can always find a combination that is sufficiently improbable to be statistically significant. Tampering with the classes is tantamount to cheating since wholly illusory values of P can be obtained (see Lewis and Burke, 1949, for further discussion).

Most statisticans take a position intermediate between these extremes. Some recommend, for instance, that the classes or cells should be determined without reference to the data, and changed only if the expected frequencies are found to be too small (see below) and then only as a last resort if no better solution can be found.

The problem of delimiting the classes cannot yet be regarded as solved. If combining classes is permissible provided the combinations are logical, how does one set about judging whether the combinations are logical? For instance, would it be right to combine classes Ai and Aii in Table 15.16? There are plainly arguments for and against. If combining classes is wrong, then what is there to stop the dishonest research worker, who has failed to get a significant value of P, juggling with his classes to make P significant? Provided the classes are combined in a logical manner the fact that they are combined will not be apparent.

4. Minimum expected frequencies. Pearson's chi-square provides a good approximation to the multinomial distribution only for large samples. Moreover, it is only for large samples that the P values associated with Pearson's chi-square approximate closely to the P values tabulated for the continuous chi-square. Only with samples of infinite size are the P values identical.

How large ought a sample to be to allow a test based on Pearson's chi-square? In the past it was customary to recommend that the expected frequency in each class or cell should be at least 5, and preferably 10. The numbers 5 and 10 appear to have been arrived at somewhat arbitrarily. It is now realised that tests using Pearson's chi-square will often give probabilities that are substantially correct even if one of the expected frequencies is as low as 1. Cochran (1954) suggests that if the degrees of freedom are larger than 1, the probabilities will be tolerably accurate provided 80% of the classes or cells have expected frequencies greater than 5, and no class or cell has an expected frequency of less than 1. Essentially similar views have been expressed by other recent writers, such as Lancaster (1969).

If the expected frequencies fail to meet Cochran's criteria, the obvious remedy is to increase the size of the sample (or samples). But often this is not practical. One alternative solution if there are enough classes or cells, is to combine them in such a way as to eliminate the expected frequencies that are too small. However, combining classes or cells introduces difficulties of a different nature. As explained previously, it undermines the test procedure and is best avoided if at all possible. Rather than combine classes or cells, it is often better to leave them alone and risk getting an inaccurate value of P.

When there is only one degree of freedom, classes or cells cannot be combined without making it impossible to calculate Pearson's chi-square. In the case of a 1 × 2 contingency table the value of P can be found using the binomial test discussed in the next chapter. This test yields exact values of P, regardless of the size of the expected frequencies.

In the case of a 2 × 2 table the exact value of P depends on whether the row and column totals are random or fixed (unless n is so large that the chi-square approximation holds). Methods of calculating P exactly are long and tedious (for details, see Conover, 1971). Fortunately, some thought has been given to the production of tables so that the value of P can be looked up directly.

Case 1: both the row totals and the column totals are free to vary randomly. If n is 10 or less, Table K may be used to find the exact value of P. This table lists all possible values of Pearson's chi-square together with the corresponding exact values of P. For $n > 10$ upper-tail values of P for Pearson's chi-square are closely approximated by the upper tail values for the continuous chi-square. Hence Table J can be used for significance testing for $n > 10$ without appreciable loss of accuracy.

Case 2: the row totals are fixed, but the column totals are free to vary (or vice versa). It appears that no tables have been prepared as yet. However, values of P are intermediate between those for case 1 and case 3. Thus an approximate idea of whether P is significant or not can be obtained by consulting the tables for both case 1 and case 3.

Case 3: both the row totals and the column totals are fixed. Tables are available which list individual cell frequencies corresponding to selected values of P (e.g. 0.05) for values of n up to 100 (Finney, *et al.*, 1963; Bennett and Horst, 1966). The tables, which do not require Pearson's chi-square to be calculated, are too bulky to reproduce in this book. They are one-tailed (unlike Tables J and K), but can be readily adapted for two-tailed tests.

If the foregoing seems altogether too complicated, there is a simpler approach due to Yates (1934). It consists of applying a "correction" to the formula for Pearson's chi-square. 0.5 is added to each observed frequency that is smaller than the corresponding expected frequency and 0.5 is subtracted from each observed frequency that is larger than the expected frequency. This is equivalent to subtracting 0.5 from $|f - F|$, the absolute difference between each observed frequency and the expected frequency. The formula for Pearson's chi-square becomes

$$\chi^2 = \Sigma \left[\frac{(|f - F| - 0.5)^2}{F} \right]$$

Because the correction reduces the size of the difference between the observed and expected frequencies, it reduces the value of Pearson's chi-square and hence the value of P (which is found from Table J in the usual way).

TABLE 15.17 - HYPOTHETICAL DATA TO ILLUSTRATE THE USE OF YATES' CORRECTION

Observed frequencies

	A	*Not A*	*Totals*
B	7	1	8
Not B	1	1	2
Totals	8	2	10

Expected frequencies

	A	*Not A*	*Totals*
B	6.4	1.6	8
Not B	1.6	0.4	2
Totals	8	2	10

Yates' correction suffers from the disadvantage that it tends to overcorrect, changing P from too small to too large. As an illustration, consider the hypothetical data shown in Table 15.17.

Pearson's chi-square calculated in the usual way is

$$\frac{(7 - 6.4)^2}{6.4} + \frac{(1 - 1.6)^2}{1.6} + \frac{(1 - 1.6)^2}{1.6} + \frac{(1 - 0.4)^2}{0.4} = 1.406$$

The value of P according to Table J is 0.24. Applying Yates' correction, Pearson's chi-square becomes

$$\frac{[|7 - 6.4| - 0.5]^2}{6.4} + \frac{[|1 - 1.6| - 0.5]^2}{1.6} + \frac{[|1 - 1.6| - 0.5]^2}{1.6} +$$

$$\frac{[|1 - 0.4| - 0.5]^2}{0.4} = 0.039$$

The value of P obtained from Table J is 0.85. The true value of P, however, is only 0.36, assuming the marginal totals to be random (Table K).

It should be remarked that Yates' correction does not always produce such bad results. If n is fairly large, it can produce a very good approximation to the exact value of P. An illustration is provided by Kendall and Stuart (1961, pp. 555-556). Nevertheless, most statisticians nowadays advise against the use of Yates' correction, since its effects are unpredictable.

5. Percentage frequencies. Tests based on Pearson's chi-square are designed to deal with absolute frequencies and not percentages. The value of Pearson's chi-square increases with increasing sample size (degrees of freedom) even if the percentage in each class remains the same. This is demonstrated in Table 15.18.

TABLE 15.18 - DEPENDENCE OF PEARSON'S CHI-SQUARE ON SIZE OF SAMPLE

Classes of A	*Classes of B* b_1	b_2	*Totals*	*Classes of A*	*Classes of B* b_1	b_2	*Totals*
a_1	12	4	16	a_1	120	40	160
a_2	8	24	32	a_2	80	240	320
Totals	20	28	n = 48	Totals	200	280	480
Value of X^2 = 10.97 with df = 1				Value of X^2 = 109.7 with df = 1			

When calculating Pearson's chi-square percentages must always be converted to absolute frequencies unless the sample size happens to be exactly 100. If the sample size is less than 100, and percentages are used to calculate Pearson's chi-square, the value of P will be too low with the possible result that the null hypothesis is rejected incorrectly. If the sample size is more than 100, substituting percentages for frequencies will have the opposite effect. The value of P will be too high, and consequently significant differences between the observed and expected frequencies may be overlooked.

6. Ordered classes. Tests based on Pearson's chi-square do not consider whether the classes or cells fall into any definite order. The same value of Pearson's chi-square is obtainable whatever the ordering. In many geographical problems the classes or cells are clearly ordered, consisting of distance zones, time bands, etc. Consider, for instance, the sample frequencies set out in Table 15.19.

TABLE 15.19 - LENGTH OF RESIDENCE IN PRESENT PROPERTY: HOUSEHOLDS IN SLUM AND TWILIGHT DISTRICTS, LEEDS

Location of householder	Length of residence				
	Under 4 years	*5 - 9 years*	*10 - 19 years*	*20 years and over*	*Total*
Slum district	94	41	41	84	260
Twilight district	217	93	97	266	673

The value of Pearson's chi-square is 4.194 with df = 3. P works out at about 0.25, suggesting that the apparent differences in length of residence between slum and twilight areas are due to chance sampling. But, although Pearson's chi-square can be calculated, it does not provide a very satisfactory test of significance since it ignores the order of occurrence of the classes. The data do in fact suggest that length of residence varies with district. In each of the first three classes the percentage of slum householders is greater than the percentage of twilight householders; in the final class the situation is reversed, the percentage of twilight householders exceeding the percentage of slum householders. It is interesting to note that if the first three classes are combined (0 to 20 years) the value of Pearson's chi-square becomes 4.167 which with df = 1 is significant at the 5% level.

Tests using Pearson's chi-square cannot be modified to take account of the ordering of the classes. Alternative tests that consider ordering exist, but they are not entirely satisfactory. The best known are the Kolmogorov-Smirnov tests (see Bradley, 1968, and Conover, 1971, for details), which can be used in a variety of one- and two-sample situations, but lack overall sensitivity (Slakter, 1965), and often give higher values of P than Pearson's chi-square despite allowing for the ordering of the classes.

Pearson's chi-square is often used to test *goodness-of-fit*. The measurements in a sample are grouped into classes and the frequencies are compared with the expected frequencies obtained by fitting a mathematical distribution to the sample, such as the normal curve. The classes are ordered, but there is only one sample, not two as in Table 15.19. The same test procedure is followed as in the standard one-sample test described earlier in the chapter, except that the degrees of freedom are reduced. The degrees of freedom are the number of classes minus the number of sample "characteristics" used to calculate the expected frequencies. For example, in calculating the expected frequencies corresponding to a normal distribution three characteristics are taken into account - the sample size, mean, and standard deviation or variance. With j classes the degrees of freedom are therefore (j - 3). In calculating the frequencies corresponding to a Poisson distribution, the sample size is considered and the mean, λ. Hence, with j classes, the degrees of freedom are (j - 2).

Cochran (1952, p.329) suggests that if Pearson's chi-square is used to test the goodness-of-fit of hump-backed curves such as the normal distribution, little inaccuracy is introduced at the 5% level of significance even if two expected frequencies are as low as 1.

As an example consider the calculation of Pearson's chi-square for the 75 distances between nearest-neighbour towns and villages in Iowa. In Chapter 9 the distances were grouped into 15 classes to give observed frequencies that closely resembled the expected frequencies obtained by fitting a normal curve with the same mean and standard deviation as the distances (see Table 9.2). In Table 15.20 the number of classes has been reduced to 11 in order to increase the expected frequencies to the minimum level required for a chi-square test.

TABLE 15.20 - CALCULATION OF PEARSON'S CHI-SQUARE FOR THE NEAREST-NEIGHBOUR DATA

Distance to nearest neighbour	*Observed frequency f*	*Expected frequency F*	*(f - F)*	*$(f - F)^2/F$*
Under 3 km	1	1.748	-0.748	0.320
3 - 3.9 km	4	2.962	+1.038	0.364
4 - 4.9 km	5	5.895	-0.895	0.136
5 - 5.9 km	10	9.562	+0.438	0.020
6 - 6.9 km	16	12.622	+3.378	0.904
7 - 7.9 km	12	13.552	-1.552	0.178
8 - 8.9 km	11	11.842	-0.842	0.060
9 - 9.9 km	8	8.415	-0.415	0.020
10 - 10.9 km	3	4.868	-1.868	0.717
11 - 11.9 km	3	2.295	+0.705	0.217
12 km or more	2	1.238	+0.762	0.469
	75	74.999	0.001	χ^2 = 3.405

Assume as a null hypothesis that the 75 measurements are a sample from a normally distributed population, and that the differences between the observed and expected frequencies are therefore due to chance sampling. As Table 15.20 shows, the value of Pearson's chi-square is 3.405 approximately. Since the expected frequencies have been calculated taking (1) the sample size, (2) the sample mean, and (3) the sample standard deviation into account, the degrees of freedom are (11 - 3) = 8. Referring to Table J, P lies between 0.90 and 0.95. The high value of P is consistent with the hypothesis that the distances are approximately normally distributed.

16 *The Binomial Test and Lilliefors' Test

The two tests described in this chapter are specialised and need not be studied in detail until they are required in a research investigation. The reader who is short of time may wish to omit this chapter and proceed to the next.

THE BINOMIAL TEST

As explained in the last chapter, the multinomial distribution is very awkward to manipulate mathematically. If there are only two classes, however, it simplifies to the binomial distribution which can be manipulated with comparative ease.

The *binomial test* may be used in place of Pearson's chi-square test in order to determine whether the frequencies or proportions in a sample divided into two classes differ significantly from the frequencies or proportions expected under some hypothesis about the population. One advantage of the binomial test is that it yields an exact value of P, not an approximation as does Pearson's chi-square. Another advantage is that the expected frequencies can be small.

TO TEST WHETHER A SAMPLE DIVIDED INTO TWO CLASSES DIFFERS SIGNIFICANTLY FROM A POPULATION EITHER IN PROPORTIONS OR RELATIVE FREQUENCIES

THEORY: Suppose there exists a population of N items, Np of which belong to one class and Nq = N - Np to another. The probability of obtaining exactly X items of the first class in a random sample consisting of n items is given by the binomial term

$$\frac{n!}{X!(n-X)!}\, p^X q^{n-X}$$

where p represents the probability that an item drawn from the population will belong to the first class, and q = 1 - p is the probability that it will belong to the second. The derivation of this formula is given in Chapter 8. In the present chapter n, the sample size, replaces k, the symbol for the number of trials.

The probability of obtaining at least X items is the probability of obtaining X, plus the probability of obtaining X + 1, plus the probability of obtaining X + 2, etc. In symbols, it is

$$\frac{n!}{X!(n-X)!}\, p^X q^{n-X} + \frac{n!}{(X+1)!(n-X-1)!}\, p^{X+1} q^{n-X-1}$$

$$+ \frac{n!}{(X+2)!(n-X-2)!} p^{X+2} q^{n-X-2} + \ldots + p^n$$

Similarly the probability of obtaining X or fewer items is

$$\frac{n!}{X!(n-X)!} p^X q^{n-X} + \frac{n!}{(X-1)!(n-X+1)!} p^{X-1} q^{n-X+1}$$
$$+ \frac{n!}{(X-2)!(n-X+2)!} p^{X-2} q^{n-X+2} + \ldots + q^n$$

THE DATA

A random sample of n items is drawn from some population. The items are then grouped into two mutually exclusive classes.

Let p denote the hypothesised proportion of items that belongs to a given class in the population. Then np is the expected number ("expected frequency") of items that belongs to the same class in the sample. Let X denote the actual number ("observed frequency") of items that belongs to the class in the sample.

PROCEDURE

1. Set up the null hypothesis: chance alone is responsible for the differences between the observed and expected frequencies. The differences, in other words, are not statistically significant.

2. Select one of the two classes making up the sample. Determine X, the number of items in that class, i.e. the observed frequency.

3. Determine the probability P of obtaining the observed value of X, or one even more extreme, through chance sampling. If X exceeds np (the frequency suggested by the null hypothesis) use the formula

$$P = 2\left[\frac{n!}{X!(n-X)!} p^X q^{n-X} + \frac{n!}{(X+1)!(n-X-1)!} p^{X+1} q^{n-X-1} + \frac{n!}{(X+2)!(n-X-2)!} p^{X+2} q^{n-X-2} + \ldots + p^n\right]$$

If X is smaller than np, use the formula

$$P = 2\left[\frac{n!}{X!(n-X)!} p^X q^{n-X} + \frac{n!}{(X-1)!(n-X+1)!} p^{X-1} q^{n-X+1} + \frac{n!}{(X-2)!(n-X+2)!} p^{X-2} q^{n-X+2} + \ldots + q^n\right]$$

These formulas are liable to lead to heavy arithmetic, especially if n is at all large. If n is large, and p is not too near 0 or 1, the value of P can be estimated using the normal distribution as an approximation to the binomial. The method has already been discussed in Chapter 11. Let $\hat{p}$ denote the sample proportion X/n. If the population is infinite, or if the sampling is with individual replacements the standard error of the sample proportion is given by

$$\sigma_{\hat{p}} = \sqrt{\frac{pq}{n}} = \sqrt{\frac{p(1-p)}{n}}$$

If the sample is drawn without individual replacements from a population of finite size

$$\sigma_{\hat{p}} = \left[\sqrt{\frac{p(1-p)}{n}}\right]\left[\sqrt{\frac{N-n}{N-1}}\right]$$

The next step is to calculate Z

$$Z = \frac{|\hat{p}-p| - \frac{1}{2n}}{\sigma_{\hat{p}}} \quad \text{or} \quad \frac{|X-pn| - 0.5}{n\sigma_{\hat{p}}}$$

The probability of obtaining the calculated value of Z, or one larger, may be found by referring to Table B.

The procedures outlined above yield values of P appropriate for a two-tailed test. For a one-tailed test the values must be halved.

Values of P associated with the binomial distribution need not be calculated: they can be looked up directly in tables. Particularly useful are the comprehensive tables issued by the Staff of the Computation Laboratory, Harvard University (1955). Also recommended are the tables prepared by the National Bureau of Standards (1949), the Ordnance Corps (1952) and Weintraub (1963). The tables only set out one tail of the binomial distribution. For two-tailed tests the values of P given by the tables must be doubled.

4. Decide whether to retain or reject the null hypothesis. If P is low, serious doubt is cast on the null hypothesis. Reject the null if P is equal to or less than whatever level of significance is judged appropriate e.g. 0.05 or 0.01. Conclude that the observed and expected frequencies differ by more than would ordinarily occur as a result of chance sampling.

If P exceeds the level of significance, reserve judgement, i.e. regard the null hypothesis as possibly, but not necessarily, correct.

EXAMPLE

TABLE 16.1 - NUMBERS OF ROMAN SETTLEMENTS IN ESSEX ON TWO TYPES OF SOIL

	Sandy or gravelly soils	*Chalk soils*	*Total*
Area of Essex in sq km	400	103	503
Numbers of Roman settlements	89	4	93

Table 16.1 repeats part of Table 15.3. It shows that sandy or gravelly soils account for 400/503 = 0.79523 of the total area under study while chalk soils account for only 103/503 = 0.20477 of this area. If the 93 Roman settlements are located randomly (and thus constitute a random sample of the infinite number of possible locations on sandy, gravelly or chalk soils in Essex) some 93(0.79523) = 73.95626 ought on average to be located on sandy or gravelly soils and 93(0.20477) = 19.04374 ought on average to be

located on chalk soils. These expected (or most probable) numbers of settlements are rather different from the actual numbers, but the differences might be thought to be simply a matter of chance. If chance is indeed responsible, the probability of obtaining the sample, or one departing further from expectation, can be calculated using either the binomial or Pearson's chi-square test. The binomial test furnishes an exact value of P and Pearson's chi-square an approximation. According to the binomial formula

$$P = 2\left[\frac{93}{89!4!}p^{89}q^{4} + \frac{93}{90!3!}p^{90}q^{3} + \frac{93}{91!2!}p^{91}q^{2} + \frac{93}{92!1!}p^{92}q^{1} + p^{93}\right]$$

where p = 0.79523 and q = 0.20477.

Remembering that p^{89} = antilog (89 log p) and q^4 = antilog (4 log q), etc.

$$\begin{aligned} P &= 2\,[2919735\ (1.30247\times10^{-9})\ (1.75824\times10^{-3}) + 129766 \\ &\quad (1.10734\times10^{-9})\ (8.58633\times10^{-3}) + 4278\ (8.80588\times10^{-10}) \\ &\quad (0.04193) + 93\ (7.00268\times10^{-10})\ (0.20477) \\ &\quad + (5.56873\times10^{-10})] \\ &= 2\,[(7.14840\times10^{-6}) + (1.23381\times10^{-6}) + (1.57962\times10^{-7}) \\ &\quad + (1.33357\times10^{-8}) + (5.56874\times10^{-10})] \\ &= 0.17\times10^{-4} \text{ approximately.} \end{aligned}$$

Hence, with random sampling, there are about 17 chances in a million of obtaining the observed sample, or one departing even more noticeably from expectation. It can be confidently assumed that the Roman settlements are located non-randomly. The differences between the observed and expected numbers of settlements are unlikely to be due solely to chance.

Pearson's chi-square yields a value of P with much less bother, although the value is only approximate

$$\chi^2 = \sum\left[\frac{(f-F)^2}{F}\right] = \frac{(89-73.95626)^2}{73.95626} + \frac{(4-19.04374)^2}{19.04374} = 14.944$$

There is one degree of freedom. Referring to Table J, Row 1, P is found to be less than 0.001, the smallest probability listed in the table. Extrapolation of the tabled values suggests that P is about 0.0001. This is rather more than the true value of 0.000017. The discrepancy arises because the chi-squared distribution is a poor approximation to the binomial if, as in the present example, one of the expected frequencies is small.

ASSUMPTIONS AND LIMITATIONS

The binomial test is often very simple to carry out. It should be used, however, only to compare a sample with a population. To compare a sample with another sample, a different test is needed such as one based on Pearson's chi-square.

The binomial test requires that the items in the sample fall into one of two mutually exclusive classes. The test is inapplicable if there are more than two classes, or if the classes overlap.

The probability p of an item falling into the first class must be constant (i.e. the same for all items) as must be the probability of it falling into the second class. If p or q vary from one item to another the test is strictly speaking invalid. Ideally the population must be infinite in size, or the sampling be carried out with individual replacements. However, if the sample is only a small fraction of the population (say 10% or less) the test can be relied upon to yield reasonably accurate results even though the sampling is without replacements. It should be noted also that if the Z approximation is used the formula for $\sigma_{\hat{p}}$ can be adjusted to allow for sampling without replacements.

LILLIEFORS' TEST

This test checks whether sample measurements closely approximate to a normal distribution. Unlike the test based on Pearson's chi-square discussed in the last chapter it does not require the measurements to be grouped into classes, which wastes information. It is a specialised version of a test known as the Kolmogorov test or Kolmogorov-Smirnov one-sample test (see Bradley, 1968, and Conover, 1971, for details), which allows sample measurements to be compared with any theoretical distribution. The Kolmogorov test is not included in this book because it has little practical use. The theoretical distribution must be specified entirely independently of the sample measurements, but in most research investigations this is impossible. Although Lilliefors' test can be used only to compare a sample with a normal population, it is more practical than the Kolmogorov test because it allows the researcher to select a normal distribution with the same mean and standard deviation as the sample.

TO TEST WHETHER A SAMPLE IS APPROXIMATELY NORMALLY DISTRIBUTED

1. Set up the null hypothesis: the population from which the sample is drawn can be considered to be normally distributed.

2. For each value of X (or measurement) in the sample, count the number of items having this value or a smaller value. Divide every number by n, the sample size, to obtain the cumulative probability distribution of the X values.

3. Calculate the cumulative probability distribution of the normal curve with the same mean and standard deviation as the sample.

4. Find D, the maximum difference between the cumulative probabilities of the two distributions.

5. Decide whether a one- or two-tailed test is appropriate. Use a one-tailed test if, prior to sampling, there could be complete certainty of the direction that D would take in the absence of sampling error (in other words, certainty that the cumulative probability of the normal distribution would exceed that of the sample, or certainty that the reverse would be the case). Use a two-tailed test, if, prior to sampling, there could be no certainty about the direction of D.

6. Refer to Table L to find the value of D that must be equalled or exceeded for P to attain the value shown at the head of the table.

The table values of P are for a two-tailed test. For a one-tailed test they should be halved.

7. Decide whether to retain or reject the null hypothesis. Retain the null if the calculated value of D is less than the value shown in the table for P = 0.05, or whatever level of significance is judged appropriate. Conclude that the sample distribution does not deviate significantly from the normal at the chosen level of significance. Reject the null if the calculated value of D is equal to or greater than the tabulated value. Conclude that the sample distribution deviates significantly from the normal at the chosen level of significance.

EXAMPLE

In the last chapter the grouped distances between the 75 towns and villages in Iowa and their nearest neighbours were shown to be approximately normally distributed. Lilliefors' test will now be applied to the original data (Table 5.1) to provide an alternative and potentially more sensitive test.

TABLE 16.2 - CUMULATIVE PROBABILITY DISTRIBUTIONS FOR THE NEAREST-NEIGHBOUR DISTANCES

Nearest-neighbour distances	*Cumulative probability*	*Nearest-neighbour distances*	*Cumulative probability*
2.7	1/75	7.2,7.2	40/75
3.4	2/75	7.3	41/75
3.5	3/75	7.4,7.4,7.4	44/75
3.6	4/75	7.5	45/75
3.9	5/75	7.8	46/75
4.2	6/75	7.9,7.9	48/75
4.4,4.4	8/75	8.0,8.0	50/75
4.8	9/75	8.3	51/75
4.9	10/75	8.4	52/75
5.1	11/75	8.5,8.5	54/75
5.2,5.2	13/75	8.6,8.6	56/75
5.4,5.4,5.4	16/75	8.8,8.8	58/75
5.6	17/75	8.9	59/75
5.7	18/75	9.0	60/75
5.8	19/75	9.3	61/75
5.9	20/75	9.4,9.4	63/75
6.0,6.0	22/75	9.5,9.5	65/75
6.2,6.2	24/75	9.7	66/75
6.3	25/75	9.9	67/75
6.4,6.4	27/75	10.0	68/75
6.5	28/75	10.6	69/75
6.6,6.6,6.6	31/75	10.7	70/75
6.7,6.7,6.7	34/75	11.0	71/75
6.9,6.9	36/75	11.6,11.6	73/75
7.0	37/75	12.5	74/75
7.1	38/75	12.7	75/75

1. Null hypothesis: nearest-neighbour distances in Iowa are normally distributed.

2. The distances between the 75 towns and villages and their nearest neighbours are listed in order of increasing size in Table 16.2 together with the corresponding cumulative probabilities. The latter are plotted in Fig. 16.1 as the line marked S(X). Because of its graphical form S(X) can be called a *step function*. For each unique distance (value of X) in the sample there is a step in the graph 1/75 units high. Where two or more X values are the same the graph rises in a step that is the appropriate multiple of 1/75.

3. The smooth curve F(X) represents the cumulative probability distribution of a normal curve having the same mean and variance as the sample. F(X) is taken from Fig. 9.12 which was plotted by cumulating the probabilities (areas) listed under the heading A in Table 9.2.

4. The maximum vertical distance between S(X) and F(X) is seen from Fig. 16.1 to occur just to the right of X = 7.5 km. At this point the value of S(X) is 0.60 and the value of F(X) is approximately 0.53. The value of D is therefore 0.07 approximately.

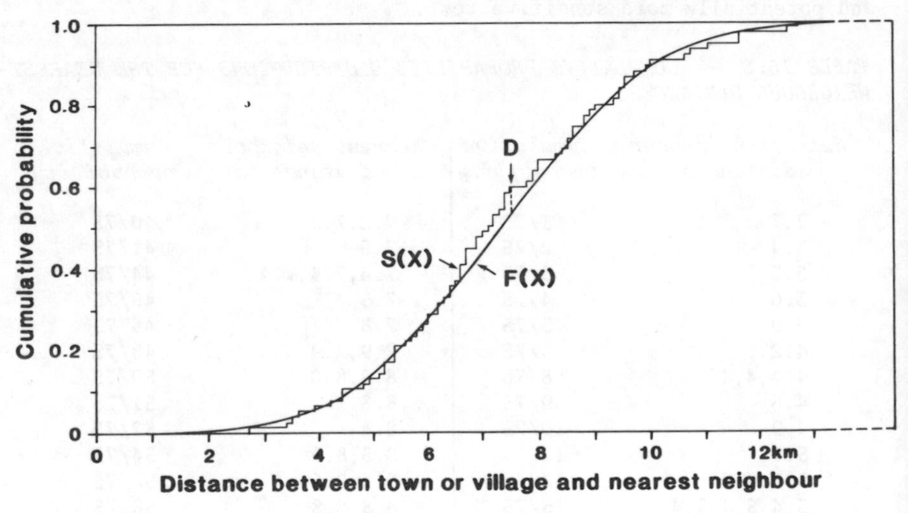

Fig. 16.1 - Graphs of S(X) and F(X) for the nearest-neighbour distances in Iowa.

The exact value of D could, of course, be found by direct calculation. However, the graphical approach is simpler and reasonably accurate.

5. According to Table L the value of D corresponding to P = 0.05 for a two-tailed test is $0.866/\sqrt{n}$. In the present instance n = 75 and the 0.05 value of D is therefore $0.866/\sqrt{75} = 0.102$. Because the observed value of D is less than the tabled value the required value of P is greater than 0.05. Notice also that the observed value of D is less than $0.736/\sqrt{75} = 0.085$, which is the tabled value for P = 0.10. The value of P corresponding to the observed value of D is evidently greater than 0.10.

6. The high value of P indicates that the null hypothesis must be retained. Retention of the null hypothesis, however, is not proof that the nearest-neighbour distances are distributed normally. Indeed, since the normal distribution tails off to infinity in both directions it is clear that the distances cannot be exactly normally distributed. All the same, the high value of P does show that the normal distribution is a very reasonable approximation to the observed distribution. The differences that occur are no more than might have been expected through chance sampling of a perfectly normal population.

ASSUMPTIONS AND LIMITATIONS

Lilliefors' test assumes that the sample is drawn randomly from a population of infinite size. If the population is finite, as in the preceding example, the calculations may be in error, though not necessarily by very much, unless the population happens to be extremely small. Another point to remember is that the variable under study is assumed to be continuous. A few ties are tolerable, but strictly speaking ties ought not to occur. If the variable is discontinuous, the test tends to be somewhat biased in favour of the null hypothesis.

Like other statistical tests, Lilliefors' test assumes that the values of the items making up the sample are independent of one another. If there is appreciable autocorrelation, the results may be seriously in error.

Since no real population is ever precisely normal, there might seem to be no need for Lilliefors' test. The test is bound to indicate non-normality provided a large enough sample is drawn. However, the test is useful for checking whether a sampled population is sufficiently normal to allow t and other tests to be used with reasonable accuracy.

17 The Product-Moment Coefficient of Correlation

INTRODUCTION

The values of one variable are often found to vary in accordance with the values of another. Yields of crops, for instance, vary with rainfall. The numbers of shops in different towns vary according to the population sizes of the towns. The amounts spent on education in developing countries vary with the amounts spent on medical services. Such co-variation between variables, or *correlation* as it is called, may be purely accidental, but in general it is the result of a causal relationship. There are three ways in which one variable may be causally related to another:

1. Simple dependence. One variable may be cause, the other effect. This is the simplest kind of relationship, and the one most commonly selected for investigation. The values of one variable determine at least in part the values of the other. For instance, the amount of rainfall greatly influences the yields of crops. The relationship is entirely one way: variations in rainfall cause variations in crop yield, but not vice versa, at least not to any important extent.

2. Interdependence. The variables may mutually interact so that it is impossible to say which is cause and which is effect. An interaction exists between the population size of towns and the number of shops that are located within them. The larger the population of the town, the greater the number of shops that can be run at a profit. Whereas a small town can support only a restricted number of shops, a large town can justify an extensive number. It would be an over-simplification, however, to regard population size purely as cause and number of shops purely as effect. As the number of shops increases, so the population size of the town naturally tends to grow. Conversely, if some shops disappear, then those people who were dependent on these shops for their livelihood or the services the shops provided may find themselves forced to move elsewhere. Thus, although population size undoubtedly controls the number of shops, the number of shops also controls population size. The relationship is two way.

3. Common dependence. Both variables may be effects of some other variable or group of variables. For example, the amount of money spent on education in different countries shows a relationship with the amount spent on medical services. This merely reflects the fact that both variables are dependent on national affluence. Rich countries can afford to spend large amounts on education and medical services; poor countries cannot; but there is no simple cause-effect

relationship between spending on education and spending on medical services. Nor is there any mutual interaction between the two variables, except that money spent on education cannot then be spent on medical services and vice versa.

The relationship between population size and number of shops is also, in part, one of dependence on other variables. The spacing of towns; their relation to the transport network; national economic policies; natural resources - all these variables and more exercise an influence on population size and number of shops.

The statistical techniques discussed in this chapter are concerned with describing relationships in numerical terms. They can be applied to any pair of variables, not just variables that are causally related. The techniques are worth applying, however, only if a causal relationship exists, or can be presumed to exist. They are irrelevant if the relationship is purely accidental.

It must be stressed that the techniques do not help one to distinguish between cause and effect. The unravelling of causal connections is an exercise in logic, rather than statistics, and as such lies outside the scope of this book. This is not to deny in any way the importance of studying causality. One ought not to apply the techniques described here without first giving detailed thought to the nature of the relationship, and the purpose of studying it.

USE OF SYMBOLS

The letters X and Y are commonly used in statistical operations to symbolise two variables that are related. If there is simple dependence between the variables, then it is the causal (or *independent*) variable that is labelled X, and the variable that is the effect (the *dependent* variable) that is labelled Y. This convention is arbitrary, but it helps to make clear the nature of the link between the variables.

Where an interaction exists between two variables, or there is common dependence, the same letters X and Y are used to symbolise the variables, but the letters are used interchangeably. It does not matter which variable is labelled X and which Y.

BIVARIATE SAMPLES AND POPULATIONS

Most of the samples and populations discussed in previous chapters have consisted of measurements on a single variable.[1] They may be described as *univariate*. Each item is represented by only a single measurement, or at the most a pair of measurements relating to the same variable.

By comparing two or more univariate samples (or populations), one may be able to establish whether two variables are related. Suppose, for instance, one were to try to decide whether wheat yields in different parts of Britain are related to rainfall. It would be possible to calculate average yields for counties with (i) low, (ii)

[1]The exceptions have consisted of counts representing attributes or pseudo-attributes (Chapters 15 and 16).

moderate, and (iii) high rainfall. If the average for (iii) exceeded the average for (ii) and the average for (ii) exceeded the average for (i), this would suggest that yields are related to rainfall (or possibly some other factor that varies from county to county in accordance with rainfall).

This kind of approach has the obvious disadvantage that only one variable (in the present instance, wheat yields) is precisely measured; the other (rainfall) is merely categorised, i.e. treated as an attribute. The data may suggest that the variables are related; what they cannot clearly indicate is the extent of the relationship. Without further information one cannot decide whether the values of the two variables correspond closely, or show only limited accordance.

The extent to which one variable is related to another can only be judged successfully if a sample or population is available consisting of both variables. Each item must be measured (or ranked) twice, once in respect of each of the variables under study. In other words, each measurement (or rank) on one variable must be paired with a measurement (or rank) on the other variable. A population or sample of this type is described as *bivariate*.

To revert to the question of wheat yields: for each county one might choose to determine both the average yields and the average rainfall. From the paired values it would be possible to judge the extent to which the two variables are related. Obviously there would be no point in determining average yields for one set of counties, and rainfall for another set. The measurements must be paired if the extent of the relationship between the variables is to be established.

Ordinarily, the two variables forming a bivariate sample or population are so distinct that they require different units of measurement. Crop yields and rainfall are a case in point. Sometimes, however, the two variables resemble each other so closely that they are measurable with the same units of measurement. If, for example, one wished to study the relationship between the width and depth of a river, one might choose a number of locations between the source and the mouth of the river in order to make measurements. Although both width and depth could be measured in metres, they would nevertheless remain distinct variables.

A sample or population made up of two variables sharing the same unit of measurement must be carefully distinguished from a sample or population of the type discussed in Chapter 13, made up of pairs of measurements relating to the same variable (slope angles before and after a storm, areas measured in two different ways, etc.). In the case of a sample or population of the latter type, it is meaningful to consider the differences between the measurements forming each pair. Where two variables are involved, however, very different techniques are called for as explained in this and following chapters. These techniques can be applied to pairs of measurements on the same variable, but usually it is more relevant to calculate the differences as discussed in Chapter 13.

SCATTER DIAGRAMS

Bivariate data can be represented graphically as a *scatter diagram* (Fig. 17.1). Two scales are shown, crossing at right angles. One is horizontal and is numbered from left to right. It is called the *X-axis* and is used for depicting the variable X. The other scale, which is vertical, is numbered upwards. Called the *Y-axis*, it is used for depicting the variable Y. If there is simple dependence between the variables, it is the independent variable that is depicted on the horizontal axis and the dependent variable on the vertical axis. The intersection of the two axes, known as the *origin*, represents the zero values of both variables. Each item making up the sample or population is plotted on the scatter diagram as a point or cross. Thus, if there are a hundred items making up a particular sample, the scatter diagram shows a hundred points or crosses.

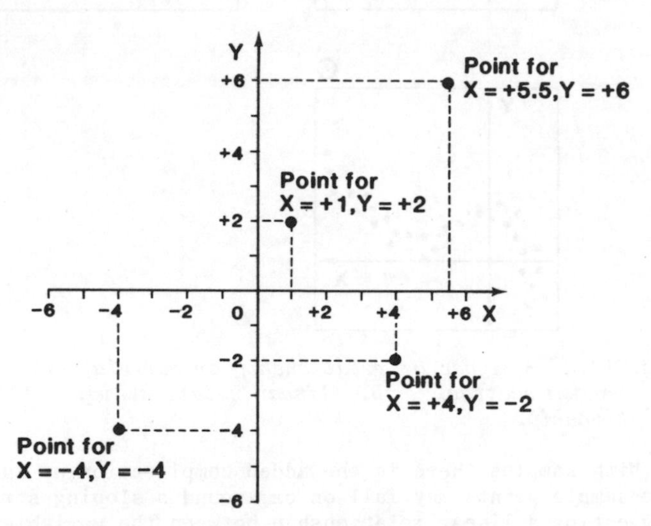

Fig. 17.1 - Scatter diagram showing 4 points (pairs of values of X and Y).

The method of positioning the points is shown in the figure. To position a point on the scatter diagram to represent a pair of values (X, Y), first go X units along the horizontal axis in the appropriate direction from the origin (right for positive values of X, left for negative values). Then go Y units upwards if Y is positive, or Y units downwards if Y is negative. X and Y are called the *co-ordinates* of the point.

Only that portion of a scatter diagram need be constructed that is necessary to show the data in hand. If all the values of X and Y are positive, for instance, only the upper right portion of the diagram is needed. Note also that it is permissible to use different scales on the horizontal and vertical axes.

A scatter diagram is useful in that it gives an immediate visual impression of the "form" of the relationship between two variables. Consider in the first place a bivariate population. A random scatter of points indicates that the two variables are *unrelated* or *uncorrelated* (Fig. 17.2A). If the points fall on or around a sloping straight line, a *linear* relationship or correlation is said to exist (Fig. 17.2B). The relationship is said to be curvilinear if the points fall on or around a curving line (Fig. 17.2C).

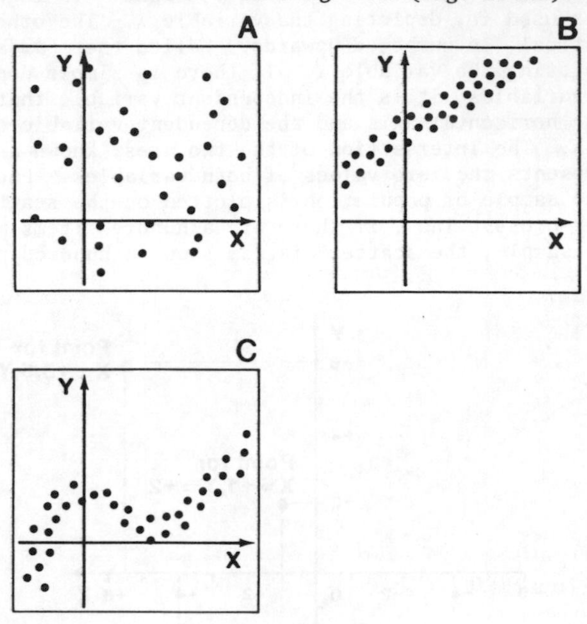

Fig. 17.2 – Types of relationship or correlation.
A. random scatter. B. linear relationship. C. curvilinear relationship.

With samples there is the added complication of sampling error. The sample points may fall on or around a sloping straight line suggesting a linear relationship between the variables when in fact there is no relationship in the parent population. Likewise the sample points may be randomly scattered even though something of a linear relationship exists in the parent population. A curvilinear relationship in a population may appear as a linear relationship in a sample, and vice versa. The reliability of a sample depends as always on its size. Large samples tend to be more representative of the population from which they are drawn than small samples.

This chapter will consider only populations and samples where the variables are linearly related, or else unrelated. Curvilinear relationships will be considered in Volume 2.

TYPES OF LINEAR RELATIONSHIP OR CORRELATION

A number of imaginary populations are shown in Fig. 17.3 illustrating various types of linear correlation or lack of correlation. In Fig. 17.3A all the points on the scatter diagram lie exactly on a sloping straight line. The correlation can be described as *perfect*. The values of X correspond exactly to the values of Y, making it possible to predict from one variable to the other with complete accuracy.

Perfect correlation rarely, if ever, occurs. It can exist only if both variables have identical distributions and Z values (where Z is the deviation of each X or Y value from the mean, divided by the standard deviation). The slightest discrepancy renders the correlation less than perfect. The data points will then not fall exactly on a sloping line.

If the points on the scatter diagram are highly concentrated around a sloping line, the variables are said to be *closely*, *strongly* or *highly* correlated (Fig. 17.3B). If the points only very roughly approximate to a sloping line, the correlation is said to be *weak* or *low* (Fig. 17.3C). As mentioned already, when the points are scattered at random (Fig. 17.3D and 17.2A), the correlation is zero

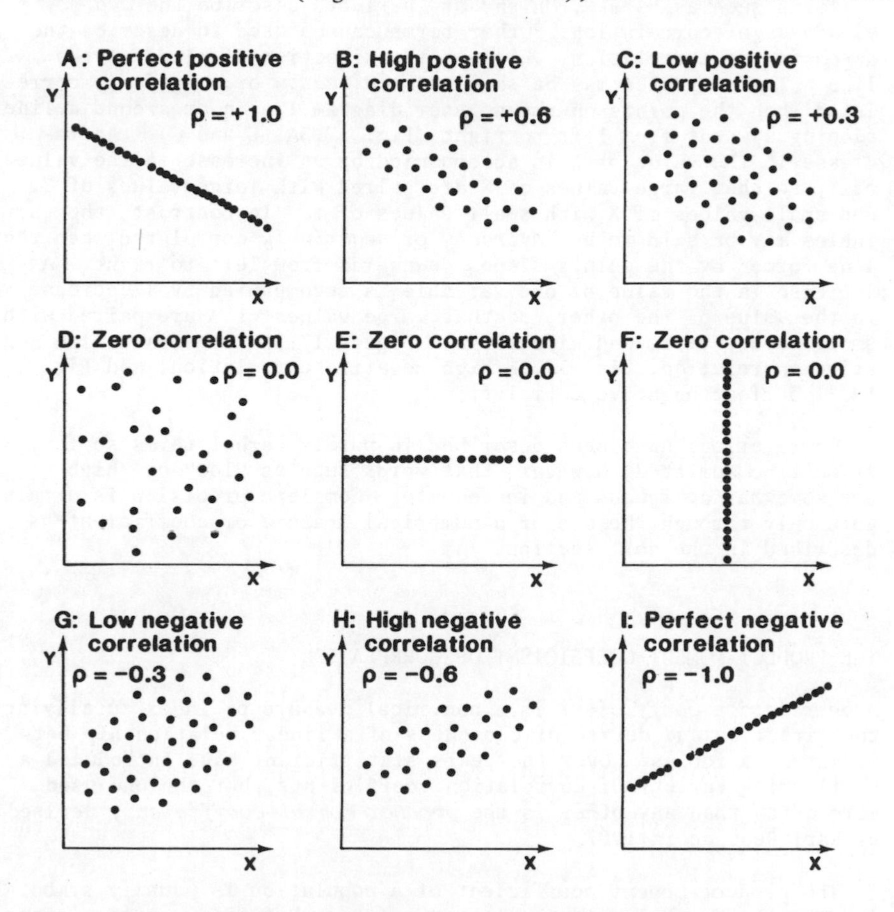

Fig. 17.3 – Scatter diagrams for a number of artificial populations to illustrate variations in the value of the product-moment coefficient.

A. Perfect positive correlation. B. High positive correlation.
C. Low positive correlation. D. Zero correlation.
E. Zero correlation. F. Zero correlation.
G. Low negative correlation. H. High negative correlation.
I. Perfect negative correlation.

and the values of X and Y vary in a totally independent and unpredictable manner.

In Fig. 17.3E all the points fall on a horizontal line and in Fig. 17.3F on a vertical line. In both instances the correlation is zero, not perfect as one might imagine at first glance. The essential point to note is that neither line slopes. The values of X in Fig. 17.3E vary independently of the value of Y which is fixed. Figure 17.3F represents the reverse situation: the value of Y is free to vary but the value of X is fixed. Correlation requires that the data points approximate to a sloping line; the line must be neither horizontal nor vertical.

Terms such as "weak", "high" or "perfect" describe the *degree of closeness* of correlation. Other terms can be used to describe the *direction* of correlation, which is the direction of slope of the line. The variables may be said to be *directly* or *positively* correlated when the points on the scatter diagram lie on or around a line sloping upwards from left to right (Fig. 17.3A, B and C). An increase in the value of X is accompanied by an increase in the value of Y, so that large values of X are paired with large values of Y, and small values of X with small values of Y. In contrast, the variables may be said to be *inversely* or *negatively* correlated when the line formed by the points slopes downwards from left to right. An increase in the value of one variable is accompanied by a decrease in the value of the other, so that large values of X are paired with small values of Y, and vice versa. Figure 17.3G illustrates low negative correlation, Fig. 17.3H high negative correlation, and Fig. 17.3I perfect negative correlation.

Correlations have been described in purely verbal terms so far. It must be admitted, however, that words such as "low" or "high" are somewhat ambiguous and inadequate. Complete precision is attainable only through the use of a numerical measure or coefficient as described in the next section.

THE PRODUCT-MOMENT COEFFICIENT OF CORRELATION

A *correlation coefficient* is a numerical measure or index specifying the direction and degree of closeness of a linear relationship between two variables. Over the years statisticians have introduced a bewildering variety of correlation coefficients, but the one used more often than any other is the *product-moment coefficient,* devised by Karl Pearson in 1907.

The product-moment coefficient of a population is usually symbolised by the Greek lower-case letter ρ *(rho)*, but a few statisticians prefer the Roman capital R. It is defined by the formula

$$\rho = \frac{\frac{1}{N}\Sigma[(X - \mu_x)(Y - \mu_y)]}{\sqrt{\frac{\Sigma(X - \mu_x)^2}{N}}\sqrt{\frac{\Sigma(Y - \mu_y)^2}{N}}}$$

where N is the number of items in the population, i.e. the number of pairs of values of X and Y, μ_x is the mean of the X values and μ_y is the mean of the Y values.

The Roman lower-case letter r is fairly universally used to symbolise the product-moment coefficient of a sample. The formula used to define r is similar to that used to define ρ save that n - 1 appears in place of N, where n is the number of items in the sample, and a different symbolism is employed for the means, $\bar{X}$ replacing μ_x as the mean of the X values and $\bar{Y}$ replacing μ_y as the mean of the Y values.

$$r = \frac{\frac{1}{n-1}[\Sigma(X - \bar{X})(Y - \bar{Y})]}{\sqrt{\frac{\Sigma(X - \bar{X})^2}{n-1}}\sqrt{\frac{\Sigma(Y - \bar{Y})^2}{n-1}}}$$

The numerator (top line) on the right-hand side of the formulas for ρ and r is called the *covariance* of X and Y. The denominator (bottom line) is the product of the standard deviations of X and Y.

The equations for ρ and r can be rewritten so as to eliminate all reference to the number of items

$$\rho = \frac{\Sigma[(X - \mu_x)(Y - \mu_y)]}{\sqrt{\Sigma(X - \mu_x)^2}\sqrt{\Sigma(Y - \mu_y)^2}} \text{ and } r = \frac{\Sigma[(X - \bar{X})(Y - \bar{Y})]}{\sqrt{\Sigma(X - \bar{X})^2}\sqrt{\Sigma(Y - \bar{Y})^2}}$$

Because of the way it is defined the product-moment coefficient cannot be less than -1 or greater than +1. A minus sign indicates an inverse or negative relationship between X and Y, a plus sign a direct or positive relationship. If there is no linear relationship between X and Y, the coefficient is zero. The more perfect the linear relationship, the more the coefficient deviates from zero.

The set of scatter diagrams shown in Fig. 17.3 illustrates the range of possible values of the correlation coefficient. In Fig. 17.3A all the points fall on a straight line sloping upwards from left to right. The correlation coefficient is +1, indicating perfect positive correlation. In Fig. 17.3B, the points approximate closely to a straight line, forming a narrow band or ellipse. The correlation is high but not perfect (ρ = +0.6). In Fig. 17.3C the points are still more spread out. The correlation is fairly low (ρ = +0.3).

In Fig. 17.3D the points are scattered randomly all over the diagram. They form a diffuse cloud with no obvious boundary. There is no linear relationship or correlation between the two variables, and hence the value of ρ is zero. In Fig. 17.3E the points fall on a horizontal straight line and in Fig. 17.3F on a vertical straight line. Again there is no relationship between the two variables and the product-moment coefficient is zero.

The remaining diagrams illustrate negative correlation. The points in Fig. 17.3G are well scattered and the coefficient is accordingly fairly low ($\rho = -0.3$). In Fig. 17.3H the points are again concentrated into a narrow band or ellipse. The coefficient is high ($\rho = -0.6$). Figure 17.3I demonstrates perfect negative correlation. The points all fall on a straight line sloping downwards from left to right. The coefficient attains its lower limit of -1.

The product-moment coefficient of correlation is an elegant and effective measure of linear relationship. Because the formulas for ρ and r are "symmetrical" as regards X and Y, it does not matter which variable is labelled X and which Y. Reversing the labels leaves the sign and absolute value of the coefficient unchanged. Note also that the units of measurement are unimportant, provided the scales of measurement are arithmetical. A correlation between, say, distance and time remains the same whether the units are kilometres and hours or altometres (1.0×10^{-18} of a metre) and nanoseconds (1.0×10^{-9} of a second). The correlation is invariant, however, only under changes of scale that involve addition, subtraction, multiplication or division. The coefficient is changed if the units of measurement are converted to logarithms or square roots.

The product-moment coefficient is useful only as a measure of *linear* relationships. It does not provide a meaningful measure of curvilinear relationships. In Fig. 17.4, for example, the pairs of values of X and Y plot as a parabolic curve. There is a perfect relationship between the two variables, but it is non-linear. The product-moment coefficient is exactly zero, despite the fact that the relationship is perfect in a curvilinear sense.

Methods of measuring curvilinear relationships are discussed in the companion volume.

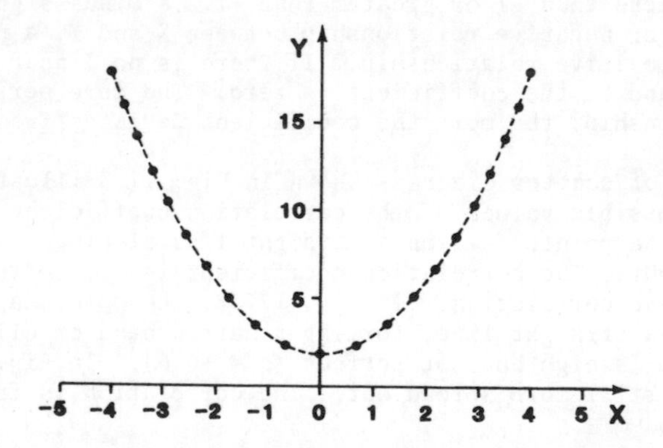

Fig. 17.4 - An example of a curvilinear relationship yielding a zero coefficient of linear correlation. The points are equally spaced along a parabola.

THE CALCULATION OF THE PRODUCT-MOMENT COEFFICIENT FROM UNGROUPED DATA

The definitional formulas given in the last section are not very convenient for calculating the product-moment coefficient unless the values of X and Y happen to be integers (whole numbers) and the means are also integers. If the values of X and Y are decimal numbers, or if the means are decimal numbers, the calculation of the coefficient can be greatly shortened using the algebraically equivalent formulas

$$\rho = \frac{N\Sigma XY - \Sigma X\Sigma Y}{\sqrt{[N\Sigma X^2 - (\Sigma X)^2]}\sqrt{[N\Sigma Y^2 - (\Sigma Y)^2]}}$$

and $$r = \frac{n\Sigma XY - \Sigma X\Sigma Y}{\sqrt{[n\Sigma X^2 - (\Sigma X)^2]}\sqrt{[n\Sigma Y^2 - (\Sigma Y)^2]}}$$

Some pocket calculators have a key that will give the standard deviation immediately a set of values is entered. If such a calculator is available, it is quickest to use the formulas

$$\rho = \frac{N\Sigma XY - \Sigma X\Sigma Y}{N^2\sigma_x\sigma_y} \quad \text{and} \quad r = \frac{n\Sigma XY - \Sigma X\Sigma Y}{n(n - 1)s_x s_y}$$

where σ_x or s_x is the standard deviation of the X values and σ_y or s_y the standard deviation of the Y values.

The most advanced pocket calculators have a key that will give the product-moment coefficient immediately a set of X and Y values is entered. They save a great deal of time, but are rather expensive.

CALCULATION OF THE CORRELATION COEFFICIENT: EXAMPLE (1)

Oxley (1974) studied the amounts of sediment and dissolved solids carried by two small streams draining into Llyn Ebyr, near Llanidloes in Central Wales. A recording instrument installed beside each stream provided continuous data on discharge (the volume of water passing the recorder in unit time). Every fortnight, Oxley visited the streams and collected some water for laboratory analysis. He was concerned to measure both the amount of sediment suspended in the water and the concentration of dissolved solids in order to see how these varied with discharge.

Table 17.1 records the discharge of one of the streams and the concentration of dissolved solids in the water for all the visits made in 1971 when the level of the stream was rising (as distinct from holding steady or falling). The discharges are listed in order of increasing magnitude rather than order of occurrence. It is immediately apparent that low concentrations of dissolved solids tend to be associated with low discharges, and high concentrations with high discharges. Figure 17.5 shows the data plotted as a scatter diagram. All 11 points approximate closely to a straight line sloping upwards from left to right. There is seemingly a high degree of positive correlation between the variables.

TABLE 17.1 - DISCHARGE AND CONCENTRATIONS OF DISSOLVED SOLIDS DURING RISING STAGES OF A SMALL STREAM

Date of visit (1971)	*Discharge of stream (litres/sec)* X	*Concentration of dissolved solids (mg/litre)* Y	X^2	Y^2	XY
June 3	0.4	34.2	0.16	1169.64	13.68
Oct. 7	2.2	40.4	4.84	1632.16	88.88
March 11	2.3	38.3	5.29	1466.89	88.09
Nov. 4	3.9	39.4	15.21	1552.36	153.66
Dec. 16	4.3	44.8	18.49	2007.04	192.64
Nov. 18	4.4	47.0	19.36	2209.00	206.80
Feb. 25	5.1	42.6	26.01	1814.76	217.26
Jan. 28	8.0	45.6	64.00	2079.36	364.80
Jan. 14	8.1	52.8	65.61	2787.84	427.68
Aug. 26	8.3	48.7	68.89	2371.69	404.21
Dec. 30	9.7	50.5	94.09	2550.25	489.85
	56.7	484.3	381.95	21640.99	2647.55

Source: Oxley (1974, p.146). The values of X and Y are taken from a graph and may not be entirely accurate. No dates are mentioned by Oxley, and no reliance should be placed on the imaginary dates given here; they are introduced simply to make clear the basis of the pairing of the X and Y values. The stream in question has no official name, but Oxley refers to it as "Ebyr North".

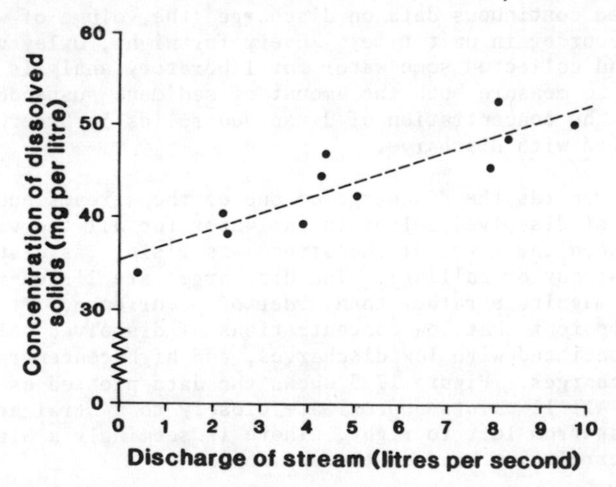

Fig. 17.5 - Amounts of dissolved solids in the water of a small stream plotted against the discharge of the stream. Data for rising stages only. The broken line represents the regression of dissolved solids on discharge (for explanation see Chapter 19). Redrawn from Oxley (1974).

The formula for Pearson's coefficient of correlation is

$$r = \frac{n\Sigma XY - \Sigma X\Sigma Y}{\sqrt{[n\Sigma X^2 - (\Sigma X)^2]}\sqrt{[n\Sigma Y^2 - (\Sigma Y)^2]}}$$

There are 11 pairs of values of X and Y, hence n = 11. Substituting the column totals shown in Table 17.1 gives

$$r = \frac{11(2647.55) - (56.7)(484.3)}{\sqrt{[11(381.95) - (56.7)^2]}\sqrt{[11(21,640.99) - (484.3)^2]}}$$

$$= \frac{29,123.03 - 27,459.81}{\sqrt{[986.56]}\sqrt{[3504.4]}} = \frac{1663.24}{\sqrt{3,457,300.864}}$$

$$= 0.89451\ldots \text{ or } 0.89 \text{ approximately.}$$

The size of the correlation coefficient suggests but does not prove that the concentration of dissolved solids is highly correlated with discharge. One cannot be wholly sure that a high correlation exists, because the coefficient is calculated from a sample and not a population. The correlation coefficient r of a sample is always an uncertain guide to the correlation coefficient ρ of the population from which the sample is taken, particularly if the sample is small. In the present instance the sample comprises only 11 pairs of values of X and Y. One must accept that with more data a somewhat different coefficient of correlation might be obtained. There is even a possibility that the correlation coefficient of the parent population is zero. In other words, the concentration of dissolved solids may be totally unrelated to the discharge, despite the high correlation observed in the sample. This question of sampling error will be explored in a later section. It will be shown that the concentration of dissolved solids is almost certainly positively correlated with discharge even if there is some sampling error. The possibility that the correlation is really zero will be shown to be so remote that it need not be seriously considered.

The discharge of the stream presumably controls the concentration of dissolved solids and not vice versa. For this reason the discharge in Table 17.1 is labelled X and the concentration Y. However, it makes no difference to the calculation of the correlation coefficient which variable is labelled X. The same value would be obtained if the concentration were to be labelled X and the discharge Y.

THE CALCULATION OF THE CORRELATION COEFFICIENT: EXAMPLE (2)

As has already been mentioned, the population size of a town partly determines the number of shops the town can support. Likewise the number of business concerns determines in part the population size of the town. The two variables interact, and hence are correlated.

Table 17.2 presents some data for a random sample of 22 towns in south-east England, with populations from 10,000 to 30,000. Although population size is labelled X and number of shops Y, the labelling is purely arbitrary. It would have been just as appropriate to have labelled population size Y and number of shops X.

TABLE 17.2 - NUMBERS OF RETAIL SHOPS IN 22 TOWNS IN SOUTH-EAST ENGLAND

Name of town	Population size (1971) X	Number of retail shops Y	X^2	Y^2	XY
1 Henley	11402	156	130005604	24336	1778712
2 Epping	11681	99	136445761	9801	1156419
3 Marlow	11706	115	137030436	13225	1346190
4 Hythe	11949	160	142778601	25600	1911840
5 Frinton	12431	195	154529761	38025	2424045
6 Haslemere	13252	164	175615504	26896	2173328
7 Lewes	14015	206	196420225	42436	2887090
8 Harwich	14892	170	221771664	28900	2531640
9 Seaford	16196	168	262310416	28224	2720928
10 East Grinstead	18569	198	344807761	39204	3676662
11 Littlehampton	18621	186	346741641	34596	3463506
12 Godalming	18634	198	347225956	39204	3689532
13 Shoreham	18804	146	353590416	21316	2745384
14 Cowes	18895	223	357021025	49729	4213585
15 Broadstairs	19996	246	399840016	60516	4919016
16 Bishop's Stortford	22084	193	487703056	37249	4262212
17 Ryde	23171	304	536895241	92416	7043984
18 Newbury	23696	246	561500416	60516	5829216
19 Herne Bay	25117	323	630863689	104329	8112791
20 Andover	25538	195	652189444	38025	4979910
21 Rayleigh	26265	175	689850225	30625	4596375
22 Hitchin	28680	293	822542400	85849	8403240
	405594	4359	8087679258	931017	84865605

Source: Report on the Census of Distribution and other services; Part 8, Area Tables London and South-East Region. Department of Industry.

A fairly high degree of correlation is obviously present in the sample data. The smaller towns tend to have relatively few shops while the larger towns tend to have very many. Figure 17.6 shows the data plotted as a scatter diagram. An overall linear arrangement or pattern is apparent. The plotted points all lie within a band sloping upwards from left to right which suggests a positive correlation between the variables. Substituting the sums of squares and products shown in the table gives

$$r = \frac{22(84,865,605) - (405,594)(4359)}{\sqrt{[22(8,087,679,258) - (405,594)^2]}\sqrt{[22(931,017) - (4359)^2]}}$$

$$= \frac{1,867,043,310 - 1,767,984,246}{\sqrt{[13,422,450,840]}\sqrt{[1,481,493]}}$$

$$= 0.70247 \text{ or } 0.70 \text{ approximately.}$$

The size of the coefficient suggests that the variables are fairly closely correlated. One must not jump too readily to conclusions,

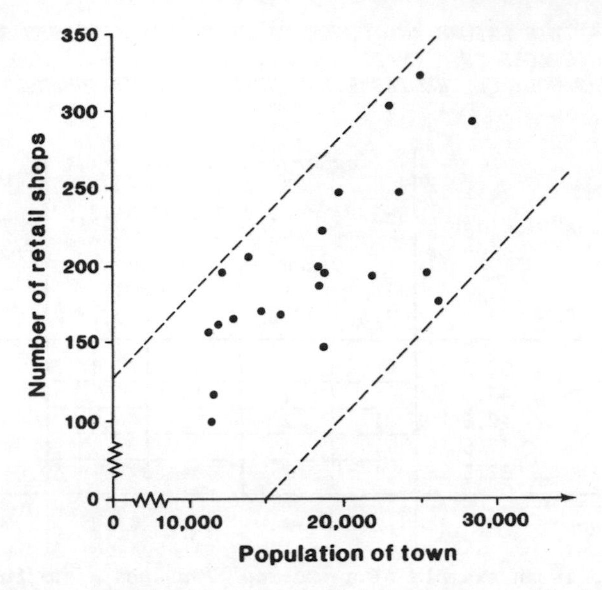

Fig. 17.6 - Population size and number of retail shops, 22 small towns in South-East England. The vertical scale has been broken between 0 and 100 and the horizontal scale between 0 and 10,000 so as to save space.

however. The correlation coefficient of the sample of towns need not be the same as the correlation coefficient of all the towns in south-east England with populations of 10,000 to 30,000, the statistical population from which the sample is drawn. It is just conceivable that the sample is wholly misleading and the number of shops is actually uncorrelated with the population size of towns in south-east England. In other words, despite the high value of r, there is a remote possibility that ρ is zero. This question of the reliability of the correlation coefficient will be examined in a later section.

*ESTIMATING THE CORRELATION COEFFICIENT FROM GROUPED DATA

Writers of books and articles often choose not to list X and Y values individually, but group them into classes. This saves space and also makes the values easier to comprehend. However, a reader anxious to calculate the product-moment coefficient is presented with a problem. Without knowing the individual values of X and Y he cannot calculate the coefficient exactly, though he can estimate it as will now be demonstrated.

Table 17.3 presents in grouped form the data for the Welsh stream already listed in Table 17.1. The classes along the left-hand side of the table are arranged in ascending order of magnitude when read from bottom to top. The classes are arranged like this in order to make the table more like a scatter diagram. The band of frequencies sloping upwards from left to right suggests that X and Y are positively correlated. A slope downwards from left to right would suggest a negative correlation.

TABLE 17.3 - DISCHARGES AND CONCENTRATIONS OF DISSOLVED SOLIDS DURING RISING STAGES OF A SMALL STREAM. DATA GROUPED INTO CLASSES: FIGURES IN THE BODY OF THE TABLE ARE NUMBERS OF MEASUREMENTS OR FREQUENCIES

Dissolved solids (mg/litre) Y	*Class mid-value m_y*	*Discharge (litres/sec) X*				*Total frequency*
		0-2.9'	*3-5.9'*	*6-8.9'*	*9-11.9'*	
		Class mid-value m_x				
		1.5	*4.5*	*7.5*	*10.5*	
50-54.9'	52.5			1	1	2
45-49.9'	47.5		1	2		3
40-44.9'	42.5	1	2			3
35-39.9'	37.5	1	1			2
30-34.9'	32.5	1				1
Total frequency		3	4	3	1	11 = n

Table 17.3 is an example of a *two-way frequency distribution* or *correlation table*. It differs from a contingency table in having ordered classes.

Because the frequencies are low the correlation coefficient is fairly easy to estimate. The frequencies in most correlation tables are larger than those shown here, and the task of estimating the correlation coefficient tends to become more difficult. Table 17.3 sacrifices realism in the interests of simplicity.

It will be recalled that the formulas for the correlation coefficient for ungrouped data are

$$\rho = \frac{N\Sigma XY - \Sigma X\Sigma Y}{\sqrt{[N\Sigma X^2 - (\Sigma X)^2]}\sqrt{[N\Sigma Y^2 - (\Sigma Y)^2]}}$$

and
$$r = \frac{n\Sigma XY - \Sigma X\Sigma Y}{\sqrt{[n\Sigma X^2 - (\Sigma X)^2]}\sqrt{[n\Sigma Y^2 - (\Sigma Y)^2]}}$$

In the case of grouped data all the summation terms have to be estimated, their precise values being unknown. To begin with the numerator: the sum of products, ΣXY, can be estimated by computing $\Sigma(fm_x m_y)$, where f is the frequency in each cell in the body of the table, m_x is the mid-value of the cell on the X scale, and m_y is the mid-value on the Y scale. ΣX can be estimated by obtaining $\Sigma(f_x m_x)$, where f_x is the frequency in each class of X (column of cells in Table 17.3) and m_x is the mid-value of the class on the X scale. ΣY can likewise be estimated by computing $\Sigma(f_y m_y)$, where f_y is the frequency in each class of Y (row of cells in Table 17.3) and m_y is the mid-value of the class on the Y scale.

Starting with the top row of Table 17.3 and working downwards from left to right, $\Sigma(fm_x m_y) = 1(7.5)(52.5) + 1(10.5)(52.5) + 1(4.5)(47.5) + 2(7.5)(47.5) + 1(1.5)(42.5) + 2(4.5)(42.5) + 1(1.5)(37.5) + 1(4.5)(37.5) + 1(1.5)(32.5) = 2591.25$. Also $\Sigma(f_x m_x) = 3(1.5) + 4(4.5) + 3(7.5) + 1(10.5) = 55.5$ and $\Sigma(f_y m_y) = 2(52.5) + 3(47.5) + 3(42.5) + 2(37.5) + 1(32.5) = 482.5$. Hence the numerator of r can be estimated as $11(2591.25) - 55.5\,(482.5) = 1725$.

Turning now to the denominator, the sum of squares, ΣX^2, can be estimated by calculating $\Sigma(f_x m_x{}^2)$, where $m_x{}^2$ is the square of the mid-value of each class of X. An estimate of $(\Sigma X)^2$ is $[\Sigma(f_x m_x)]^2$. For the stream data $\Sigma(f_x m_x{}^2) = 3(1.5)^2 + 4(4.5)^2 + 3(7.5)^2 + 1(10.5)^2 = 366.75$. $[\Sigma(f_x m_x)]^2 = (55.5)^2 = 3080.25$. Hence the first half of the denominator can be estimated to be $\sqrt{[11(366.75) - 3080.25]} = \sqrt{(954)}$. An estimate of ΣY^2 is $\Sigma(f_y m_y{}^2)$, where $m_y{}^2$ is the square of the mid-value of each class of Y. $(\Sigma Y)^2$ can be estimated by calculating $[\Sigma(f_y m_y)]^2$. In the case of the stream data $\Sigma(f_y m_y{}^2) = 2(52.5)^2 + 3(47.5)^2 + 3(42.5)^2 + 2(37.5)^2 + 1(32.5)^2 = 21{,}568.75$. $[\Sigma(f_y m_y)]^2 = (482.5)^2 = 232{,}806.25$. An estimate of the second half of the denominator of r is, therefore, $\sqrt{[11(21{,}568.75) - 232{,}806.25]} = \sqrt{(4450)}$.

Putting numerator and denominator together, the value of r is estimated to be

$$r = \frac{1725}{\sqrt{(954)}\sqrt{(4450)}} = 0.84 \text{ approximately.}$$

This estimate is 6 per cent smaller than the true value. The estimate of the correlation coefficient from grouped data always tends to produce under-estimates, the effect being most marked if n is small and if the grouping is coarse. No correction factor has been devised that has won general acceptance, however.

The procedure just described for estimating the correlation coefficient can lead to rather tiresome arithmetic if the frequencies are at all large. If the class intervals happen to be equal as in Table 17.3, the arithmetic can be drastically shortened by making use of assumed means. The estimate of the correlation coefficient of a population is

$$\rho = \frac{N\Sigma(FD_x D_y) - \Sigma(F_x D_x)\Sigma(F_y D_y)}{\sqrt{\{N\Sigma(F_x D_x{}^2) - [\Sigma(F_x D_x)]^2\}}\sqrt{\{N\Sigma(F_y D_y{}^2) - [\Sigma(F_y D_y)]^2\}}}$$

and for a sample

$$r = \frac{n\Sigma(fd_x d_y) - \Sigma(f_x d_x)\Sigma(f_y d_y)}{\sqrt{\{n\Sigma(f_x d_x{}^2) - [\Sigma(f_x d_x)]^2\}}\sqrt{\{n\Sigma(f_y d_y{}^2) - [\Sigma(f_y d_y)]^2\}}}$$

where D_x (or d_x) is the deviation of a cell or class from the assumed mean of X, and D_y (or d_y) is the deviation from the assumed mean of Y.

TABLE 17.4 - ESTIMATION OF THE CORRELATION COEFFICIENT USING ASSUMED MEANS

Dissolved solids (mg/litre) Y	*Discharge (litres/sec) X* 0-2.9'	3-5.9'	6-8.9'	9-11.9'	f_y	d_y	f_yd_y	$f_yd_y^2$
50-54.9'			1 **2**	1 **4**	2	+2	+4	+8
45-49.9'		1 **0**	2 **2**		3	+1	+3	+3
40-44.9'	1 **0**	2 **0**			2	0	0	0
35-39.9'	1 **1**	1 **0**			2	-1	-2	+2
30-34.9'	1 **2**				1	-2	-2	+4
f_x	3	4	3	1	$\Sigma f_y = \Sigma f_x$ =n=11		$\Sigma(f_yd_y)$ =+3	$\Sigma(f_yd_y^2)$ =17
d_x	-1	0	+1	+2				
f_xd_x	-3	0	3	2	$\Sigma(f_xd_x)$=2			
$f_xd_x^2$	+3	0	+3	+4	$\Sigma(f_xd_x^2)$ =10			$\Sigma(fd_xd_y)$ =**11**

The calculations for the stream data are set out in Table 17.4. The mean of X is assumed to be 4.5 litres/sec and the mean of Y 42.5 mg/litre. Figures in bold face in each cell are values of fd_xd_y. They are obtained by multiplying the frequency of the cell (shown in ordinary type) by d_x and d_y. The summation $\Sigma(fd_xd_y)$ is shown in bold face in the lower right corner of the table: it happens to have the same value as n, but this is coincidence.

Substituting the appropriate totals in the formula for r gives

$$r = \frac{11(11) - 2(3)}{\sqrt{[11(10) - (2)^2]}\sqrt{[11(17) - (3)^2]}} = 0.84 \text{ approximately.}$$

It is no coincidence that this estimate is the same as that obtained earlier: the two methods of estimation can be shown to be equivalent algebraically.

THE BIVARIATE NORMAL DISTRIBUTION

The correlation coefficient of a sample, like any other statistic, is subject to sampling error. It cannot be relied upon to be precisely the same as the correlation coefficient of the population from which the sample is taken, even if calculated from individual values of X and Y rather than grouped values. If the sample is large, the difference between the sample coefficient and the population coefficient is likely to be small, possibly not even worth bothering about. But if the sample is small, the difference may well be substantial. The sample may suggest, for example, that there is quite a close correlation between X and Y even though no correlation exists in the parent population. Or the sample may suggest that there is no correlation when in fact there is quite a close correlation.

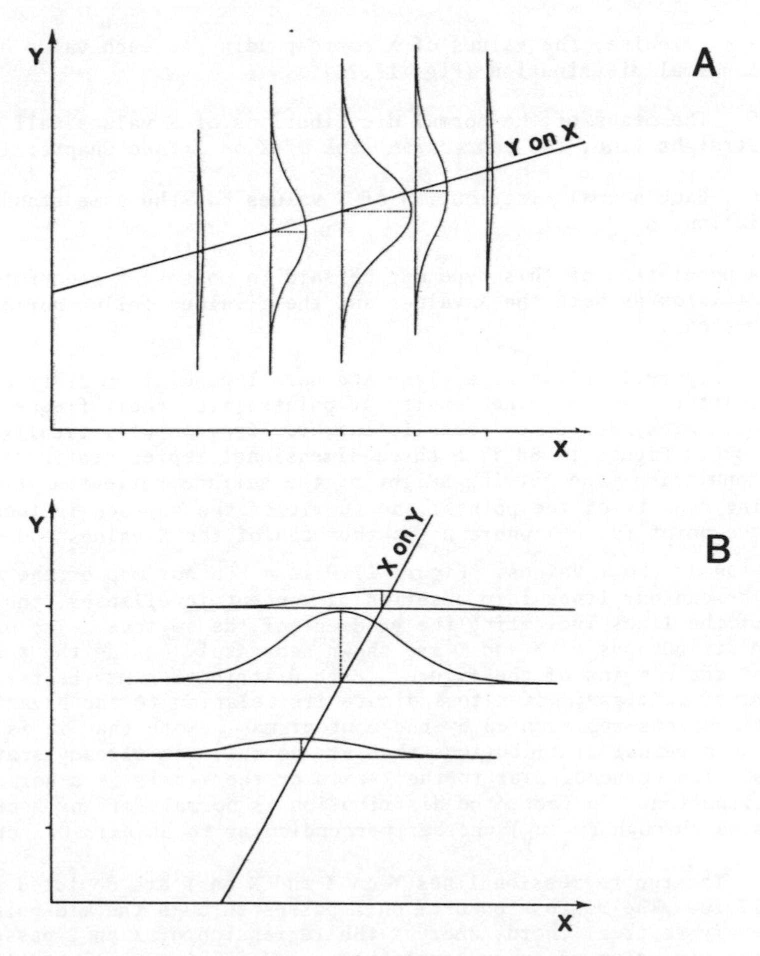

Fig. 17.7A - Bivariate normal distribution. The normal distribution of Y values is shown for five selected values of X.

Fig. 17.7B - Bivariate normal distribution. The normal distribution of X values is shown for three selected values of Y.

It is not just sample size that determines the amount of sampling error associated with r. The value of ρ is important, and also the manner in which the X and Y values are distributed in the population. Somewhat unrealistically, statisticians have largely restricted their investigations of sampling error to populations where

1. The values of Y corresponding to each value of X form a normal distribution (Fig. 17.7A).

2. The means of the normal distributions of Y values fall on a straight line. This line is referred to as the *regression line of Y on X* (see Chapter 19).

3. Each normal distribution of Y values has the same standard deviation, $\sigma_{y.x}$.

4. Likewise, the values of X corresponding to each value of Y form a normal distribution (Fig. 17.7B).

5. The means of the normal distributions of X values fall on a straight line, the *regression line of X on Y* (see Chapter 19).

6. Each normal distribution of X values has the same standard deviation, $\sigma_{x.y}$.

A population of this type may be said to possess a *bivariate distribution* as both the X values and the Y values follow normal distributions.

Figure 17.8A shows a bivariate normal population plotted as a scatter diagram. The density of points (i.e. their frequency per unit area) decreases steadily outwards from an elliptically shaped core. Figure 17.8B is a three-dimensional representation of the same population, the varying height of the surface reflecting the varying density of the points: the summit of the surface is located at the point (μ_x, μ_y) where μ_x is the mean of the X values and μ_y the mean of the Y values. Figure 17.9 is a contour map of the surface. The contour lines form a series of concentric ellipses, the spacing of the lines indicating the gradient of the surface. The univariate distributions of X and Y are shown separately beside the X and Y axes at the margins of the figure. Each distribution may be termed a *marginal distribution* to indicate its relation to the bivariate distributions represented by the contour map. Note that it is not just the marginal distributions that are normal. As already stated, each section perpendicular to the X-axis or the Y-axis is a normal distribution. In fact, the distribution is normal for any section passing through (μ_x, μ_y) whether perpendicular to an axis or not.

The two regression lines Y on X and X on Y are depicted in Fig. 17.10. The regression of Y on X passes through the mid-point of every vertical chord, whereas the regression of X on Y passes through the mid-point of every horizontal chord. This is illustrated for the two chords shown by the broken lines. Note also that the major axis of the ellipses formed by the contour lines does not bisect the angle between the two regression lines, but lies closer to the regression Y on X. If the ellipses were inclined at 45°, the major axis would exactly bisect the angle between the regression lines. If the ellipses were more steeply inclined, the major axis would lie closer to the regression of X on Y.

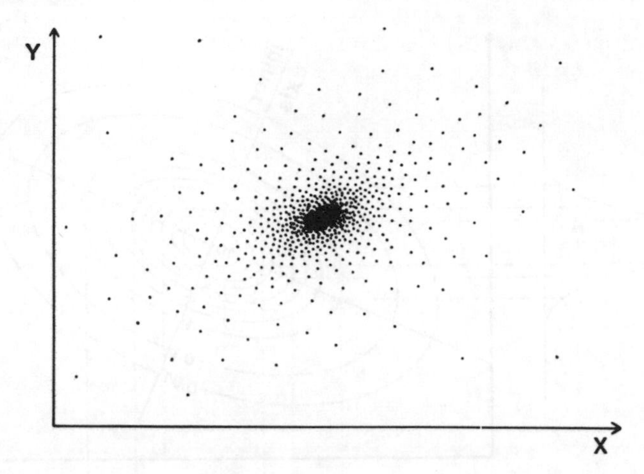

Fig. 17.8A - Scatter diagram showing a bivariate normal population. Each dot represents a member of the population.

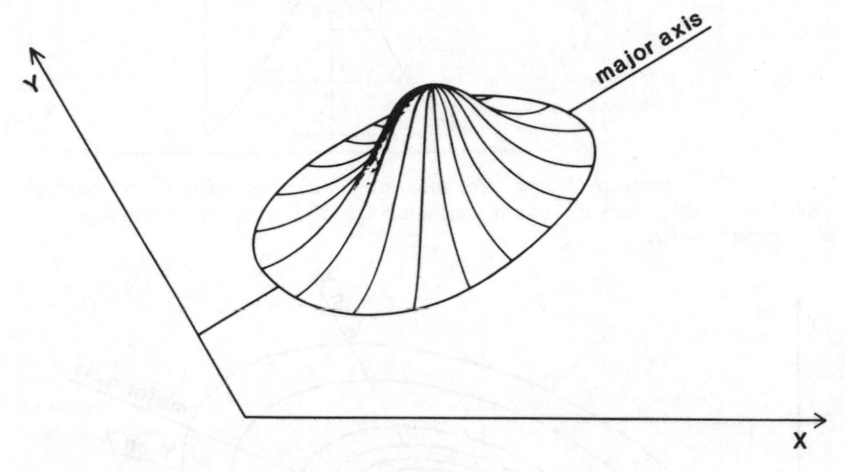

Fig. 17.8B - Three-dimensional representation of a bivariate normal population

The bivariate normal distribution is a mathematical model like the Poisson or binomial distributions. It is not, perhaps, a very realistic model. No real population is precisely bivariate normal, and, although some populations approximate fairly closely to the model, they are the exceptions rather than the rule. Most populations do not begin to approach bivariate normality unless the scale of measurement of one or both variables is altered, for instance, by taking logarithms. It would be idle to pretend that there are not other bivariate distributions that conform more closely with reality. They do not provide a convenient starting point, however, from which to derive the sampling distribution r. The reason why statisticians have paid so much attention to the bivariate normal distribution is that it is relatively easy to manipulate mathematically. Other distributions, although more realistic, are far less tractable.

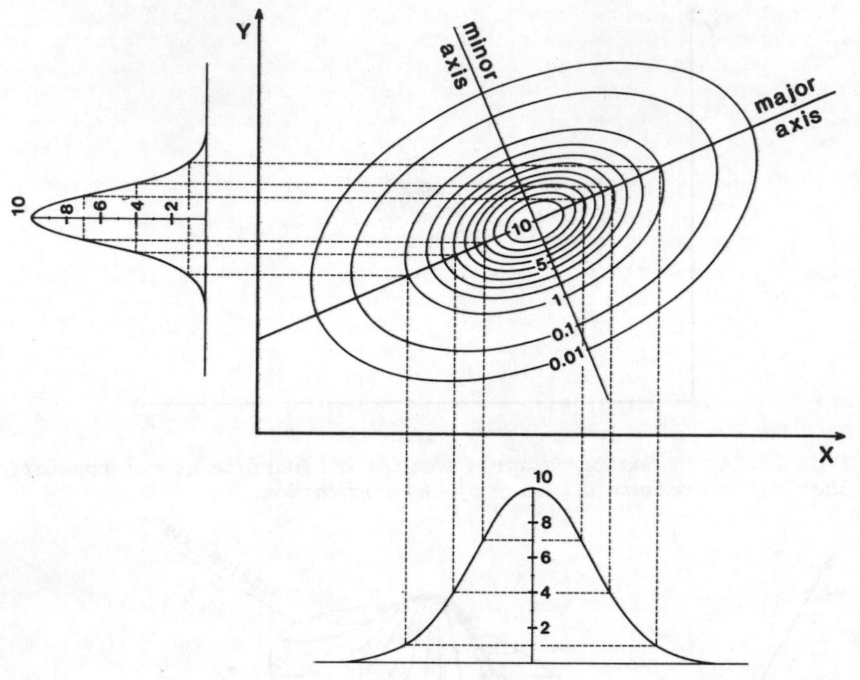

Fig. 17.9 - Contour lines for the surface of a bivariate normal population. The normal distributions at the margins represent X and Y separately.

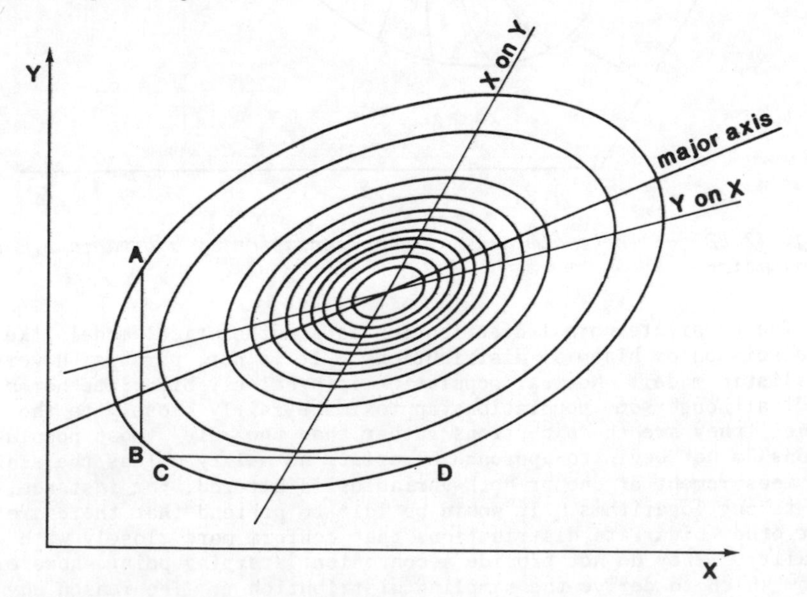

Fig. 17.10 - Bivariate normal population with regressions of Y on X and X on Y.

THE SAMPLING DISTRIBUTION OF r

Suppose an infinite number of samples of size n are drawn with individual replacements from a population with a bivariate normal distribution. Also suppose that the X values of the different items in the population are independent of each other, as are the Y values for any given X. What will be the shape of the sampling distribution of the values of r for the different samples? The answer is that the shape varies according to the value of ρ and the value of n.

If ρ is zero (that is to say, if X and Y are uncorrelated) the distribution of r is symmetrical around a mean of zero (Fig. 17.11). The distribution tends to normality as n increases, but only slowly with the result that for small values of n the distribution diverges considerably from normality. When n is 3 the distribution is U-shaped, becoming rectangular when n is 4. A central peak is present only when n is 5 or more. Even when n is as large as 50 slight deviations from normality persist especially in the tails of the distribution, though this is not apparent in Fig. 17.11. Exact normality is attained only when n is infinite.

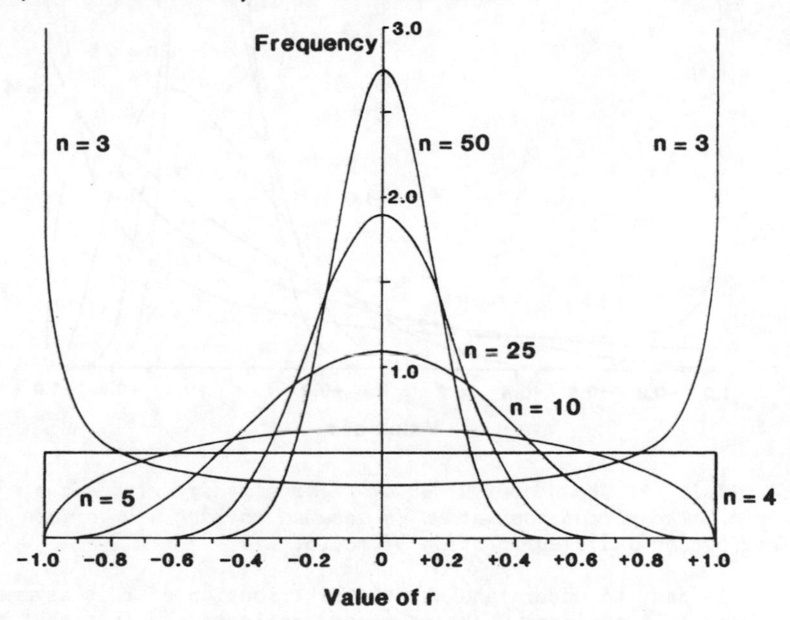

Fig. 17.11 - The distribution of r for six different sizes of sample. The parent population is assumed to have a bivariate normal distribution and zero correlation.

If ρ is not zero (that is to say, if X and Y are correlated) the distribution of r again tends to normality as n increases, though more slowly than where ρ is zero. For all finite values of n the distribution is asymmetrical (Fig. 17.12), the degree of asymmetry increasing the more ρ deviates from zero. When n is 3, the distribution takes the form of a lop-sided U. The larger side is on the right if ρ is positive, and on the left if ρ is negative. When n is 4 the distribution of r is J-shaped for positive values of ρ and reversed J-shaped for negative values. The steepness increases the more ρ deviates from zero. When n is 5 or more the distribution is unimodal and skewed. The skew is to the left if ρ is positive and to the right if ρ is negative. The amount of skew increases rapidly as ρ approaches + or -1, and decreases slowly with the size of n. When n is infinite the distribution of r is exactly normal.

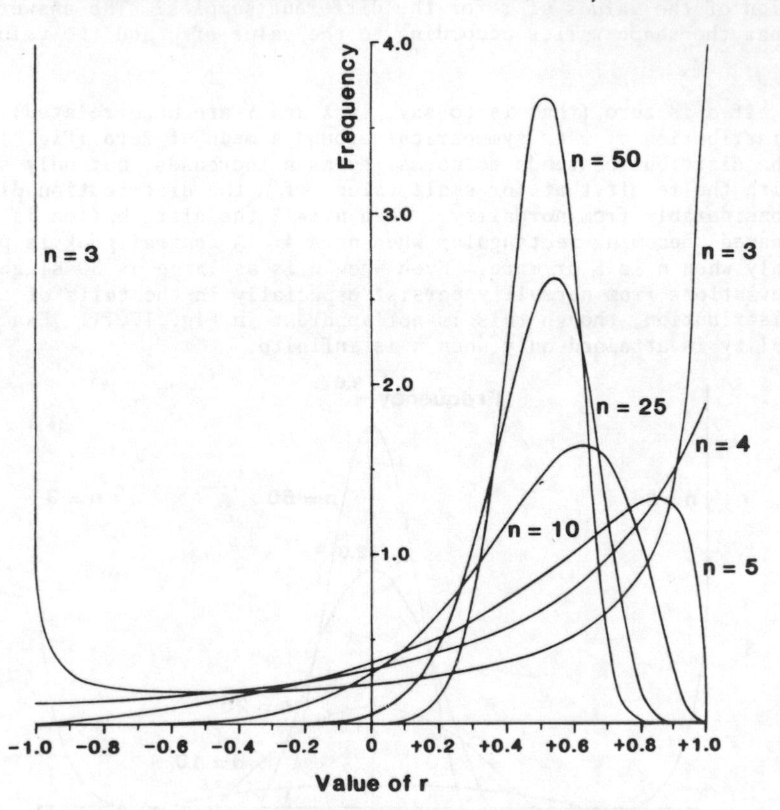

Fig. 17.12 - The distribution of r for six different sizes of sample. The parent population is assumed to have a bivariate normal distribution and a correlation of + 0.5.

It is easy to understand why the distribution of r is asymmetrical when ρ is not zero. The asymmetry reflects the fact that the upper limit of r (r = +1) and the lower limit (r = -1) are different distances from ρ. The value of r is free to deviate more in one direction than in the other. If ρ is +0.8, for example, r can deviate by only 0.2 in an upward direction as against 1.8 in a downward direction. Since r is as likely to deviate in an upward direction as in a downward direction, the distribution of r is necessarily asymmetrical - the more so when n is small, because the sampling error then tends to be large.

The standard error of r (that is to say, the standard deviation of the sampling distribution of r) decreases with increasing n. Consequently the asymmetry of the sampling distribution also decreases. If n is large and ρ is not too near -1 or +1, the standard error is approximately equal to $(1 - \rho^2)/\sqrt{(n - 1)}$. In the case of samples of size 50 drawn from a bivariate normal population with a correlation coefficient of +0.5, the standard error is approximately 0.11. The distribution of the values of r is not far removed from normal. Approximately 95 per cent of the samples yield values of r lying within two standard errors of the value of ρ, that is to say between +0.29 and +0.71. The distribution of r is virtually symmetrical because neither the upper limit of +1 nor the lower limit of -1 constitutes an effective restriction on the fluctuation of the values of r.

The mean of the sampling distribution of r is the same as the population mean ρ only if ρ is 0, +1 or -1. If ρ is any other value, the mean of the sampling distribution is slightly biased in the direction of zero. The bias decreases as n increases and can be safely ignored if n is large. If n is small, the bias may need to be taken into account in estimating the value of ρ from a value of r. Olkin and Pratt (1958) have shown that for a bivariate normal population an unbiased estimate of ρ is given by $\hat{\rho}$ where

$$\hat{\rho} = r\left[1 + \frac{1 - r^2}{2(n - 2)} + \frac{9(1 - r^2)}{8n(n - 2)} + \ldots\right]$$

The first two terms are usually all that need to be calculated to obtain a reasonably unbiased estimate of ρ.

The distribution of r for any given ρ and n can be written as an equation (see Kendall and Stuart, 1963, vol. 1, pp. 385-387). Unfortunately, the equation is complicated and awkward to use, unless ρ happens to be zero. If ρ is zero, the distribution of r is related to that of Student's t. For any n, r is distributed as $\sqrt{[t/(t^2 + n - 2)]}$, t having (n - 2) degrees of freedom. This result holds good even if X is not normally distributed but only Y. If ρ is not zero, however, the distribution of r bears no simple relation to that of t.

TO TEST WHETHER THE CORRELATION COEFFICIENT OF A POPULATION IS ZERO

THE DATA

The population under study consists of paired values of two variables X and Y. It has an unknown correlation coefficient ρ. For any X the values of Y form a normal distribution.

The correlation coefficient of a sample obtained from the population has a value r which is not zero. It is desired to test the hypothesis that in the population there is no correlation between X and Y, in other words that $\rho = 0$.

PROCEDURE

1. Set up the null hypothesis: there is no correlation between the two variables in the population, i.e. $\rho = 0$. Sampling error is alone responsible for r not being zero.

2. EITHER calculate Student's t where $t = [|r|\sqrt{(n - 2)}]/\sqrt{(1 - r^2)}$

and enter Table E with (n - 2) degrees of freedom to find P, the probability of obtaining the calculated value of t or one greater through chance sampling. P is the probability of obtaining a correlation coefficient equal to r, or one deviating from zero by a greater amount than r.

OR, without calculating t, enter Table M to find whether P is greater than, equal to, or less than the 0.05 and 0.01 levels of significance. Table M is computed from Table E using the formula given above.

Tables E and M are designed for two-tailed tests. For one-tailed tests the tabled probabilities should be halved.

3. Decide whether to retain or reject the null hypothesis. If P is low serious doubt is cast on the null hypothesis. Reject the null if P is equal to or less than whatever level of significance is judged appropriate, e.g. 0.05 or 0.01. Conclude that the deviation of r from zero is significantly greater than that which would ordinarily occur as a result of sampling error given a value of ρ of zero. Assume that the two variables are indeed correlated.

If P is greater than the level of significance reserve judgement, i.e. regard the null hypothesis as possibly though not necessarily correct.

EXAMPLE (1)

In the case of the stream data, r = 0.89451 and n = 11. Is the value of r statistically significant?

The concentrations of dissolved solids are not precisely normally distributed, and therefore the following test is only approximate. Another difficulty is that the sample is not truly random.

1. Null hypothesis: the observed correlation results from chance sampling of a population in which the correlation coefficient ρ is actually zero.

2. $$t = \frac{0.89451\sqrt{(11 - 2)}}{\sqrt{[1 - (0.89451)^2]}} = \frac{0.89451\sqrt{(9)}}{\sqrt{(0.19985)}} = 6.00 \text{ approximately.}$$

Entering Table E with 9 degrees of freedom P is seen to be approximately 0.0002. In other words there are about two chances in ten thousand of obtaining such a high value of r or one higher still as a result of sampling error.

A less precise value of P can be obtained using Table M. The 0.05 level of r for 9 degrees of freedom is 0.602, and the 0.01 level is 0.735. The value of P associated with the sample value of r is therefore less than 0.01. The table does not allow a more accurate fixing.

3. Because the value of P is very low the null hypothesis can be rejected. There would appear to be a positive correlation between the concentration of dissolved solids and the discharge.

The sign of the correlation is somewhat difficult to explain. Increasing discharge on most streams is associated with decreasing concentrations of dissolved solids. In other words, there is a negative correlation. Water finds its way into the streams as surface runoff (overland flow) after rains, as throughflow in the soil layer, and as groundwater discharge from rocks outcropping along the stream channel. The groundwater results from vertical percolation of water into the rocks from the soil above. In theory, the groundwater ought to contain more dissolved solids than either the surface runoff or the throughflow since it is in contact with particles of soil and rock for longer periods of time. Surface runoff and throughflow reach streams comparatively quickly, and have less time to dissolve solids. Following rain, stream levels rise because of the addition of quantities of surface runoff and throughflow to the groundwater. The high discharges are usually associated with low concentrations of dissolved solids because the solute-rich groundwater is diluted. Low discharges between storms are derived largely or entirely from groundwater and tend to be associated with relatively high concentration of dissolved solids. Oxley's observations are at variance with this general relationship and this makes them all the more interesting. For some reason the correlation between the concentration of dissolved solids and the discharge is positive only when the stream is rising, i.e. when the discharge is increasing. Oxley showed that when the level of the stream is steady or falling, the correlation is negative just as one would expect.

EXAMPLE (2)

The correlation between the population sizes of the 22 towns in south-east England and the numbers of shops in each town is 0.70247. Is the correlation statistically significant?

The value of Student's t is

$$t = \frac{0.70247\sqrt{(22 - 2)}}{\sqrt{[1 - (0.71812)^2]}} = \frac{0.70247\sqrt{20}}{\sqrt{0.48430}} = 4.51 \text{ approximately.}$$

There are 22 - 2 = 20 degrees of freedom. The value of P according to Table E is less than 0.0005. In other words, there is only a remote possibility that the true value of the correlation coefficient is zero.

ASSUMPTIONS AND LIMITATIONS

1. The test assumes that the sampling is with individual replacements or the population is infinite in size. If the population is finite in size and the sampling is without replacements, the value of P will be too high, thereby biasing the test in favour of the null hypothesis. The amount of bias will not usually be worth bothering about, however, if the sample constitutes less than about ten per cent of the population.

2. The test assumes that the Y values corresponding to each value of X in the population are normally distributed with the same standard deviation $\sigma_{y.x}$. Until recently this assumption was generally

thought not to be an important practical limitation. E.S. Pearson (1929) in a pioneer paper studied samples of size 20 and 30 drawn from two populations, each a mixture of three bivariate normal distributions. Although the populations departed greatly from bivariate normality, Pearson found that the effect on the sampling distribution of r was not serious. He concluded that "the normal bivariate surface can be mutilated and distorted to a remarkable degree without affecting the frequency distribution of r" to an appreciable extent. Later statisticians experimented with other non-normal distributions (exponential, χ^2, etc.) and obtained additional evidence to suggest that the distribution of r is fairly robust in the face of parental non-normality (see Kowalski, 1972, for a useful historical survey). It was recognised that in certain circumstances the distribution of r can depart greatly from that for the normal case, but these circumstances were held to be wholly exceptional. The distribution of r was considered to be remarkably insensitive to non-normality provided the value of ρ is close to zero. No one worried much about non-normality for a test of the hypothesis that $\rho = 0$ even for a sample as small as 12 or so. Any error in the value of P was believed to be sufficiently small to ignore.

This easy-going optimism has been replaced by a mood of distinct caution, even pessimism, following the appearance of a detailed paper by Kowalski (1972). Kowalski experimented with a great variety of non-normal populations, using a high speed computer to derive the sampling distribution of r. He discovered that the robustness of r is something of a myth. It is true that the distribution of r is fairly insensitive to certain types of non-normality, but nevertheless there exist many instances, even when $\rho = 0$, when the distribution of r in samples from non-normal populations departs substantially from the distribution suggested by normal theory. This is the case even for large sample sizes. The test $\rho = 0$ may give results that are seriously in error if the distribution of the Y values for each value of X is not approximately normal. Since one can never be sure with small samples how far the assumption of normality is justified, one must not place too much reliance on the result of the test.

If the sample is small, and it is clear that the Y values depart so greatly from normality that the test cannot be trusted, either of two alternatives can be followed. One is to transform the Y values so that they approximate to a normal distribution. The other is to obtain a randomisation distribution of r by pairing every X value in the sample with every Y value, for example by keeping the X values fixed and permuting the Y values. Under the null hypothesis all n! possible sets of pairings of X and Y are equiprobable. The probability P is k/n! where k is the number of sets of pairings that yield a value of r equal to or greater than the sample value.

Enumeration of the n! sets of pairings is very time-consuming if n is large. The randomisation is undoubtedly best carried out using a computer.

3. The X values of the different items in the population must be independent of each other, and the Y values for a given X must also be independent. Slight autocorrelation can be ignored, but major autocorrelation will cause serious errors in the value of P. If the

autocorrelation is positive, the value of P will be too low.

TO TEST WHETHER THE CORRELATION COEFFICIENT OF A POPULATION IS SOME SPECIFIED VALUE OTHER THAN ZERO

THEORY

The test just described examines the hypothesis that ρ is zero. Occasionally it is necessary to examine the hypothesis that ρ is some specified value other than zero. A different test is needed because the sampling distribution of r is not symmetrical and is consequently not related to the t-distribution. It is necessary to transform r into a new quantity, Fisher's $\mathbf{Z}$. This $\mathbf{Z}$ has nothing to do with the more familiar Z, $Z = |X - \mu|/\sigma$, hence the bold type. $\mathbf{Z}$ is defined as follows

$$\mathbf{Z} = \frac{1}{2}\log_e(1 + r) - \frac{1}{2}\log_e(1 - r) = \frac{1}{2}\log_e[(1 + r)/(1 - r)]$$

Unlike r, $\mathbf{Z}$ can take on any value from minus infinity to plus infinity. As n increases, the sampling distribution of $\mathbf{Z}$ tends to normality much more quickly than r. Unless n is very small, the sampling distribution of $\mathbf{Z}$ can be assumed to be approximately normal around a mean of $\mathbf{Z}_\rho = \frac{1}{2}\log_e[(1 + \rho)/(1 - \rho)]$. The standard deviation (standard error) of $\mathbf{Z}$ is $\sigma_{\mathbf{Z}} = 1/\sqrt{(n - 3)}$ approximately.

PROCEDURE

1. Set up the null hypothesis: the correlation coefficient of the population is ρ, ρ being a specified value other than zero. The value of r differs from ρ purely as a result of sampling error.

2. Calculate $\mathbf{Z} = \frac{1}{2}\log_e[(1 + r)/(1 - r)]$, $\mathbf{Z}_\rho = \frac{1}{2}\log_e[(1 + \rho)/(1 - \rho)]$ and $\sigma_{\mathbf{Z}} = 1/\sqrt{(n - 3)}$.

3. Calculate $Z = |\mathbf{Z} - \mathbf{Z}_\rho|/\sigma_{\mathbf{Z}}$

4. Consult Table B ("the Z table") to find P, the probability of obtaining a difference between r and ρ equal to or greater than the difference actually observed. The value of P listed in the table is appropriate for a two-tailed test. For a one-tailed test it should be halved.

5. Decide whether to retain or reject the null hypothesis. If P is low serious doubt is cast on the null hypothesis. Reject the null if P is equal to or less than whatever level of significance is judged appropriate, e.g. 0.05 or 0.01. Conclude that the deviation of r from zero is significantly greater than that which would ordinarily occur as a result of sampling error given a bivariate normal population with a correlation coefficient ρ as specified. Assume that the true value of the correlation coefficient lies nearer to r_s than the specified value ρ.

If P is greater than the level of significance reserve judgement, i.e. regard the null hypothesis as possibly, though not necessarily, correct.

EXAMPLE

The value of r for the Welsh stream is +0.89451, n being 11. Does this value of r differ significantly from a value of ρ of, say, 0.75?

$$\mathbf{Z} = \tfrac{1}{2}\log_e[(1 + 0.89451)/(1 - 0.89451)] = 1.44405$$

$$\mathbf{Z}_\rho = \tfrac{1}{2}\log_e[(1 + 0.75)/(1 - 0.75)] = 0.97296$$

$$\sigma_{\mathbf{Z}} = \frac{1}{\sqrt{(11 - 3)}} = 0.35355$$

$$Z = \frac{1.44405 - 0.97296}{0.35355} = 1.332 \text{ to 3 decimal places.}$$

Table B shows that a Z value equal to or greater than 1.332 has a probability of 0.1828. Hence the hypothesis that ρ is +0.75 is not discredited. The difference between the observed r and the hypothetical ρ is not statistically significant.

ASSUMPTIONS AND LIMITATIONS

The test assumes that the sampling is with individual replacements or else that the population is infinite in size. Provided the sample is small, however, no great inaccuracy will result if the population is finite in size and the sampling is without replacements.

The test also assumes that the values of X are independent of each other, and that the values of Y for a given X are independent. Serious errors may result if major autocorrelation is present. In theory, the population must have a bivariate normal distribution. It appears that the test is fairly insensitive to certain types of non-normality, in particular, skewness; nevertheless considerable errors are possible in certain circumstances (see Gayen, 1951, and Kowalski, 1972, for further information).

A further assumption is that $\mathbf{Z}$ is normally distributed. If n is small, the distribution of $\mathbf{Z}$ will only be very approximately normal and an incorrect value of P will be obtained.

PROCEDURE

TO SET CONFIDENCE LIMITS AROUND THE POPULATION CORRELATION COEFFICIENT ρ

1. Convert r to $\mathbf{Z}$ using the formula $\mathbf{Z} = \frac{1}{2}\log_e[(1 + r)/(1 - r)]$.

2. Calculate $\sigma_{\mathbf{Z}}$ where $\sigma_{\mathbf{Z}} = 1/\sqrt{(n - 3)}$.

3. Calculate the confidence limits L_1 and L_2 around the transformed value of ρ. The 95% confidence limits are $\mathbf{Z} - 1.960\ \sigma_{\mathbf{Z}}$ and $\mathbf{Z} + 1.960\ \sigma_{\mathbf{Z}}$ and the 99% confidence limits are $\mathbf{Z} - 2.576\ \sigma_{\mathbf{Z}}$ and $\mathbf{Z} + 2.576\ \sigma_{\mathbf{Z}}$.

4. Find the limits around the value of ρ corresponding to the limits around the transformed value of ρ calculated in step 3. The limits are $(e^{2L_1} - 1)/(e^{2L_1} + 1)$ and $(e^{2L_2} - 1)/(e^{2L_2} + 1)$. Snedecor and Cochran (1967, pp. 558 and 559) provide useful tables for converting r or ρ to Fisher's **Z** and converting Fisher's **Z** back to r or ρ.

EXAMPLE

For the stream data, $r = 0.89451$. This corresponds to a value of **Z** of $\frac{1}{2}\log_e (1.89451/0.10549) = 1.44405$. The standard error $\sigma_{\mathbf{Z}} = 1\sqrt{(11 - 3)} = 0.35355$. The 95% confidence limits for the transformed value of ρ are therefore $1.44405 - 1.960\ (0.35355)$ and $1.44405 + 1.960\ (0.35355) = 0.75109$ and 2.13701.

The values of ρ corresponding to the 95% limits of the transformed ρ are $[e^{2(0.75109)} - 1]/[e^{2(0.75109)} + 1]$ and $[e^{2(2.13701)} - 1]/[e^{2(2.13701)} + 1]$ or $3.49147/5.49147$ and $70.80973/72.80973$ or 0.64 and 0.97 approximately.

The limits are fairly widely spaced because the sample is small. They are not equidistant either side of r because the sampling distribution of r is skewed. The limits are calculated on the assumption that the population is bivariate normal and that the size of sample is sufficiently large for the sampling distribution of **Z** to be normal. Neither of these assumptions is really justified, and hence the limits are only approximate.

TO TEST THE HYPOTHESIS THAT THE CORRELATION COEFFICIENTS OF TWO SAMPLES DIFFER SIGNIFICANTLY[1]

THE DATA

Two samples are drawn independently from different populations. One sample of size n_1 has a correlation coefficient r_1; the other sample of size n_2 has a correlation coefficient r_2. It is desired to test the hypothesis that the two samples are drawn from populations with the same correlation coefficient ρ.

PROCEDURE

1. Set up the null hypothesis: the difference between the two correlation coefficients r_1 and r_2 is due entirely to sampling error. The two samples are drawn from populations with the same correlation coefficient ρ. Both r_1 and r_2 are estimates of ρ.

2. Convert each r to **Z** using the formula $\mathbf{Z} = \frac{1}{2}\log_e[(1 + r)/(1 - r)]$.

3. Find the absolute difference in the **Z** values, namely $|\mathbf{Z}_1 - \mathbf{Z}_2|$.

---

[1] A more elaborate test for three or more samples is available: see Snedecor and Cochran (1967, pp. 186-187).

4. Calculate the standard error of this difference, $\sigma_{\mathbf{z}_1-\mathbf{z}_2}$

$$\sigma_{\mathbf{z}_1-\mathbf{z}_2} = \sqrt{\left[\left(\frac{1}{n_1 - 3}\right) + \left(\frac{1}{n_2 - 3}\right)\right]}$$

5. Calculate Z, the ratio of the difference in the **Z** values to the standard error of the difference.

$$Z = \frac{|\mathbf{Z}_1 - \mathbf{Z}_2|}{\sigma_{\mathbf{z}_1-\mathbf{z}_2}}$$

6. From Table B ("the Z table") find P, the probability of obtaining the calculated difference, or one greater, through chance sampling. The value of P listed in the table is appropriate for a two-tailed test. For a one-tailed test, it should be halved.

7. Decide whether to retain or reject the null hypothesis. If P is low, serious doubt is cast on the null hypothesis. Reject the null if P is equal to or less than whatever level of significance is judged appropriate, e.g. 0.05 or 0.01. Conclude that the difference between the sample correlation coefficients is more than would ordinarily occur if the samples were drawn from populations with identical correlation coefficients.

If P is greater than the level of significance, reserve judgement i.e. regard the null hypothesis as possibly, though not necessarily, correct.

EXAMPLE

The value of r for the Welsh stream is +0.89 (to 2 decimal places) and n is 11. The value of r for a neighbouring stream investigated by Oxley (1974) at the same time is +0.83, n being 12. Are the two correlation coefficients significantly different?

1. Null hypothesis: the two samples are drawn from populations with the same correlation coefficient. In other words the correlation between the concentration of dissolved solids and the discharge is the same for the two streams. The correlation coefficients of the two samples differ purely because of chance.

2. $\mathbf{Z}_1 = \frac{1}{2}\log_e (1 + 0.89)/(1 - 0.89) = 1.42193$ approximately.

$\mathbf{Z}_2 = \frac{1}{2}\log_e (1 + 0.83)/(1 - 0.83) = 1.18814$ approximately.

3. $|\mathbf{Z}_1 - \mathbf{Z}_2| = 0.23379$.

4. $\sigma_{\mathbf{z}_1-\mathbf{z}_2} = \frac{1}{(11 - 3)} + \frac{1}{(12 - 3)} = 0.48591$.

5. $Z = \frac{0.23379}{0.48591} = 0.48$ approximately.

6. P = 0.6312.

7. Because the value of P is so high the null hypothesis cannot be rejected. The correlation between the concentration of dissolved solids and the discharge is possibly different for the two streams but the small size of the samples does not allow any definite conclusion.

If the null hypothesis is correct, both correlation coefficients are estimates of a common population coefficient ρ. An estimate of ρ can be obtained by combining the samples. It works out at +0.86.

18 The Interpretation of Correlation Coefficients

Correlation coefficients can be all too readily misinterpreted. Before drawing a conclusion the student or research worker should check through the catalogue of possible misunderstandings set out below.

1. SAMPLING ERROR

Correlation coefficients calculated from sample data are subject to chance fluctuations. The tests discussed in the last chapter provide objective ways of examining this problem if the sampling is random. They are inappropriate if the X and Y values are selected non-randomly (or are autocorrelated, see below). Unfortunately, because it is often difficult or impossible to carry out random sampling, less effective types of sampling are often substituted. The interpretation of correlation coefficients based on non-random samples is fraught with difficulty. How much sampling error is involved is largely a matter of guesswork.

2. LINEAR VERSUS CURVILINEAR RELATIONSHIPS

As mentioned in Chapter 17, correlation coefficients measure only linear correlation. They do not provide an effective measure of curvilinear relationship. If the correlation coefficient of a population is zero, one must not assume that the two variables in question (X and Y) are unrelated: there may be a very strong curvilinear relationship. If the population coefficient is zero, all one can say is that there is no linear relationship.

Some samples are so small that one cannot be sure whether a relationship is linear or curvilinear. The data points are so scattered it is impossible to say whether they are best described by a line or a curve. Some statisticians argue that there is no point in calculating a coefficient of correlation if it is unclear whether the relationship in question is linear. Others take the view that in the absence of clear evidence of curvilinearity it is reasonable to calculate a coefficient of correlation, even if the sample contains only a few items and they only roughly approximate to a straight line.

Measures of curvilinear relationship are considered in the companion volume.

3. CONFOUNDING

When one calculates a coefficient of correlation between two variables X and Y one must always bear in mind the possibility that the correlation is badly distorted by other, unexamined variables. One may wish to believe for the sake of simplicity that the unexamined variables are random in their effects and thus of no great consequence but often the opposite is true: the effects of the unexamined variables may be far from random, and the relationship between X and Y may be so systematically distorted as to be seriously misleading. The relationship may be observed to be curvilinear, when in fact it would be linear if one could only eliminate the unwanted distortion caused by the unexamined variables. Or the relationship may be observed to be linear, although in the absence of the unexamined variables it would actually be curvilinear. A further possibility is that the sign of the observed correlation is misleading and that the opposite sign would occur if one could remove the influence of the unexamined variables.

The problem one faces of trying to disentangle the relationship of the examined variables from their relationships with the unexamined variables can be called the *problem of confounding*, since the separate effects of the variables are easily confused or confounded. It is a problem that should be avoided or minimised, if at all possible, by careful collection of data. The obvious strategy in the case of a laboratory experiment is to try and hold all the variables constant, except X and Y. If there are variables that cannot be held constant, it may be possible to ensure that their effects on X and Y are random and thus relatively uncomplicated. Consider, for instance, an experiment in which sandstone blocks are alternately dried and soaked in natural brines of varying salt concentration. Suppose that it is desired to establish the degree of correlation between the durability of the sandstone (Y) and the salt concentration of the brines (X). In the interest of simplicity it would be best to try to hold as many other variables constant as possible. Thus, the blocks ought all to be cut to the same shape and size and soaked for the same length of time. It would be impossible, of course, to ensure that all the blocks were identical in lithology but one could minimise the unwanted effects of variations in grain-size, porosity, etc. by randomly assigning the blocks to the different brines. In this way one could hope to establish the correlation between X and Y with minimal inaccuracies caused by confounding.

Field investigations cannot be as successfully controlled as laboratory experiments, and the collection of data is inevitably rather more haphazard. Sometimes, however, one can arrange to stratify a random sample so as to reduce the risk of confounding. It may be known, for instance, in advance of sampling that the relation between X and Y is likely to be severely distorted by variations in a third variable (say, Z). One may be able to take advantage of this knowledge by stratifying the sampling so that pairs of values of X and Y are collected for selected values of Z. This would permit a study of the way in which the correlation between X and Y varies with the value of Z. Another approach that can be applied in both field and laboratory investigations is to make measurements of not only X and Y but all other variables that might systematically distort the relationship. The mass of data so collected can then be used to calculate coefficients of multiple and partial correlation as described in the companion volume.

4. THE RANGE OF X VALUES

The size of a correlation coefficient depends in part on the range of X values under consideration. When X is allowed to range only from 0 to 10 the correlation will not usually be as high as when X is allowed to range from 0 to 20 or 0 to 100. This is illustrated in Fig. 18.1. The ten open circles are the same distances from the sloping line as the ten solid circles. They form precisely the same pattern. Although the correlation coefficient for each set of circles is 0.79, that for both sets is 0.93. Doubling the range of X increases the size of r by 15% in this instance. Trebling the range would increase it still further.

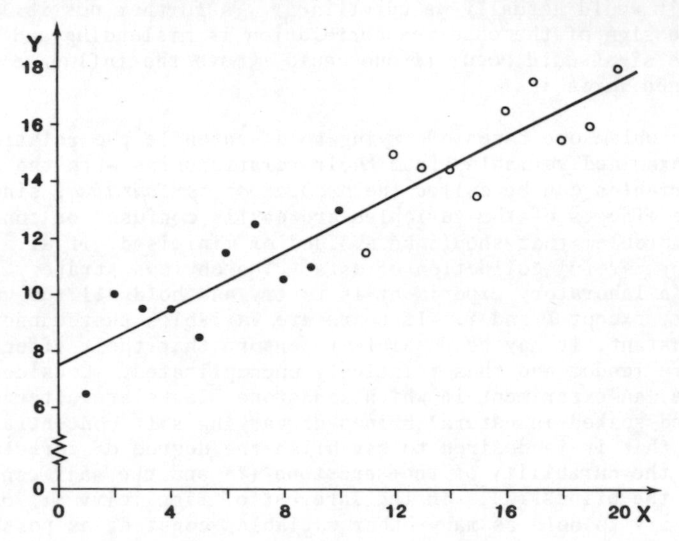

Fig. 18.1 - Diagram to illustrate the effect of the range of X values on the size of the correlation coefficient. For explanation see text.

The random sample of towns in south-east England with populations between 10,000 and 30,000 suggested that the number of shops (Y) is moderately correlated with population size (X). The value of r was found to be +0.70. If towns ranging from 10,000 to 100,000 had been sampled, a much higher coefficient would almost certainly have been obtained. Conversely, had the sampling been restricted to towns ranging from 10,000 to 15,000, the coefficient would almost certainly have been rather lower. The size of the coefficient in part reflects the range of town sizes selected for investigation.

In laboratory experiments it is common practice to select certain values of X because of their simplicity or convenience and to allow only the values of Y to vary randomly. The rate of solution (Y) of limestone blocks in dilute acids at different temperatures might be measured at X = 0, 5, 10, 15 and 20°C. Or the spacing (Y) of ripples of sand in a wind tunnel might be measured when the wind velocity (X) attains 20, 30, 40 and 50 km per hour. Correlation coefficients derived from such experiments are doubly difficult to interpret because the size of the coefficients depends partly on the range of X values and partly on which X values are selected within the range. Observe also that no valid tests of significance can be carried out because the sampling is non-random.

5. AUTOCORRELATION

The items that make up some samples and populations are geographical or historical in character: they are different places, areas, events or periods of time. The values of such items are seldom distributed randomly in time or space but are *autocorrelated*. The autocorrelation is said to be *positive* if the values of items that are close together are more alike than the values of items that are far apart. If the reverse condition occurs - in other words, if the values of items that are close together are less alike than the values of items that are far apart - the autocorrelation is said to be *negative*. Positive autocorrelation is common, but negative autocorrelation is only rarely met with.

Correlation coefficients calculated from autocorrelated values of X and Y have the same values that they would have if there was no autocorrelation. No correction factor has to be applied to the coefficients to allow for the autocorrelation. The coefficients of random samples provide unbiased estimates of the coefficients of the parent populations irrespective of any autocorrelation. Unfortunately, the formulas for the standard error given in previous chapters are invalid if autocorrelation is present. In the case of positive autocorrelation the true standard error is larger than that suggested by the formulas, which means that the significance tests that have been described give P values that are too low. If the positive autocorrelation is strong, the tests may seriously over-state the significance of the correlation coefficient. A test could suggest, for example, that a coefficient is significant at the 0.001 level when it is not significant even at the 0.05 level. The confidence limits for the coefficient may be far too close together because the standard error is seriously under-estimated.[1]

The formulas for the standard error can be modified to allow for autocorrelation, but only with considerable difficulty. At present there is no simple way of eliminating the influence of autocorrelation from significance tests and from calculations of confidence limits (Cliff and Ord, 1981). In general, it seems wise not to run a test if the sample values of X or Y show appreciable autocorrelation. Methods of detecting and measuring autocorrelation in samples will be discussed in the companion volume.

6. NONSENSE CORRELATIONS

A correlation between two variables X and Y does not necessarily indicate that X causes Y, or that Y causes X, even when sampling error is not involved. X and Y may interact so that both are simultaneously cause and effect. Or X and Y may both be effects of another variable or group of variables. A further possibility is that the correlation has no physical significance whatsoever, but represents an accidental parallelism or opposition of two variables in time or space. The annual birth rate in the Netherlands has declined steadily in recent years. In 1955, for example, the birth rate stood at 21.3 births per 1000 population. By 1975 the rate was down to 13.0, almost half the level of 20 years before. Even more

[1]Negative autocorrelation introduces errors of the opposite kind. The tests understate the significance and the confidence limits are set too far apart.

dramatic has been the decline in the fortunes of the White Stork (see Table 18.1 and Fig. 18.2). At the turn of the century the Stork was widespread if not abundant in the Netherlands. Its bulky nests were a familiar sight on roofs and chimneys in many villages, particularly in the east. Now Storks have all but disappeared. In 1955 57 pairs of Storks bred (as against 312 in 1939). In 1975 only 9 pairs were left, despite desperate efforts by conservationists, including the erection of platforms to serve as nesting sites.

TABLE 18.1 - THE WHITE STORK, CICONIA CICONIA, IN THE NETHERLANDS

Year	*Number of pairs of Storks*	*Birth rate per 1000 population (human)*	*Year*	*Number of pairs of Storks*	*Birth rate per 1000 population (human)*
1955	57	21.3	1966	28	19.2
1956	63	21.3	1967	19	18.9
1957	73	21.2	1968	19	18.6
1958	56	21.2	1969	19	19.2
1959	50	21.4	1970	14	18.3
1960	48	20.8	1971	15	17.2
1961	46	21.3	1972	9	16.1
1962	50	20.9	1973	6	14.5
1963	33	20.9	1974	8	13.8
1964	29	20.7	1975	9	13.0
1965	32	19.9			

Sources: Lack, 1966, Jonkers, 1976, and E.E.C. Publications.

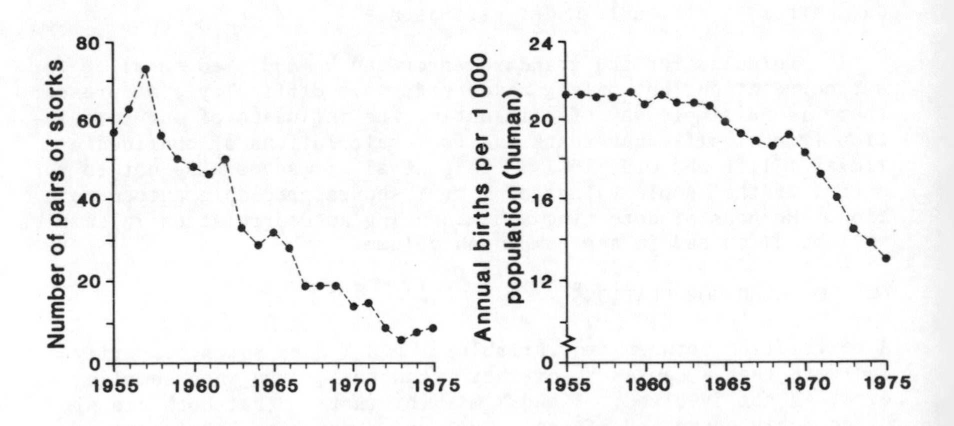

Fig. 18.2 - Nonsense correlation of the annual birth rate in the Netherlands with the number of pairs of Storks.

The decline in the annual birth rate matches the decline in the number of Storks with remarkable accuracy. The correlation works out at 0.83, which is much too large a figure to be due to sampling error. The correlation has to be accepted as statistically significant,[1] even though the variables do not seem to be causally related in any way. The birth rate in the Netherlands has declined in response to cultural and economic pressures. Improved methods of contraception and the increased availability of abortion have undoubtedly played a part. The decline in the numbers of Storks would appear to have had very different causes. As Lack (1966, p.313) wrote: "Part, but probably only part, of the widespread decrease is attributable to human factors, especially to the drainage of damp ground where the Storks find most of their food, also to the erection of cables, into which they fly, and perhaps to the poisoning of locusts in Africa and to increased shooting of migrants around the Mediterranean and in some parts of tropical Africa. There were, however, marked fluctuations in Europe in the nineteenth century and the first part of the twentieth century ..., so long-term causes are probably involved as well".

If there is a causal connection between the birth rate in the Netherlands and the number of Storks it can only be a very tenuous one (unless the old tradition is true that Storks bring babies). In all probability the two variables are wholly independent, but because they vary with time in a fairly similar manner they are strongly correlated. Although statistically significant, the correlation is seemingly without physical significance and may be looked upon as an example of a *nonsense correlation*. This term was coined by Yule (1926) to describe an accidental correlation between two independent variables that happen to vary systematically over time or space. Experience has shown that geographical data are just as susceptible to producing nonsense correlations as data for periods of time. For instance, nations that have the highest public expenditure on libraries have the highest death rates from cancer. Nations that spend little on libraries experience fewer deaths from cancer. Yet reading books does not cause cancer, and fear of cancer is not the reason why money is spent on libraries. It would seem to be largely or wholly accidental that a correlation exists between library expenditure and the incidence of cancer.

A disquieting feature of nonsense correlations is that they cannot be proved to be nonsense. A correlation that at first glance appears to be nonsense may in fact be meaningful, because an unexpected causal relationship may exist. Just because one cannot think of any rational explanation for a particular correlation this does not make it nonsense. What cannot be explained immediately is not necessarily inexplicable.

There is a real danger that only the most outlandish and ludicrous correlations will be correctly identified as nonsense correlations, while those that fit existing theories or prejudices will be automatically assumed to be meaningful. Some correspondence in *The Times* on the subject of the relationship between money supply and inflation

[1] This statement is perhaps too dogmatic since the data do not constitute a random sample and are autocorrelated. No test of significance can be legitimately applied. Nevertheless, the notion that the correlation is a mere accident of sampling is scarcely credible given the size of the coefficient. Were the sample random and were autocorrelation absent the value of P would be less than 0.000005 ($t = 6.40$ with 19 df).

illustrates the point perfectly. Professor Lord Kaldor (March 31, 1977) asserted that "there is no historical evidence whatever" to show that excess money supply leads to inflation. This view was promptly challenged by Professor Mills (April 4th) who produced figures to show that increases in the money supply during the years 1965-1973 were matched by subsequent increases in prices. Professor Mills obtained a correlation of +0.85 between the money supply in each year and the increase in prices two years later. This result would be significant at the 1% level if the sampling were properly random and there were no autocorrelation. Professor Lord Kaldor (April 6th) remained unimpressed. He argued that the correlation was partly or wholly accidental, there being no good evidence to suggest that the supply of money actually causes inflation. In his view the high correlation was to a large extent the consequence of rising world commodity prices.

Dr. Llewellyn and Mr. Witcomb (April 6th) calculated a correlation of -0.87 between the numbers of cases of dysentery in Scotland each year and increases in prices one year later, asserting light-heartedly that this proved that Scottish dysentery held prices down. Now this correlation is so absurd that it can be safely assumed to be nonsense, but the correlation noted by Professor Mills is not so easily dismissed. While many economists would agree with Professor Lord Kaldor that excess money supply does not feed inflation to an appreciable extent, and would therefore dismiss the observed correlation as largely if not entirely nonsense, others would claim that excess money supply has a major effect on inflation, and would accept the correlation as evidence of the correctness of this claim. One man's nonsense correlation is all too often another man's meaningful relationship.

7. SPURIOUS CORRELATION

Variables that are defined in such a way as to overlap will automatically be correlated. The size of the correlation will reflect the degree of overlap. If the overlap is total, in other words if the variables are synonymous, the correlation will be perfect. If the variables only overlap to a slight extent, the resulting correlation will be low.

A correlation that is due solely to the presence of an overlap is said to be *spurious*.[1] It is implicit in the definition of the variables and has no deeper significance. If the variables are redefined so as to eliminate the overlap, the correlation disappears.

As an example consider the three variables A, B and C shown in Table 18.2. The variables are largely if not wholly random, the values being taken from a table of supposedly random digits. However, variable C has been allowed to assume a greater range of values than variables A and B. The coefficient of variation of C is 9.0 whereas the coefficient of A is only 0.8, and that of B only 0.7

---

[1]The term "spurious correlation" is perhaps unfortunate since it might mistakenly be thought to be synonymous with the term "nonsense correlation" which means something entirely different. It is too firmly entrenched in the statistical literature, however, to be replaceable at this stage.

(the coefficient of variation is the standard deviation divided by the mean).

For all practical purposes A, B and C may be regarded as uncorrelated. The coefficients of correlation are all close to zero (r for A and B = 0.01, r for A and C = -0.07 and r for B and C = +0.05), which is not surprising given that the variables are supposed to be random.

In Table 18.2 A, B and C have been combined to create two additional variables X = A/C and Y = B/C. Since A, B and C are uncorrelated, one might expect their derivatives X and Y to be uncorrelated. Actually they are very highly correlated (r = +0.95). Correlation is present because the variables are defined in such a way that they overlap. They share a *common element*: the variable C which appears in both their formulas. It is the variation in C which sets up the correlation. If C were a constant (i.e. the same for all items), X and Y would be uncorrelated. Dividing by C would change the scale of A and B but it would not introduce any correlation.

TABLE 18.2 - NINE HYPOTHETICAL VARIABLES

ITEM	*VARIABLE* *A*	*B*	*C*	*X=A/C*	*Y=B/C*	*X'=AC*	*Y'=BC*	*X"=A+B*	*X'''=A-B*
a	8	6	-4	-2.0000	-1.5000	-32	-24	14	2
b	4	4	+64	+0.0625	+0.0625	+256	+256	8	0
c	6	1	-87	-0.0680	-0.0115	-522	-87	7	5
d	4	7	+62	+0.0645	+0.1129	+248	434	11	-3
e	1	3	-60	-0.0167	-0.0500	-60	-180	4	-2
f	0	9	-22	0.0000	-0.4091	0	-198	9	-9
g	8	6	+55	+0.1455	+0.1091	+440	+330	14	2
h	4	1	+26	+0.1538	+0.0385	+104	+26	5	3
i	3	9	-69	-0.0435	-0.1304	-207	-621	12	-6
j	5	3	-91	-0.0769	-0.0330	-455	-273	8	2
k	1	2	-13	+0.0769	-0.1538	-13	-26	3	-1
l	2	8	+49	+0.0408	+0.1633	+98	+392	10	-6
m	3	2	+90	+0.0333	+0.0222	+270	+180	5	1
n	1	5	+69	+0.0145	+0.0725	+69	+345	6	-4
o	0	1	+33	0.0000	+0.0303	0	33	1	-1

By substituting X = A/C and Y = B/C in the formula for r one can prove that the value of r depends on the size of the coefficients of variation of A, B and C. The value of r is approximately $C_C^2/\sqrt{[(C_A^2 + C_C^2)(C_B^2 + C_C^2)]}$ where C_A, C_B and C_C are the coefficients of variation of A, B and C respectively (for a proof see Pearson, 1897). If the coefficients of variation are equal, the correlation is 0.50. If the coefficient of C is twice that of A and B, the correlation is 0.80. The correlation is 0.90 if the coefficient of C is three times that of A and B. In the present instance, the coefficient is more than three times that of A or B and consequently the correlation exceeds 0.90.

Spurious correlations are developed not only between ratios but also between products that share a common element. The correlation between X' = AC and Y' = BC in Table 18.2 is 0.69, the variable C once again providing the common element: this time as a multiplier and not as a divisor. Spurious correlations can also be obtained between products and ratios that share a common element. The variables X = A/C and Y' = BC in Table 18.2 yield a correlation of 0.14. Spurious correlations are also developed between sums of variables and the individual variables contributing to the sums. The correlation between X" = A + B and B in Table 18.2 is +0.75 and that between X''' = A - B and B is -0.75.

A correlation between two variables sharing a common element may be partly spurious and partly real. It may be larger, in other words, than the correlation that would occur if the variables were random.[1] One must be careful not to assume that every correlation involving a common element is completely spurious. Equally, one must take care to avoid the opposite mistake of thinking that a correlation has some physical significance when in fact it is completely spurious. Appearances can be deceptive, as the following examples demonstrate.

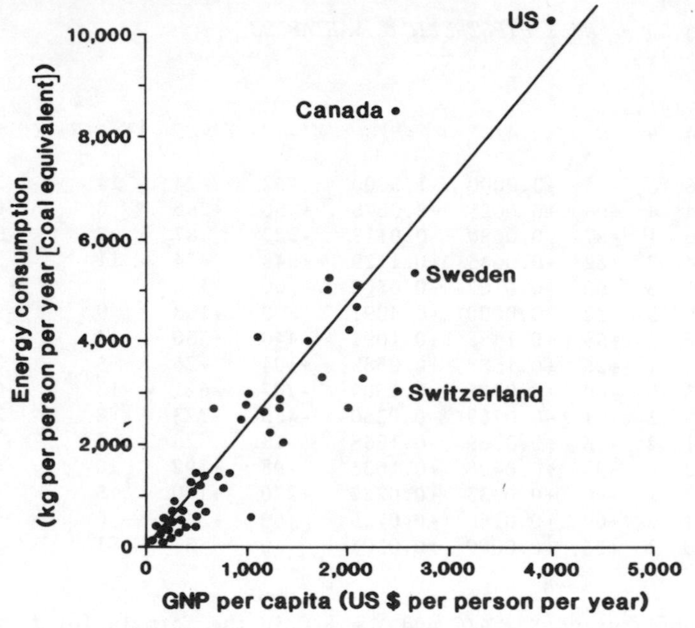

Fig. 18.3 - World Energy consumption and GNP per capita. Redrawn from Meadows et al. (1972).

Figure 18.3 is taken from *The Limits to Growth* by Meadows *et al.* (1972), the influential report on world population and resources commissioned by the Club of Rome. Each of the 50 dots on the graph represents a different country. The variables depicted are total output per capita (GNP per capita) and energy consumption per capita. Although the 50 countries consume very different amounts of energy,

---

[1] Although the correlation is usually larger, it can be smaller. Thus, if the spurious correlation is small and negative in sign and the underlying or real correlation is large and positive, the result may be a weakly positive correlation.

energy consumption per capita correlates very closely with total output per capita. The value of Pearson's coefficient of correlation would seem to be about +0.92. (Meadows *et al.* do not calculate the coefficient, and since they do not provide any raw data, the value of the coefficient cannot be determined precisely. One can measure the co-ordinates of each dot in Fig. 18.3, however, and from the measurements estimate the value to be 0.92). The high degree of correlation depicted in Fig. 18.3 is perhaps surprising. Considering that countries differ greatly in climate, local fuel prices, and emphasis on heavy industry, one might have expected a much lower correlation. Just why is the correlation coefficient so high?

It is important not to overlook the fact that the per capita figures for output and energy consumption are derived by dividing national totals by population size. Because the two variables share the same denominator (population size), the observed correlation may be partly or wholly spurious. In order to make sense of Fig. 18.3 one must try to determine the amount of spurious correlation. If the correlation is entirely spurious, then Fig. 18.3 is seriously misleading.

As already mentioned, if A, B and C are random variables, and the ratio A/C is correlated with the ratio B/C, the value of Pearson's coefficient is approximately $C_C^2/\sqrt{[(C_A^2 + C_C^2)(C_B^2 + C_C^2)]}$ where C_A, C_B and C_C are the coefficients of variation of A, B and C respectively. In the present instance, C_C is the coefficient of variation of population size. C_A and C_B are the coefficients of variation of GNP and energy consumption. The exact values of C_C, C_A and C_B cannot be calculated since Meadows *et al.* fail to provide the necessary raw data. However, by taking a fresh sample of countries around the world, including those named in Fig. 18.3, C_C can be estimated to be 1.9, C_A to be 6.1 and C_B to be 3.1. Substituting these values in the formula for r gives

$$r = \frac{(1.9)^2}{\sqrt{[(6.1)^2 + (1.9)^2][(3.1)^2 + (1.9)^2]}} = \frac{3.61}{\sqrt{[(40.82)(13.22)]}}$$

$$= 0.16 \text{ approximately.}$$

The amount of spurious correlation is low because both C_A and C_B are appreciably larger than C_C. Figure 18.3 is therefore not as deceptive as one might imagine. Only (0.16/0.92) × 100 = 17% cf the observed correlation is spurious. 83% has some physical significance, unless it be due to sampling error. This latter possibility can be checked by running a significance test. With 50 - 2 = 48 degrees of freedom, the value of r = 0.92 - 0.16 = 0.76 is significant at the 1% level. It is thus almost certain that part of the observed correlation is real. 95% confidence limits for the part that is real can be calculated in the usual way (see Chapter 17) and are 0.61 to 0.86.

A further example of the danger of spurious correlation may prove helpful. In recent years British industry has tended to become much more widely dispersed. There has been a steady decline in the number of manufacturing companies and jobs in inner-city areas and conurbations, contrasting with an increase in numbers in smaller industrial areas. As Keeble (1978) has pointed out, a variety of economic and social forces seem to have been at work to produce these changes. Conurbations and inner-city areas suffer from dilapidated factory premises and lack space for modern manufacturing industry. With their ageing housing stock and increased traffic congestion they are not considered attractive places in which to live. Industrialists and workers have tended to move into the suburbs and rural areas in search of more modern houses with gardens and garages. A further factor has been the substantial improvement in road communications, mostly outside urban areas, which has conferred greater locational freedom and flexibility on many firms. Central and local government planning policies have also helped to disperse industry away from inner-city areas and conurbations.

Regional changes in manufacturing jobs are summarised in Table 18.3. Regions with the biggest concentrations of industries declined markedly between 1965 and 1975, while regions with the smallest concentrations exhibited modest growth. There is an almost perfect inverse correlation between the number of manufacturing jobs in a region in 1965 and the changes in the succeeding ten years. The greater the number of jobs, the greater the proportion of employment lost. The value of Pearson's coefficient is -0.985. The degree of correlation would be astonishing were it not for the fact that the variables overlap. One variable can be written C and the other A - C, where C is the 1965 figure for each region and A the 1975 figure. Because C is a common element, some degree of correlation is bound to occur, even if C and A are random variables. The correlation, in other words, is only partly real.

Substituting X = A - C and Y = C in the formula for r, one obtains the approximation

$$r \simeq -\frac{1}{\sqrt{(1 + K^2)}}$$

where K is the ratio of the standard deviation of C to that of A. This is the amount of spurious correlation to be expected if A and B are random variables. Referring to Table 18.3 the standard deviation for 1965 is found to be 618.539 and that for 1975 480.122. K is the ratio 618.539/480.122 = 1.28, which makes r approximately -0.613. Some 62% of the observed correlation of -0.985 is in fact spurious and of no physical significance. The "real" correlation is only 0.613 - 0.985 = -0.372, a puny figure compared with "apparent" correlation. It is not significant at the 5% level.

The dangers of spurious correlation were first pointed out by Karl Pearson in 1897, but since then the subject has attracted surprisingly little attention. Many statistics textbooks in current use make no mention of spurious correlation. Perhaps because statisticians have not issued enough warnings, geographers and other scientists have often failed to recognise spurious correlation and have been guilty of drawing incorrect inferences. Even a cursory inspection of geographical writings will yield many examples of the use of ratio correlations where the ratios share either a numerator or a denominator.

TABLE 18.3 REGIONAL CHANGES IN THE NUMBER OF MANUFACTURING JOBS 1965-75

	Manufacturing employment					
	1965		*1975*		*Change 1965-75*	
	thousands	*% U.K.*	*thousands*	*% U.K.*	*thousands*	*%*
East Anglia	167	2.0	198	2.6	-31	+18.6
Wales	311	3.6	317	4.2	+6	+2.0
South West	422	4.9	427	5.7	+5	+1.2
North	459	5.4	454	6.1	-5	-1.1
East Midlands	620	7.2	593	7.9	-27	-4.4
Northern Ireland	171	2.0	154	2.1	-17	-9.9
Scotland	725	8.5	637	8.5	-88	-12.1
West Midlands	1185	13.8	1021	13.6	-164	-13.8
Yorks and Humber	860	10.1	733	9.8	-127	-14.8
North West	1252	14.6	1042	13.9	-210	-16.8
South East	2389	27.9	1913	25.6	-476	-19.9
United Kingdom	8561	100.0	7489	100.0	-1072	-12.5

Source: *Dept. Employment Gaz.*, August, 1976, also reproduced in Keeble (1978), Table II.

Apart from Pearson, the main contributors to the theory of spurious correlation have been Reed (1921) and Benson (1965), who have developed formulas for calculating the amount of spurious correlation for a variety of ratios, products and sums. Some of the more useful formulas are set out in Table 18.4. The formulas assume that A, B and C are random variables and hence uncorrelated. As before C_A, C_B and C_C are the coefficients of variation of A, B and C, and K is the ratio of the standard deviation of C to the standard deviation of A.

8. CORRELATIONS BASED ON MEAN OR AGGREGATED VALUES

A correlation calculated from average or aggregated (grouped) values of X and Y will tend to differ from one calculated from the individual values on which the averages or aggregates are based. Usually, the correlation becomes larger when the individual values are averaged or aggregated, but it can become smaller or even change sign in certain circumstances. The more values are combined, the greater becomes the range of values over which the correlation coefficient can vary.

As a concrete illustration[1] consider the yields of wheat and main crop potatoes shown in Table 18.5. The correlation between the yields for the different counties is +0.10, indicating that there is

[1]This illustration is an updated version of one given by Yule and Kendall (1950, pp. 310-313).

TABLE 18.4 - USEFUL FORMULAS FOR CALCULATING AMOUNTS OF SPURIOUS CORRELATION

Source of spurious correlation	*Symbolism*	*Approximate formula for r*
Correlation of ratios with common numerator	X=C/A Y=C/B	$+\frac{C_C^2}{\sqrt{[(C_A^2 + C_C^2)(C_B^2 + C_C^2)]}}$
Correlation of ratios with common denominator	X=A/C Y=B/C	$+\frac{C_C^2}{\sqrt{[(C_A^2 + C_C^2)(C_B^2 + C_C^2)]}}$
Correlation of ratios where the numerator of one ratio is the denominator of the other	X=A/C Y=C/B	$-\frac{C_C^2}{\sqrt{[(C_A^2 + C_C^2)(C_B^2 + C_C^2)]}}$
Correlation of a ratio with its own numerator	X=C/A Y=C	$+\frac{C_C}{\sqrt{(C_A^2 + C_C^2)}}$
Correlation of a ratio with its own denominator	X=A/C Y=C	$-\frac{C_C}{\sqrt{(C_A^2 + C_C^2)}}$
Correlation of a ratio with the reciprocal of its numerator	X=C/A Y=1/C	$-\frac{C_C}{\sqrt{(C_A^2 + C_C^2)}}$
Correlation of products with a common variable	X=AC Y=BC	$+\frac{C_C^2}{\sqrt{[(C_A^2 + C_C^2)(C_B^2 + C_C^2)]}}$
Correlation of a product with one of the constituent variables	X=AC Y=C	$+\frac{C_C}{\sqrt{(C_A^2 + C_C^2)}}$
Correlation of a sum with one of the constituent variables	X=A+C Y=C	$+\frac{1}{\sqrt{(1 + K^2)}}$
Correlation of a difference with one of the constituent variables	X=A-C Y=C	$-\frac{1}{\sqrt{(1 + K^2)}}$

only a weak relationship between the two variables at a county level. The correlation rises to +0.46, however, if the 46 counties are grouped on an alphabetical basis into 23 pairs (Avon and Bedfordshire, Berkshire and Buckinghamshire, Cambridgeshire and Cheshire, Cleveland and Cornwall, etc.) and the correlation is calculated using the mean yields for each pair. The correlation rises still higher to +0.58 if the counties are taken in groups of four (Avon and Bedfordshire and Berkshire and Buckinghamshire, Cambridgeshire and Cheshire and Cleveland and Cornwall, etc.). It reaches +0.91 if groups of 15 are employed. Clearly, the correlation coefficient is very sensitive to changes in the size of the areas upon which it is based.

TABLE 18.5 - ESTIMATED YIELD PER HECTARE OF WHEAT AND MAINCROP POTATOES IN ENGLAND IN 1974

County	*Economic Planning Region*	*Yield in tonnes per hectare*	
		Wheat	*Maincrop potatoes*
Avon	South West	4.5	28.9
Bedfordshire	South East	5.4	32.1
Berkshire	South East	4.5	28.4
Buckinghamshire	South East	4.5	29.4
Cambridgeshire	East Anglia	5.2	31.9
Cheshire	North West	4.4	36.7
Cleveland	North	5.2	30.1
Cornwall	South West	4.6	34.4
Cumbria	North	5.4	34.4
Derbyshire	East Midlands	5.4	34.4
Devon	South West	3.8	30.6
Dorset	South West	4.3	27.1
Durham	North	5.2	29.4
East Sussex	South East	4.3	31.1
Essex	South East	5.1	34.6
Gloucestershire	South West	4.8	28.9
Greater London	South East	4.7	32.9
Greater Manchester	North West	5.2	32.6
Hampshire	South East	5.2	32.1
Hereford and Worcester	West Midlands	4.7	34.9
Hertfordshire	South East	4.8	30.6
Humberside	Yorkshire and Humberside	5.2	32.4
Isle of Wight	South East	4.8	34.6
Kent	South East	4.4	36.2
Lancashire	North West	4.8	36.7
Leicestershire	East Midlands	4.8	29.4
Lincolnshire	East Midlands	5.1	34.9
Merseyside	North West	5.2	34.1
Norfolk	East Anglia	4.8	38.4
Northamptonshire	East Midlands	5.0	31.1
Northumberland	North	5.8	33.4
North Yorkshire	Yorkshire and Humberside	5.3	33.4
Nottinghamshire	East Midlands	5.2	36.2
Oxfordshire	South East	4.5	31.6
Shropshire	West Midlands	4.7	31.6
Staffordshire	West Midlands	5.0	36.9
Somerset	South West	4.7	38.7
South Yorkshire	Yorkshire and Humberside	5.3	30.1
Suffolk	East Anglia	5.2	33.9
Surrey	South East	4.3	29.4
Tyne and Wear	North	5.4	31.6
Warwickshire	West Midlands	4.7	33.1
West Midlands	West Midlands	4.8	33.1
West Sussex	South East	4.6	41.9
West Yorkshire	Yorkshire and Humberside	5.0	29.6
Wiltshire	South West	4.6	34.6

Combining counties alphabetically (that is, randomly or semi-randomly) is not, of course, customary or very logical. In geographical research it normally makes better sense to combine areas that border one another. Random groupings of areas tend to be of little interest. When the counties shown in Table 18.5 are combined into the eight regions used for economic planning purposes (The South West, The South East, East Anglia, The East Midlands, The West Midlands, The North West, Yorkshire and Humberside, and The North) the correlation between the wheat and potato yields changes sign, becoming -0.57. It becomes still more negative (-0.92) if the regions are combined to form four provinces (Northern England, The Midlands, East Anglia and Southern England).

When X and Y values for separate periods of time are combined, the correlation is liable to change in just the same way as with areal data. A correlation based on average values for months will tend to be higher than one based on the values for individual days. If average values for years are substituted, the correlation will tend to be still further improved. With certain pairs of variables the correlation can be made to change all the way from 0 to +1 (or -1) depending upon the size of the groupings employed in its calculation. This somewhat disconcerting phenomenon may be termed the *problem of modifiable items*.[1] The items involved (counties, years, or whatever) are entirely arbitrary, and can be split or combined as desired. No correlation can be correctly interpreted unless the size of the items is taken into account. A clear distinction exists between items that are modifiable and those that are non-modifiable. People, households, individual farms or factories are separate entities that cannot be meaningfully split or combined, except in special circumstances. They are essentially non-modifiable.

The problem of modifiable items has been much discussed by statisticians and geographers (Yule and Kendall, 1950; Robinson, 1956; Thomas and Anderson, 1965; Curry, 1966; Clark and Avery, 1976; Openshaw and Taylor, 1979 and 1981; Cliff and Ord, 1981), but great uncertainty persists about what controls the variation in the correlation coefficient when the items are split or combined. There seem to be several different effects that contribute to the variation, and the interaction of these effects is still not properly understood. The first effect concerns what may be called errors. As already explained, the individual values of Y may vary to some extent because of the intervention of variables other than X that have not been examined. Unexamined variables may also cause the individual values of X to vary. It is thus possible to think of the individual values of Y (or X) as being subject to a varying amount of "error" which may be either positive or negative. If the individual values are averaged or aggregated, the positive and negative errors may cancel out to some extent, so improving the correlation. The wheat yields in one county, for example, might be increased because of favourable weather

[1]Yule and Kendall in their textbook (1950, p.312) refer to modifiable units, but the term "units" is best avoided since it tends to direct attention to the units of measurement rather than to the items being measured. Modifications to the units of measurement are not what are being considered in the present section although they can cause problems in correlation analysis.

conditions that happen to affect that county alone. The yields in a second county, however, might be depressed because of a localised outbreak of fungus disease. If the two counties are combined, the errors caused by the weather and the disease will cancel out to some degree and this will tend to improve the correlation.

A second effect of averaging or aggregating X and Y values is that the range of X values may be reduced more than the range of the Y values, which will tend to decrease the size of the correlation and thus work against the first effect. The two opposing effects may roughly balance out, leaving the correlation virtually unchanged, or one may predominate. Much depends on the detailed "structure" of the data and the way in which the items are averaged or aggregated. If positive autocorrelation is present and contiguous items are combined, the correlation coefficient will be less variable than if autocorrelation is absent or the items are grouped randomly (Openshaw and Taylor, 1979). Exactly how the autocorrelation affects the coefficient remains to be further studied. One might have expected that the errors would tend to reinforce one another when contiguous items are combined since the errors would tend to be of similar sign and amount. The autocorrelation ought, therefore, to increase the variability of the correlation coefficient. In fact, it works in reverse, which is somewhat puzzling.

9. HETEROGENEOUS POPULATIONS AND SAMPLES

Populations and samples consisting of different groups of items are said to be heterogeneous. They raise many problems in correlation analysis.

Consider, for example, the hypothetical population shown in Fig. 18.4A. The data points are so arranged that the correlation coefficient works out at exactly zero. One would ordinarily suppose that the two variables X and Y are uncorrelated. In actual fact they are highly correlated. The population is heterogeneous in that it combines two distinct groups of items (or sub-populations) identified by the open and solid circles. If the two groups of items are analysed separately, a strong correlation is apparent, which is positive in the case of the items shown by the solid circles (Fig. 18.4B) and negative in the case of the items shown by the open circles (Fig. 18.4C). The amount of positive correlation exactly balances the amount of negative correlation, and therefore, the coefficient for the population as a whole is zero. There can be no doubt, however, that the population coefficient is seriously misleading. The heterogeneity of the population effectively conceals the linear relationships that exist between X and Y.

Heterogeneous populations and samples can give rise to correlation coefficients that are not only misleadingly low, as in the case of Fig. 18.4A, but also misleadingly high. Figure 18.4D shows a hypothetical population consisting of three distinct groups of items. No correlation exists for the individual groups, but when they are combined a strong positive correlation is developed, which is in fact an illusion.

Even if only one item in a population or sample is different from the rest, the value of the correlation coefficient may be greatly affected. Figure 18.4E shows a hypothetical population that contains one markedly aberrant item. For the population as a whole the correlation coefficient is 0.80, suggesting a fairly close relationship between X and Y. If the aberrant item is excluded from the calculations, however, the correlation coefficient drops to zero. In the case of real data, it would be obviously important to check whether the aberrant item is correctly included in the population. If, in fact, it belongs to a different population, it ought not to be used in calculating the correlation coefficient. And if it is correctly included, then arguably the coefficient is not worth calculating, since its value is so greatly dependent on the single aberrant item. If the coefficient is calculated, then the reader should be warned that its value is greatly affected by the inclusion of the one aberrant item.

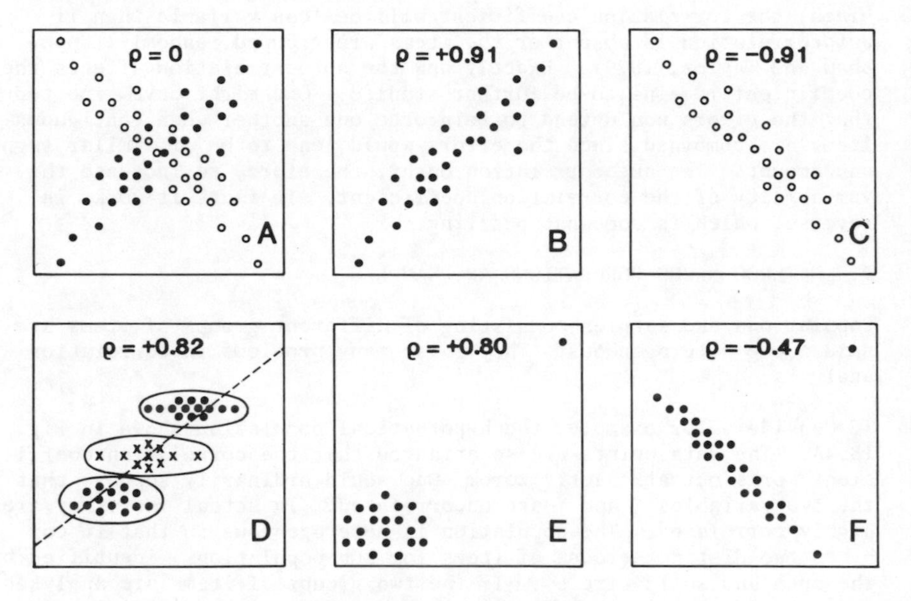

Fig. 18.4 - Four hypothetical populations exhibiting heterogeneity

A. Scatter diagram showing zero correlation. Two different groups of items have been accidentally combined and treated as one population.

B. and C. Scatter diagrams for each of the two groups of items shown in A. The group shown in B exhibits high positive correlation, the group in C high negative correlation.

D. Scatter diagram showing misleadingly high correlation resulting from the combination of three distinct groups of items in which the correlation is zero.

E. Scatter diagram with one aberrant item that is solely responsible for the correlation.

F. Scatter diagram with one aberrant item that greatly decreases the correlation.

Figure 18.4F illustrates the reverse situation. The correlation is decreased from -0.97 to -0.47 because of the presence of a single aberrant item far removed from the others on the scatter diagram. In the case of real data one would need to check whether the item is correctly included in the population.

The effects of heterogeneity are complicated enough, but it is the definition of heterogeneity that poses most difficulty in correlation analysis. In a sense all populations are heterogeneous. Human populations, for example, consist of males and females, children and adults, married and unmarried persons. A population of glaciers is likely to include some that are fast flowing, some that are slow, some that are advancing, some that are retreating. One can always ruin a perfectly good correlation by splitting a population (or sample) into a number of parts that individually show little or no correlation. The issue is not whether a particular population is heterogeneous, but whether it is sufficiently homogeneous to be treated as a single entity. Often it is not possible to arrive at a definite conclusion. This is the case in the example that follows.

Beneath many sea cliffs there is a gently sloping platform of rock that has been carved by the waves. A.J. Trenhaille (1974) visited a number of coastal sites in England and Wales where platforms are well developed, and at each site determined the tidal range and mean gradient of the platform. Table 18.6 gives the tidal range and gradient of 11 platforms cut in Liassic rocks (limestone and shale) in Yorkshire and Glamorgan. The correlation coefficient for the 4 Yorkshire platforms is 0.68 whereas that for the 7 platforms in Glamorgan is only 0.46. By contrast, the coefficient for all 11 platforms is an impressive 0.96.

TABLE 18.6 - TIDAL RANGE AND GRADIENTS OF WAVE-CUT PLATFORMS, YORKSHIRE AND GLAMORGAN

Location *1-4 Yorkshire* *5-11 Glamorgan*	*Tidal Range (m)* *X*	*Platform gradient (mins)* *Y*	X^2	Y^2	*XY*
1. Skinningrove	4.57	30	20.8849	900	137.10
2. Staithes	4.57	36	20.8849	1296	164.52
3. Robin Hood's Bay	4.69	45	21.9961	2025	211.05
4. Scarborough	5.18	45	26.8324	2025	233.10
5. Nash-St. Donats	9.14	135	83.5396	18225	1233.90
6. Tresilian	9.30	180	86.4900	32400	1674.00
7. Col-Hugh	9.45	150	89.3025	22500	1417.50
8. Monknash	9.60	120	92.1600	14400	1152.00
9. Summerhouse	10.06	180	101.2036	32400	1810.80
10. Porthkerry	10.52	180	110.6704	32400	1893.60
11. Lavernock	10.88	165	118.3744	27225	1795.20
Totals	87.96	1266	772.3388	185796	11722.77

Source: Trenhaille (1974) and personal communication.

In Fig. 18.5 the data are set out in the form of a scatter diagram. The Yorkshire and Glamorgan platforms are totally separate, both tidal range and platform gradient being very much less in Yorkshire than in Glamorgan. One could easily argue that the two counties represent distinct populations and so require separate analysis. To judge from the coefficients for each county (0.68 and 0.46) tidal range and platform gradient would appear to be no more than moderately correlated. Since the number of observations is small, sampling error is likely to be large, and hence there is a distinct possibility that the two variables are uncorrelated (neither correlation coefficient is significant at the 5% level).

One could as easily argue, however, that the platforms belong to the same population since they are all cut in Liassic rock. The size of the overall coefficient (0.96) suggests that the variables may be extremely closely correlated, though of course there is always sampling error to consider. With 9 degrees of freedom the correlation is significant at well beyond the 1% level, and the 95% confidence limits are 0.86 to 0.99.

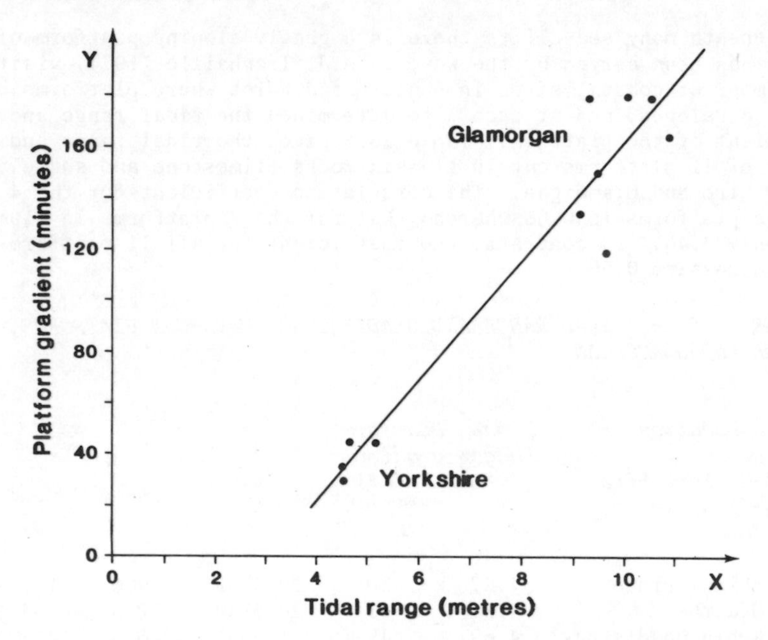

Fig. 18.5 - Tidal range and gradients of 11 wave-cut platforms

Whether the overall coefficient is accepted as a realistic estimate of the degree of correlation depends on whether one accepts the hypothesis of a single population. There is clear evidence of heterogeneity, but can it be safely ignored in calculating the correlation coefficient? Apart from running a significance test to check the hypothesis that the two samples are drawn from the same bivariate normal population, little can be done to arrive at a reliable answer. And the test is not much help:

1. Null hypothesis: the two samples are drawn from populations with the same correlation coefficient. In other words the correlation be-

tween tidal range and platform gradient is the same in the two areas. The correlation coefficients of the two samples differ purely because of chance.

2. $\mathbf{Z}_1 = \frac{1}{2} \log_e \left(\frac{1 + 0.68}{1 - 0.68}\right) = 0.829$ approximately.

$\mathbf{Z}_2 = \frac{1}{2} \log_e \left(\frac{1 + 0.46}{1 - 0.46}\right) = 0.497$ approximately.

3. $|\mathbf{Z}_1 - \mathbf{Z}_2| = 0.332.$

4. $\sigma_{\mathbf{Z}_1 - \mathbf{Z}_2} = \sqrt{\left[\frac{1}{(4 - 3)} + \frac{1}{(7 - 3)}\right]} = \sqrt{1.25} = 1.118$ approximately.

5. $Z = 0.332/1.118 = 0.30$ approximately.

6. $P = 0.76$ (Table B).

7. The null hypothesis cannot be rejected since the value of P is high. The difference between the correlation coefficients may well be due to chance sampling.

Although the significance test does not cast doubt on the null hypothesis, this does not mean that the hypothesis is proven. Since the samples are small they would have to be drawn from very different populations to have much chance of being declared significantly different. And even if the samples are drawn from the same population there is nothing to show that it is even approximately bivariate normal, so the significance test may not be valid. The degree to which the two variables are correlated is thus problematic. If the two samples are analysed separately, it is quite possible that there is no correlation. On the other hand, if the two samples are lumped together, it is possible that there is a quite high correlation.

Uncertainties of this kind are commonly encountered in research investigations and have done much to bring correlation analysis into disrepute.

10. ERRORS OF MEASUREMENT AND ESTIMATION

Correlation coefficients are usually calculated on the assumption that the measurements of X and Y are entirely accurate. The sad truth is that errors of measurement are virtually unavoidable. If the errors are random, they will tend to produce a more disorganised array of points on the scatter diagram than would be the case if there were no errors. The correlation coefficient will be reduced below its true value. Systematic errors may also reduce the correlation coefficient, although they are equally likely to increase it.

A single serious error will result in an aberrant point on the scatter diagram far removed from the other points (Fig. 18.4E and F). Usually such an error is detectable, but if overlooked it may seriously affect the value of the correlation coefficient, as discussed in the previous section.

Sometimes, it is necessary to make do with estimated values for one or both variables. The estimates may differ from the true values and render the correlation coefficient unreliable. Errors of estimation are no different in principle from errors of measurement. For example, if the errors of estimation are random, the correlation coefficient will be reduced, thus understating the degree of relationship between X and Y. Systematic errors of estimation may increase or decrease the correlation coefficient.

19 Linear Regression

INTRODUCTION

The last two chapters have discussed the calculation and interpretation of correlation coefficients, which are useful, if sometimes misleading, measures of linear relationship between variables. As already explained, the size (absolute value) of a correlation coefficient indicates how closely the data points approximate to a sloping straight line. The sign of the coefficient indicates the direction of slope of the line.

It is, of course, always important to be able to calculate the degree (and direction) of correlation between two variables. But often one is interested not so much in evaluating the correlation as in finding a line that approximates reasonably closely to the data points and provides a convenient summary or generalisation of the way the values of one variable are related to the values of the other. The line can be added to the scatter diagram to emphasise the position of the data points. It can also be written down as an equation which specifies the precise position of the line in relation to the horizontal and vertical axes.

This chapter is concerned with methods of fitting lines when one variable is cause and the other effect. Except where stated otherwise, it will be assumed that the causal variable is plotted on the horizontal axis with the label X, while the variable that is effect is plotted on the vertical axis with the label Y. Methods of fitting lines when the two variables interact, or are both dependent on another variable, or group of variables, are discussed in the companion volume.

EQUATION OF A STRAIGHT LINE

Consider the pairs of values of X and Y set out in Table 19.1 and graphed in Fig. 19.1. There is clearly a simple rule relating the value of Y to the value of X. The value of Y corresponding to each value of X is exactly 7.5 plus half X. Written in the form of an equation, the rule for obtaining Y from X is $Y = 7.5 + 0.5X$. From a mathematical point of view the equation is more interesting than the actual values of X and Y. It constitutes an abstraction or mathematical law. At the same time it provides a concise and exact description of the relationship between the two variables. The actual values hardly matter, since the equation is obeyed without exception.

As Fig. 19.1 demonstrates, the pairs of values of X and Y all fall precisely on a straight line. The line is the graphical expression of the equation $Y = 7.5 + 0.5X$. Y is the height of the

TABLE 19.1 - DATA TO ILLUSTRATE A PERFECT LINEAR RELATIONSHIP

X	1.0	2.0	3.0	4.0	5.0	6.0	7.0	8.0	9.0	10.0	11.0	12.0
Y	8.0	8.5	9.0	9.5	10.0	10.5	11.0	11.5	12.0	12.5	13.0	13.5

line at any point X on the horizontal axis. It can be seen that the line crosses the vertical axis at 7.5 units above the origin. The height of the line at this point is determined by the magnitude of the first constant on the right-hand side of the equation. Thus, when X is made equal to zero, the height of the line is

$$Y = 7.5 + 0.5(0) = 7.5$$

The multiplying constant 0.5 in the equation determines the gradient or slope of the line. An increase of one unit in X results in an increase in Y of 0.5. The multiplying constant 0.5 therefore measures the rate of change of Y with respect to X.

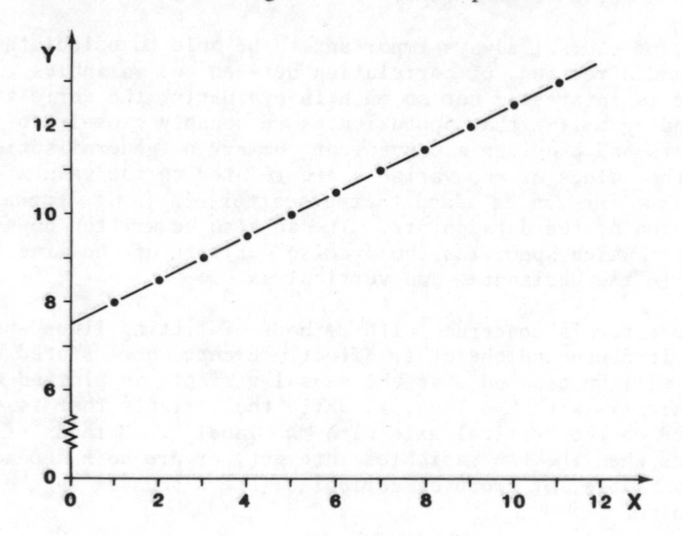

Fig. 19.1 - Plot of the (X,Y) values listed in Table 19.1. The vertical scale has been broken between 0 and 5 to save space.

This discussion of one particular equation and the corresponding straight-line graph can be readily generalised. Any equation of the type $Y = a + bX$ where a and b are numbers (*constants*) plots as a straight line on a scatter diagram. Conversely, any straight line on a scatter diagram can be represented by an equation of the type $Y = a + bX$. In every case, Y is the height of the line corresponding to a point X on the horizontal axis. The constant a determines the *Y intercept* or height of the line at $X = 0$, in other words it fixes the point at which the line cuts the vertical or Y axis. The constant b determines the *gradient* or slope of the line, in other words the increase in Y when X increases by one unit.

Figure 19.2 presents examples of straight lines with different values of a and b. If a is positive, the line intercepts the Y axis a units above the origin. If a is zero, the line passes through the origin. The line intercepts the Y axis a units below the horizontal axis if a is negative. The sign of b indicates the direction

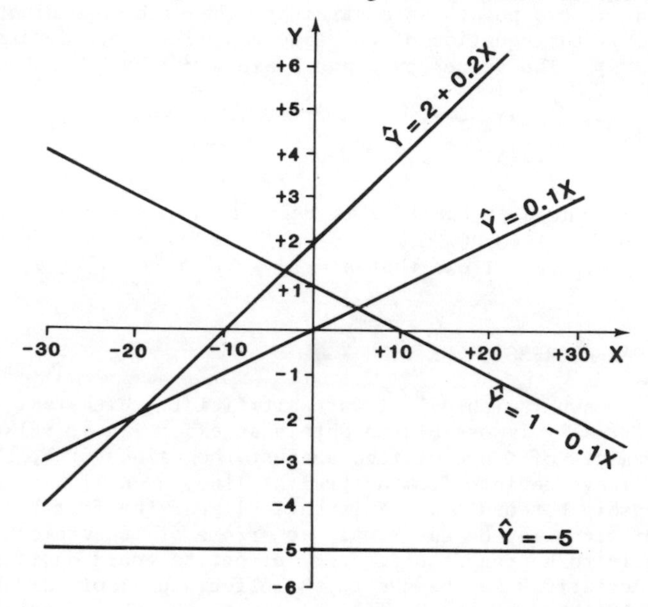

Fig. 19.2 – Some examples of straight-line equations.

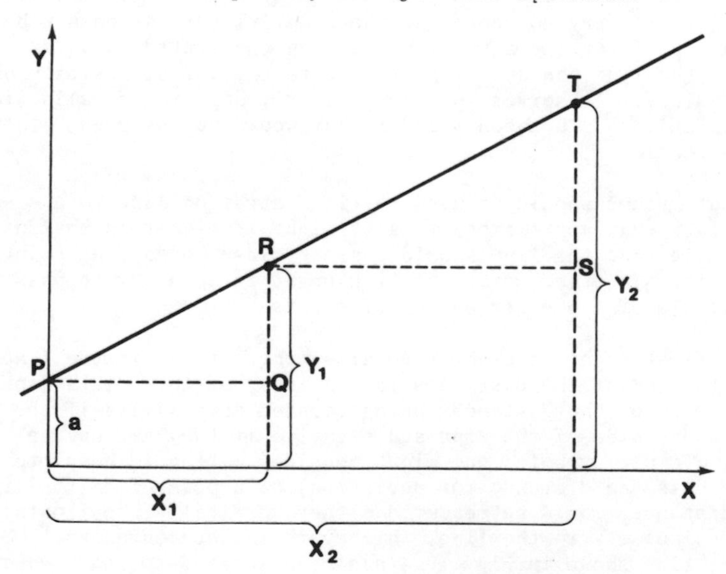

Fig. 19.3 – Diagram to show the method of calculating the equation of the straight line PRT joining two data points, R and T.

of slope of the line. If b is positive, the line slopes upwards from left to right. If b is negative, the line slopes downwards from left to right. The absolute value of b (that is, the value disregarding the sign) measures the amount of slope. If the absolute value is large, the line slopes steeply; if it is small, the slope is gentle. A zero value of b indicates a horizontal line.

Given any two points on a straight line with co-ordinates (X_1, Y_1) and (X_2, Y_2) the equation of the line can be at once determined (Fig. 19.3). The values of a and b are given by

$$a = Y_1 - X_1 \left(\frac{Y_2 - Y_1}{X_2 - X_1} \right) \quad \text{and} \quad b = \frac{Y_2 - Y_1}{X_2 - X_1}$$

The gradient b of the line is equal to TS/RS = $(Y_2 - Y_1)/(X_2 - X_1)$. It is also equal to RQ/PQ = $(Y_1 - a)/X_1$. Since $(Y_2 - Y_1)/(X_2 - X_1) = (Y_1 - a)/X_1$ it follows that $a = Y_1 - X_1[(Y_2 - Y_1)/(X_2 - X_1)]$.

THE LINEAR REGRESSION OF Y ON X

The data shown in Table 19.1 were artificial. With real variables it is virtually impossible to obtain an exact linear relationship. If the values of X are plotted against the values of Y, the points almost always deviate from a straight line, even if the cause-effect relationship between X and Y is basically of the form $Y = a + bX$. The deviations may be due simply to errors of measurement, which are liable to happen despite every effort to guard against them. Or the deviations may be due to the effects on Y of variables other than X. Either way the deviations may be thought of as chance disturbances that are of no great consequence. It is natural, therefore, to try to represent the underlying relationship between X and Y by inserting a straight line on the scatter diagram, even though the line can only approximate to the actual position of the data points. It serves as a convenient model or generalisation of the relationship between X and Y that would be observed "other things being equal".[1]

What method should be used to fit a straight line to a scatter of points that approximate to a straight line? It is obviously desirable that the line should pass as close to as many points as possible. In other words the line ought in some way to pass through the middle of the distribution of points.

At first sight it might seem all-important to fit the line so that the sum of the distances (deviations) of the points from the line is zero, the distances being counted as positive if the points are on one side of the line and negative if they are on the other side. To use symbols: one might consider making Σd equal to zero, where d is the distance (or deviation) of a point from the line. The distances could be measured either vertically, horizontally, or perpendicularly to the line, whichever is most convenient.

The line shown in Fig. 19.4 has been located so that the sum of

[1] It is assumed in this and later sections that confounding has not occurred and that the data points can be taken more or less at their face value (apart from possible unrepresentativeness due to sampling error). Any measurement errors in X are considered to be so small that they can be safely neglected. It is also assumed that the observed relationship is not due to spurious or nonsense correlation.

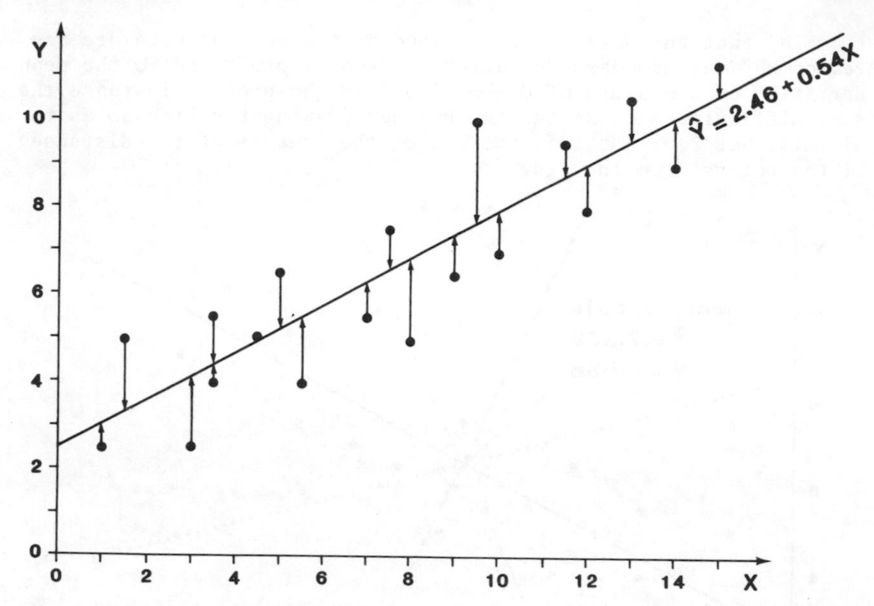

Fig. 19.4 - Vertical deviations of sample points about a fitted line. Arrows indicate deviations. One point happens to lie almost on the line and its deviation is too small to show. The line is drawn so that the algebraic sum of the vertical deviations is zero, that is to say the deviations above the line exactly balance those below.

the distances of the points from the line is zero. (Although only the vertical distances are depicted the sum is zero even if the distances are measured horizontally or perpendicularly.) There can be no doubt that the line provides a "good fit" or approximation to the data points. Unfortunately, the line is only one of an infinite number of lines that can be drawn for which the sum of the distances is zero. It can be proved that any straight line that passes through the mean of the data (i.e. the point $\overline{X}$, $\overline{Y}$ in the case of a sample, or μ_x, μ_y in the case of a population) possesses the property that $\Sigma d = 0$. In Fig. 19.5, for example, a series of straight lines are shown passing through the mean of the data, and in each case the positive distances exactly balance (cancel out) the negative distances. Only one of the lines shown provides an adequate fit to the data; the rest fail to slope in the right direction or slope too steeply.

What is clearly required is a method of fitting a straight line to a scatter of points that provides a unique solution. The best method might seem to be to minimise the sum of the distances of the points from the line. To revert to symbols: one might consider making $\Sigma|d|$ a minimum, where $|d|$ is the absolute distance of a point from the line, that is to say the distance ignoring sign. There is only one line possessing this property (Fig. 19.6), and off-hand it would appear to be entirely satisfactory. It passes closer to the data points than any other line one can construct and in this sense provides the best possible fit to the data. Unfortunately, the modulus sign $|\;|$ is difficult to manipulate algebraically. It will be recalled that the difficulties raised by the modulus sign are one reason why the standard deviation is usually preferred to the mean deviation as a measure of dispersion. In the present instance the same difficulties cause one to consider drawing the line so as to minimise not $\Sigma|d|$ but Σd^2, the sum of the squares of the distances of the points from the line.

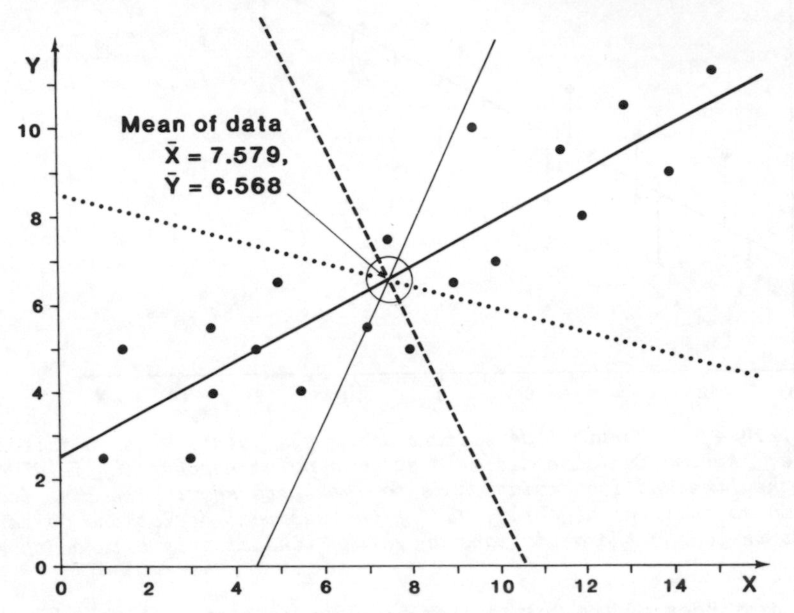

Fig. 19.5 - A selection of lines for which $\Sigma d = 0$. Any straight line that passes through the mean of a given set of data points possesses the property that the algebraic sum of the deviations is zero.

There is a further reason for not seeking to minimise $\Sigma|d|$. The line that minimises Σd^2 is one of the lines for which $\Sigma d = 0$ and it always passes through the mean of the data points which is a useful characteristic. The line that minimises $\Sigma|d|$ does not pass through the mean except in special circumstances and it possesses no particularly useful characteristics.

The best method, then, of fitting a line to a scatter of points is to make the sum of the squared distances a minimum. But in which direction should the distances be measured? The answer must depend on the nature of the causal relationship between X and Y. If X is cause and Y is effect (in other words, if X is the independent variable and Y is the dependent variable) it is clearly the vertical distances that are most relevant. If, contrary to the usual symbolism, X were to depend on Y, then it would be appropriate to consider the horizontal and not the vertical distances. This possibility will be discussed later in the chapter. For the moment it will be assumed that X causes Y and not the reverse. Other possibilities are considered in the companion volume, including minimising the sum of the squares of the perpendicular distances of the points from the line in cases where X and Y interact, or are both dependent on another variable or group of variables.

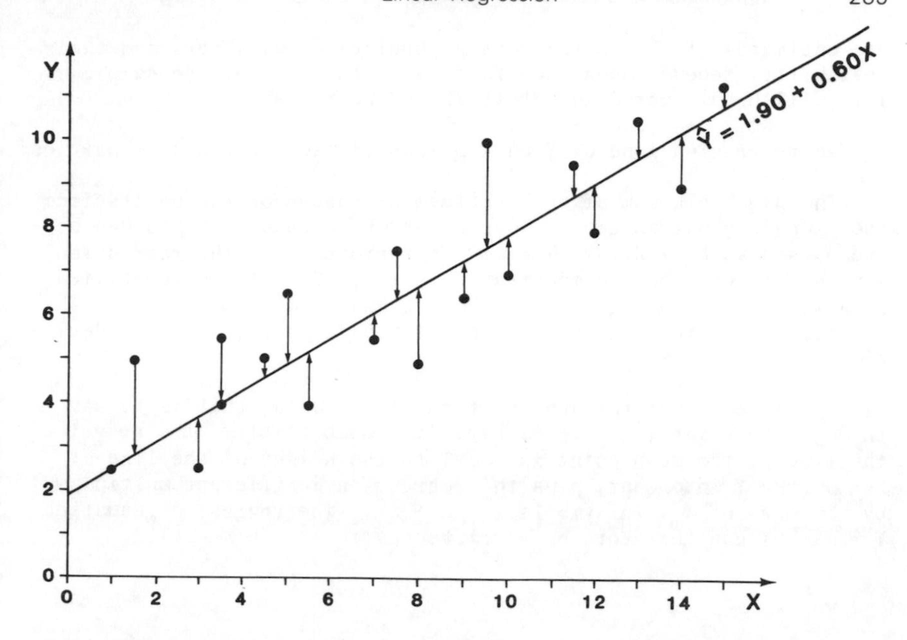

Fig. 19.6 - A straight line fitted to a scatter of points so that $\Sigma|d|$*, the sum of the vertical deviations, is a minimum.*

A line that is fitted to a set of data points so as to minimise the sum of the squares of the vertical distances is said to be fitted according to the *method of least squares*. The line is called the *regression line of Y on X* or *linear regression of Y on X*. The equation of the line is referred to as the *linear regression equation of Y on X* or, more shortly, the *regression equation of Y on X*. In the case of a sample, the equation can be written as

$$\hat{Y} = a + bX$$

where $\hat{Y}$ is the value of Y calculated from the value of X, that is to say the height of the line at the point X. The ^ or "hat" symbol is introduced to distinguish the calculated value from the actual value which may be very different. The constant a on the right-hand side of the equation is referred to as the *Y intercept* or, more simply, as the *intercept* of the regression line. The gradient b is called the *regression coefficient of Y on X*.

In the case of a population it is usual to employ a slightly different symbolism, the regression equation being written

$$\mu = \alpha + \beta X$$

where μ carries the same meaning as $\hat{Y}$ in the previous equation. Likewise α is the equivalent of a, and β is the equivalent of b. It is convenient to employ different symbols for samples and populations since samples are subject to sampling error. $\hat{Y}$, a and b are estimates of the corresponding population parameters and their reliability depends upon such factors as the size of the sample. This will be discussed in detail in the next chapter.

The regression line of Y on X possesses two important properties:

1. The algebraic sum of the vertical distances of the points from the line is equal to zero. In the case of a sample,[1] this can be symbolised as $\Sigma d = \Sigma(Y - \hat{Y}) = 0$. This property of the regression line has already been demonstrated in Fig. 19.4, the unidentified line shown there being in fact the regression line of Y on X. The positive deviations (distances) exactly balance the negative deviations.

2. The line passes through the mean of the data, that is to say through the point $(\bar{X}, \bar{Y})$. As Fig. 19.7 demonstrates, the height of the line at the mean point is equal to the height of the line at X = 0 (the Y intercept) plus the regression coefficient multiplied by the mean of X. In symbols $\bar{Y} = a + b\bar{X}$. The regression equation $\hat{Y} = a + bX$ can therefore be rewritten as $\hat{Y} = \bar{Y} + b(X - \bar{X})$.

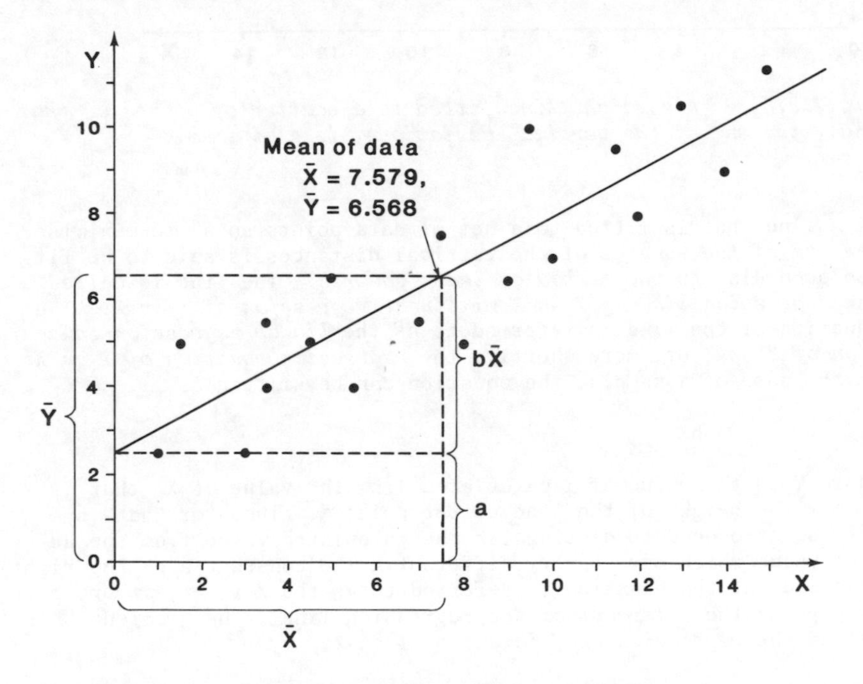

Fig. 19.7 - Diagram to show that $\overline{Y} = a + b\overline{X}$

Usually the scatter of points on a graph is too great to permit the accurate fitting of a regression line by eye. The regression line is best fitted by calculating its equation directly from the individual values of X and Y. The regression coefficient b is given

[1] To save space the symbolism adopted here and later in this section refers only to a sample. It is left to the reader to make the appropriate modifications for a population.

by the equation

$$b = \frac{n\Sigma XY - \Sigma X\Sigma Y}{n\Sigma X^2 - (\Sigma X)^2}$$

and the Y intercept (or a) by the equation

$$a = \frac{\Sigma Y}{n} - \frac{b\Sigma X}{n}$$

n being the number of pairs of values of X and Y (the number of items in the sample) and ΣXY the sum of the products of the X and Y values. The two equations follow directly from the requirement that Σd^2 be made a minimum. No assumptions about the distribution of the distances need be made.[1]

The equations for a and b can be written in several different ways. Because $\Sigma X/n = \bar{X}$ and $\Sigma Y/n = \bar{Y}$ the first equation becomes

$$a = \bar{Y} - b\bar{X}$$

while the second equation becomes

$$b = \frac{\Sigma XY - \dfrac{\Sigma X\Sigma Y}{n}}{\Sigma X^2 - \dfrac{(\Sigma X)^2}{n}} = \frac{\Sigma XY - n\bar{X}\bar{Y}}{\Sigma X^2 - n(\bar{X})^2}$$

$$\begin{aligned}\text{Now } \Sigma[(X - \bar{X})(Y - \bar{Y})] &= \Sigma XY - \bar{Y}\Sigma X - \bar{X}\Sigma Y + n\bar{X}\bar{Y} \\ &= \Sigma XY - n\bar{X}\bar{Y} - n\bar{X}\bar{Y} + n\bar{X}\bar{Y} = \Sigma XY - n\bar{X}\bar{Y}\end{aligned}$$

and $\Sigma(X - \bar{X})^2 = \Sigma X^2 - n(\bar{X})^2$

$$\text{Hence } b = \frac{\Sigma[(X - \bar{X})(Y - \bar{Y})]}{\Sigma(X - \bar{X})^2}$$

THE RELATIONSHIP BETWEEN THE CORRELATION AND REGRESSION COEFFICIENTS

The equation for the correlation coefficient closely resembles the equation for the regression coefficient b. Thus, for a sample

$$r = \frac{n\Sigma XY - \Sigma X\Sigma Y}{\sqrt{[n\Sigma X^2 - (\Sigma X)^2]}\sqrt{[n\Sigma Y^2 - (\Sigma Y)^2]}} \quad \text{and} \quad b = \frac{n\Sigma XY - \Sigma X\Sigma Y}{n\Sigma X^2 - (\Sigma X)^2}$$

---

[1] The equations can be derived using simple algebra. The equation of the line is $\hat{Y} = a + bX$, where a and b are constants that minimise the sum of the squared deviations of the points from the line. This sum is $\Sigma d^2 = \Sigma(Y - \hat{Y})^2$. Substituting $(a + bX)$ for $\hat{Y}$ in the summation gives $\Sigma(Y - a - bX)^2$, which on expansion becomes $\Sigma Y^2 - 2a\Sigma Y - 2b\Sigma XY + na^2 + 2ab\Sigma X + b^2\Sigma X^2$. The expansion can be reordered as $na^2 + 2(b\Sigma X - \Sigma Y)a + [\Sigma Y^2 - 2b\Sigma XY + b^2\Sigma X^2]$, which is a quadratic in a. The quadratic takes a minimum value when $a = -[2(b\Sigma X - \Sigma Y)]/2n = (\Sigma Y/n) - (b\Sigma X/n)$. The expansion can also be reordered as a quadratic in b, viz. $(\Sigma X^2)b^2 + 2(a\Sigma X - \Sigma XY)b + [\Sigma Y^2 - 2a\Sigma Y + na^2]$. The minimum value of this quadratic is $b = -[2(a\Sigma X - \Sigma XY)]/2\Sigma X^2 = (\Sigma XY - a\Sigma X)/\Sigma X^2$. Substituting for a in the last expression gives $b = [n\Sigma XY - \Sigma X\Sigma Y]/[n\Sigma X^2 - (\Sigma X)^2]$.

The equation for r becomes, after a little rearrangement

$$r = \frac{n\Sigma XY - \Sigma X\Sigma Y}{n\Sigma X^2 - (\Sigma X)^2} \left\{\frac{\sqrt{[n\Sigma X^2 - (\Sigma X)^2]}}{\sqrt{[n\Sigma Y^2 - (\Sigma Y)^2]}}\right\}$$

The term in braces (curly brackets) is the ratio of the standard deviation of X to that of Y, namely s_x/s_y. The equation for r can therefore be written

$$r = b\left(\frac{s_x}{s_y}\right)$$

The corresponding equation for b is

$$b = r\left(\frac{s_y}{s_x}\right)$$

If the standard deviations of X and Y are identical, then r equals b. The equation of the regression line Y on X is $\hat{Y} = a + bX$. It has already been shown that this equation is equivalent to $\hat{Y} - \overline{Y} = b(X - \overline{X})$. Substituting for b, one obtains the equation

$$\hat{Y} - \overline{Y} = r\left(\frac{s_y}{s_x}\right) (X - \overline{X})$$

In the case of a population the corresponding equations are

$$\rho = \beta\left(\frac{\sigma_x}{\sigma_y}\right), \ \beta = \rho\left(\frac{\sigma_y}{\sigma_x}\right) \ \text{and} \ \mu - \mu_y = \rho\left(\frac{\sigma_y}{\sigma_x}\right) \ (X - \mu_x)$$

where μ_x is the mean of all the X values in the population and μ_y is the mean of all the Y values.

THE CALCULATION OF THE REGRESSION EQUATION, AN EXAMPLE

The data for the discharge and solute concentration of the small stream in Central Wales (Chapter 17) will be used to illustrate the calculation of the regression equation $\hat{Y} = a + bX$. Although the data have been set out previously (Table 17.1), they are reproduced in Table 19.2 to save referring back.

From the totals shown in the table the values of a and b can be obtained as follows

$$b = \frac{n\Sigma XY - \Sigma X\Sigma Y}{n\Sigma X^2 - (\Sigma X)^2} = \frac{11(2647.55) - (56.7)(484.3)}{11(381.95) - (56.7)^2} = \frac{1663.24}{986.56}$$

$$= 1.6859 = 1.7 \text{ approximately.}$$

$$a = \frac{\Sigma Y}{n} - \frac{b\Sigma X}{n} = \frac{484.3}{11} - \left(\frac{1663.24}{986.56}\right)\left(\frac{56.7}{11}\right) = 44.02727 - 8.69004$$

$= 35.33723$ or 35.3 approximately.

Hence the regression equation is approximately $\hat{Y} = 35.3 + 1.7X$.

TABLE 19.2 - DISCHARGE AND CONCENTRATIONS OF DISSOLVED SOLIDS DURING RISING STAGES OF A SMALL STREAM

Date of visit (1971)	*Discharge of stream (litres/sec)* X	*Concentration of dissolved solids (mg/litre)* Y	X^2	Y^2	XY
June 3	0.4	34.2	0.16	1169.64	13.68
Oct. 7	2.2	40.4	4.84	1632.16	88.88
March 11	2.3	38.3	5.29	1466.89	88.09
Nov. 4	3.9	39.4	15.21	1552.36	153.66
Dec. 16	4.3	44.8	18.49	2007.04	192.64
Nov. 18	4.4	47.0	19.36	2209.00	206.80
Feb. 25	5.1	42.6	26.01	1814.76	217.26
Jan. 28	8.0	45.6	64.00	2079.36	364.80
Jan. 14	8.1	52.8	65.61	2787.84	427.68
Aug. 26	8.3	48.7	68.89	2371.69	404.21
Dec. 30	9.7	50.5	94.09	2550.25	489.85
	56.7	484.3	381.95	21640.99	2647.55

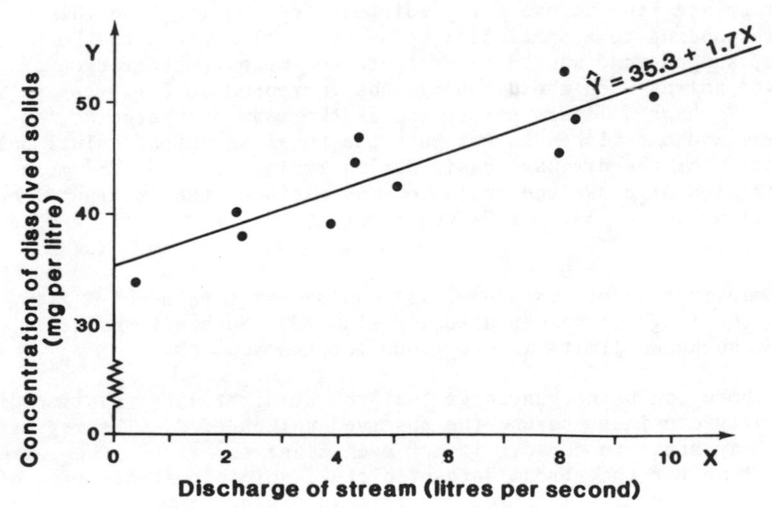

Fig. 19.8 - Linear regression of the concentration of dissolved solids on discharge for rising stages of a small stream in Central Wales

Figure 19.8 shows the regression line fitted to the data points. It can be seen that the points approximate fairly closely to the line. If it were not for chance errors of measurement, and the intrusive effects of other variables (e.g. changes in the temperature of the water, seasonal changes in the growth of plants), the points might perhaps fall exactly on a straight line. It is meaningful therefore to try to represent the "real" or underlying relationship between discharge and solute concentration by fitting a straight line to the scatter diagram even though the line is necessarily an approximation to the actual data points.

PREDICTION

Scientists are frequently concerned to predict (or estimate) the values of one variable from the values of another. The demographer attempts to predict the population in the year 2000 on the basis of current trends in the birth rate. The economic geographer is interested in predicting changes in regional employment resulting from improvements in communications, and in estimating future levels of energy consumption as standards of living increase. The climatologist who constructs a map of the variation of rainfall in a mountainous region may have measurements for only a few places. He predicts the rainfall at places where no measurements have been made from a knowledge of their altitude, it being a fact that rainfall is closely related to altitude. Likewise, the agricultural geographer who is concerned to predict average crop yields within a region may rely on the fact that yields are closely related to rainfall. Having estimated the average rainfall for farms within the region, he predicts average yields from knowledge of the way the yields are related to rainfall.

A regression line fitted to a scatter of points attempts to show the mean Y corresponding to each value of X. It is therefore the appropriate line to use for predicting (estimating) the value of Y corresponding to a particular value of X. To return to the stream data, suppose one wanted to estimate the mean concentration of dissolved solids when the discharge has increased to 5 litres per second. Perhaps 5 litres per second is the mean discharge of the stream and one wishes to estimate the total amount of solutional lowering of the drainage basin during rising stages. The mean concentration of dissolved solids on the basis of the straight line relationship $\hat{Y} = 35.3 + 1.7X$ works out at $35.3 + (1.7)(5) = 43.8$ mg per litre.

Sometimes an estimate of Y is required at a value of X lying beyond the range of values actually studied. Such *extrapolation* beyond the known limits is dangerous for two reasons:

1. There can be no guarantee that the straight-line relationship actually continues beyond the observed values of X. The regression line may start to curve. It may even cease to exist. The regression line for the stream data predicts a solute concentration of 35.3 when the discharge is zero, but this is plainly impossible. If the stream were to dry up, the concentration of dissolved solids would dwindle to zero, not 35.3. The regression line also suggests that the solute concentration increases steadily with discharge. There

must eventually come a point, however, when the concentration stops rising and levels out, or starts to fall, otherwise the stream would become a mud flow.

Problems of this sort arise repeatedly in economic and social forecasting where it is customary to plot variables against time on the X axis. One can seldom be sure how long a particular linear relationship (or trend) will continue. The economist Sir Alec Cairncross once wrote of the dangers of prediction:

> "A trend, to use the language of Gertrude Stein, is
> a trend is a trend.
> But the question is: will it bend?
> Will it alter its course,
> Through some unforeseen force,
> And come to a premature end?"

2. Even if the straight-line relationship holds good beyond the range of observed values, there is the difficulty that $\overline{Y}$ becomes less dependable as an estimator of Y the more the value of X deviates from the mean $\overline{X}$. The same is true of $\overline{Y}$ as an estimator of μ, the mean of Y at X in the parent population. These matters are taken up in detail in the next chapter.

The following section can be omitted on first reading.

*ESTIMATING REGRESSION EQUATIONS FROM GROUPED DATA

The regression coefficient can be estimated from grouped data using the formula

$$\beta = \frac{N\Sigma(FM_xM_y) - \Sigma(F_xM_x)\Sigma(F_yM_y)}{N\Sigma(F_xM_x)^2 - [\Sigma(F_xM_x)]^2}$$

for a population, or the formula

$$b = \frac{n\Sigma(fm_xm_y) - \Sigma(f_xm_x)\Sigma(f_ym_y)}{n\Sigma(f_xm_x)^2 - [\Sigma(f_xm_x)]^2}$$

for a sample. The symbols should be reasonably familiar since they have already been used in the formulas for estimating the correlation coefficient (see Chapter 17). f (or F) is the frequency in each cell of the correlation table, f_x (or F_x) is the frequency in each class of X, and f_y (or F_y) is the frequency in each class of Y. m_x (or M_x) is the mid-value of a cell or class on the X scale and m_y (or M_y) is the mid-value on the Y scale.

The Y intercept of a population regression line is given by

$$\alpha = \frac{\Sigma(F_yM_y)}{N} - \frac{\beta\Sigma(F_xM_x)}{N}$$

The corresponding formula for a sample regression line is

$$a = \frac{\Sigma(f_y m_y)}{n} - \frac{b\Sigma(f_x m_x)}{n}$$

These formulas for the regression coefficient and the Y intercept can necessitate much laborious calculation, particularly if there are many classes and the class intervals are not simple numbers. Provided the class intervals are equal, the calculations can be greatly simplified by making use of assumed means. For a population

$$\beta = \frac{i_y}{i_x}\left\{\frac{N\Sigma(FD_x D_y) - \Sigma(F_x D_x)\Sigma(F_y D_y)}{N\Sigma(F_x D_x{}^2) - [\Sigma(F_x D_x)]^2}\right\}$$

$$\alpha = \left[Y_0 + \frac{\Sigma(F_y D_y)i_y}{N}\right] - \beta\left[X_0 - \frac{\Sigma(F_x D_x)i_x}{N}\right]$$

and for a sample

$$b = \frac{i_y}{i_x}\left\{\frac{n\Sigma(fd_x d_y) - \Sigma(f_x d_x)\Sigma(f_y d_y)}{n\Sigma(f_x d_x{}^2) - [\Sigma(f_x d_x)]^2}\right\}$$

$$a = \left[Y_0 + \frac{\Sigma(f_y d_y)i_y}{n}\right] - b\left[X_0 - \frac{\Sigma(f_x d_x)i_x}{n}\right]$$

TABLE 19.3 - ESTIMATION OF b FROM GROUPED DATA

Dissolved solids (mg/litre) Y	*Discharge (litres/sec) X* 0-2.9'	3-5.9'	6-8.9'	9-11.9'	f_y	d_y	$f_y d_y$
50-54.9'			^{1} **2**	^{1} **4**	2	+2	+4
45-49.9'		^{1} **0**	^{2} **2**		3	+1	+3
40-44.9'	^{1} **0**	^{2} **0**			2	0	0
35-39.9'	^{1} **1**	^{1} **0**			2	-1	-2
30-34.9'	^{1} **2**				1	-2	-2
f_x	3	4	3	1	$\Sigma f_y = \Sigma f_x$ $= n = 11$		$\Sigma(f_y d_y)$ $= +3$
d_x	-1	0	+1	+2			
$f_x d_x$	-3	0	3	2	$\Sigma(f_x d_x)$ $= 2$		
$f_x d_x{}^2$	+3	0	+3	+4	$\Sigma(f_x d_x^2)$ $= 10$		$\Sigma(fd_x d_y)$ = **11**

where d_x (or D_x) is the deviation of a cell or class from X_0 (the assumed mean of X), d_y (or D_y) is the deviation of a cell or class from Y_0 (the assumed mean of Y), i_x is the class interval on the X scale and i_y is the class interval on the Y scale.

Although still notational monsters, these formulas are much easier to work with than the basic formulas. Table 19.3 illustrates the first steps in the estimation of b for the stream data. It closely resembles Table 17.4 which was constructed in order to estimate the correlation coefficient. The mean of X is assumed to be 4.5 litres/sec and the mean of Y 42.5 mg/litre. The value of fd_xd_y is shown in each cell in bold face and the summation fd_xd_y is shown in bold face in the lower right-hand corner of the table.

Substituting the appropriate totals in the formula for b gives

$$b = \frac{5}{3}\left[\frac{11(11) - 2(3)}{11(10) - (2)}\right] = \frac{5}{3}\left[\frac{115}{106}\right] = 1.8082\ldots$$

$$= 1.8 \text{ approximately.}$$

The value of a can be estimated as follows

$$a = \left[42.5 + \frac{3(5)}{11}\right] - 1.8082\left[4.5 - \frac{2(3)}{11}\right]$$

$$= [43.8636\ldots] - [7.1505\ldots]$$

$$= 36.7 \text{ approximately.}$$

The estimated regression equation is therefore $\hat{Y} = 36.7 + 1.8X$. The equation calculated earlier from the individual values of X and Y is $\hat{Y} = 35.3 + 1.7X$. In estimating a and b the assumption is made that the values of each variable are concentrated at the centre of each class or cell. This assumption is clearly unrealistic if the distributions of X and Y are markedly skewed and the class intervals are large. The distributions in the present instance are approximately symmetrical and therefore the estimates are close to the real values.

THE STANDARD ERROR OF ESTIMATE, $s_{y.x}$

Having calculated a regression line, it is important to observe how closely the line approximates to the data. Figure 19.9 compares two different populations that happen to have identical regression lines. In one case the points fall on or very close to the line, demonstrating that X and Y are almost exactly linearly related. In the other case the points deviate greatly from the line, demonstrating that X and Y are only very approximately linearly related. Whereas in the first case the line closely "fits" the data, in the second it "fits" only moderately well.

A simple numerical measure can be devised to express the extent to which a regression line approximates to the data from which it is

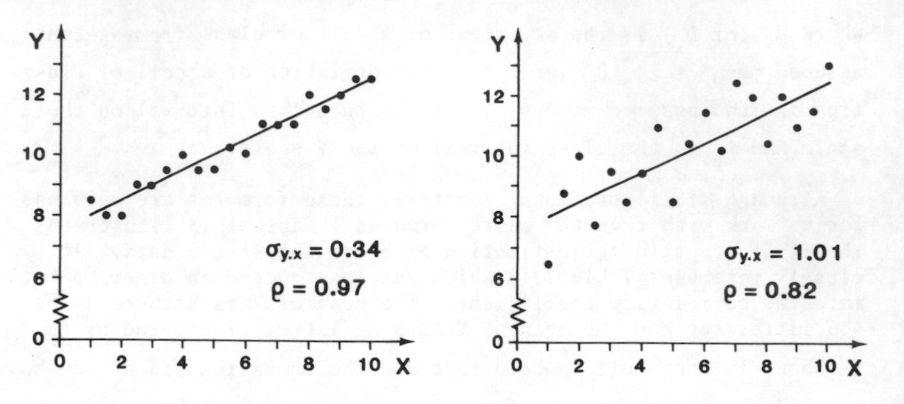

Fig. 19.9 - Two populations with identical regression lines ($\hat{Y} = 7.5 + 0.5X$) but different degrees of scatter or variability

calculated. The vertical deviation ε of a point (X,Y) from the line is equal to $(Y - \mu)$. Because ε is what is left when μ is subtracted from Y it may be called a *residual*. It is positive if the point lies above the line (μ less than Y) and negative if the point lies below (μ greater than Y). The square of the vertical deviation, $(Y - \mu)^2$ or ε^2, is positive regardless of whether the point lies above or below the line. In the case of a population the mean squared deviation or variance of the points about the regression line is given by

$$\sigma_{y.x}^2 = \frac{\Sigma(Y - \mu)^2}{N} \quad \text{or} \quad \frac{\Sigma\varepsilon^2}{N}$$

where N is the size of the population, i.e. the number of points on the scatter diagram.

The standard deviation of the points about the regression line is

$$\sigma_{y.x} = \sqrt{\frac{\Sigma(Y - \mu)^2}{N}} \quad \text{or} \quad \sqrt{\frac{\Sigma\varepsilon^2}{N}}$$

The method of calculating the equation of the regression line ensures that $\Sigma(Y - \mu)^2$ and hence $\sigma_{y.x}^2$ is a minimum. No other straight line can be fitted to the data so as to yield a smaller value of $\sigma_{y.x}^2$.

In the case of a sample of n items the quantity

$$s_{y.x}^2 = \frac{\Sigma(Y - \hat{Y})^2}{n - 2} \quad \text{or} \quad \frac{\Sigma d^2}{n - 2}$$

provides an unbiased estimate of $\sigma_{y.x}^2$. Each d in the formula is the residual or vertical deviation of a point from the sample regression line and not the population line. It may seem surprising that the denominator on the right-hand side is n - 2 and not n - 1 as in the standard formula for sample variance. However, if n - 1 is employed the result tends on average to be a slight under-estimate of $\sigma_{y.x}^2$. It is difficult to explain why division by n - 2 rather than n - 1 provides an unbiased estimate of $\sigma_{y.x}^2$ without going into rather complicated mathematics. An explanation of sorts is provided by the concept of degrees of freedom. The denominator n - 2 records the number of degrees of freedom associated with $\Sigma(Y - \hat{Y})^2$. Out of a total of n degrees of freedom two are lost because a minimum of two points is needed in order to calculate the regression equation. If a sample consists of only two points, the fitted regression line will pass exactly through both. The value of $s_{y.x}^2$ will be zero regardless of the actual value of $\sigma_{y.x}^2$. Given only two points no degrees of freedom are available for estimating $\sigma_{y.x}^2$.

The square root of $s_{y.x}^2$, namely

$$s_{y.x} = \sqrt{\frac{\Sigma(Y - \hat{Y})^2}{n - 2}}$$

is often referred to as the *standard error of estimate*. Just as $s_{y.x}^2$ provides an estimate of $\sigma_{y.x}^2$ so $s_{y.x}$ provides an estimate of $\sigma_{y.x}$.

Calculating $\sigma_{y.x}$ and $s_{y.x}$ from the individual deviations $(Y - \hat{Y})$ from regression is liable to become tedious especially if there are numerous pairs of values of X and Y. It may sometimes be simpler to use the formulas

$$\sigma_{y.x}^2 = \frac{\Sigma Y^2 - \Sigma \mu^2}{N} \quad \text{and} \quad s_{y.x}^2 = \frac{\Sigma Y^2 - \Sigma \hat{Y}^2}{n - 2}$$

The only drawback is that the predicted values (μ or $\hat{Y}$) must be squared. If the predicted values are large, finding the sum of their squares ($\Sigma\mu^2$ or $\Sigma\hat{Y}^2$) may well be difficult. Fortunately, other formulas exist that make no reference to μ or $\hat{Y}$, and these are always simple to use

$$\sigma_{y.x}^2 = \left\{\Sigma(Y - \bar{Y})^2 - \frac{[\Sigma(X - \bar{X})(Y - \bar{Y})]^2}{\Sigma(X - \bar{X})^2}\right\} \Big/ N$$

$$= \left\{\left[\Sigma Y^2 - \frac{(\Sigma Y)^2}{N}\right] - \frac{\left[\Sigma XY - \frac{\Sigma X \Sigma Y}{N}\right]^2}{\left[\Sigma X^2 - \frac{(\Sigma X)^2}{N}\right]}\right\} \Big/ N$$

or $$\frac{N\Sigma Y^2 - (\Sigma Y)^2 - \beta(N\Sigma XY - \Sigma X\Sigma Y)}{N^2} = \frac{\Sigma Y^2 - \alpha\Sigma Y - \beta\Sigma XY}{N}$$

$$s_{y.x} = \left\{\Sigma(Y - \bar{Y})^2 - \frac{[\Sigma(X - \bar{X})(Y - \bar{Y})]^2}{\Sigma(X - \bar{X})^2}\right\} \Big/ (n - 2)$$

$$= \left\{\left[\Sigma Y^2 - \frac{(\Sigma Y)^2}{n}\right] - \frac{\left[\Sigma XY - \frac{\Sigma X \Sigma Y}{n}\right]^2}{\left[\Sigma X^2 - \frac{(\Sigma X)^2}{n}\right]}\right\} \Big/ (n - 2)$$

or $$\frac{n\Sigma Y^2 - (\Sigma Y)^2 - b(n\Sigma XY - \Sigma X\Sigma Y)}{n(n-2)} = \frac{\Sigma Y^2 - a\Sigma Y - b\Sigma XY}{n-2}$$

Since all the terms on the right-hand side are calculated in the course of obtaining the equation of the regression line, little extra work is required to find $\sigma_{y.x}^2$ or $s_{y.x}^2$. Thus for the stream data

$$s_{y.x}^2 = \left\{ \left[21640.99 - \frac{(484.3)^2}{11}\right] - \frac{\left[2647.55 - \frac{(56.7)(484.3)}{11}\right]^2}{\left[381.95 - \frac{(56.7)^2}{11}\right]} \right\} \Big/ (11 - 2)$$

$$= \{318.58182 - 254.91398\}/9 = 7.0742 \text{ approximately.}$$

$$s_{y.x} = \sqrt{7.0742} = 2.6597 \text{ approximately.}$$

The same result can be obtained more laboriously by summing the squared deviations as shown in Table 19.4.

TABLE 19.4 - CALCULATION OF THE STANDARD ERROR OF ESTIMATE FOR THE STREAM DATA

Date of visit (1971)	*Discharge of stream (litres/sec)* X	*Concentration of dissolved solids (mg/litre)* Y	*Height of the regression line at X* $\hat{Y}$	*Vertical deviation or residual* $(Y - \hat{Y})$	*Square of vertical deviation* $(Y - \hat{Y})^2$
June 3	0.4	34.2	36.01159	-1.81159	3.28186
Oct. 7	2.2	40.4	39.04621	+1.35379	1.83275
March 11	2.3	38.3	39.21480	-0.91480	0.83686
Nov. 4	3.9	39.4	41.91224	-2.51224	6.31133
Dec. 16	4.3	44.8	42.58660	+2.21340	4.89916
Nov. 18	4.4	47.0	42.75519	+4.24481	18.01845
Feb. 25	5.1	42.6	43.93531	-1.33531	1.78307
Jan. 28	8.0	45.6	48.82442	-3.22442	10.39689
Jan. 14	8.1	52.8	48.99301	+3.80699	14.49317
Aug. 26	8.3	48.7	49.33019	-0.63019	0.39714
Dec. 30	9.7	50.5	51.69045	-1.19045	1.41717
	56.7	484.3	484.30001	0.0000	63.66785

$$s_{y.x}^2 = \frac{\Sigma(Y - \hat{Y})^2}{n-2} = \frac{63.66785}{9} = 7.0742 \text{ approximately.}$$

and $s_{y.x} = \sqrt{7.0742} = 2.6597$ approximately.

THE RELATIONSHIP BETWEEN THE STANDARD ERROR OF ESTIMATE AND THE CORRELATION COEFFICIENT

The standard error of estimate must not be confused with Pearson's coefficient of correlation, which measures the degree and direction of linear relationship between variables. The standard error of estimate measures only the degree of scatter of the data points about the regression line (it is a measure, in other words, of the size of the residuals). Other things being equal, it is true that the greater the scatter of points about the regression line, the less close is the linear relationship between the variables. One might imagine, therefore, that the standard error of estimate is an inverse measure of the degree of correlation, but this is only partly the case. The degree of correlation is indicated not only by the closeness with which the points approximate to a straight line, but also by the slope of the line, i.e. the rate at which Y changes with X. The standard error of estimate is independent of the slope of the regression line. In Fig. 19.10 the same horizontal regression line has been fitted to two populations of data points. Although in Fig. 19.10A the points are well scattered around the line, and in Fig. 19.10B they are positioned close to the line, in neither case is there any linear relationship between X and Y. When the value of X changes, the mean value of Y remains the same. The correlation for each population is zero, but the standard error is a positive quantity proportional to the degree of scatter (or size of the residuals).

Figure 19.11 depicts two populations with regression lines that differ markedly in slope (the regression coefficient of one population is twice that of the other). Although the slope differs, the scatter of points about each line is precisely the same. In the case of the population with the gentle slope Y increases only sluggishly with X. The degree of relationship between X and Y is less close than in the case of the population with the steep slope where Y increases rapidly as X increases. The populations display different correlations but the same standard error of estimate.

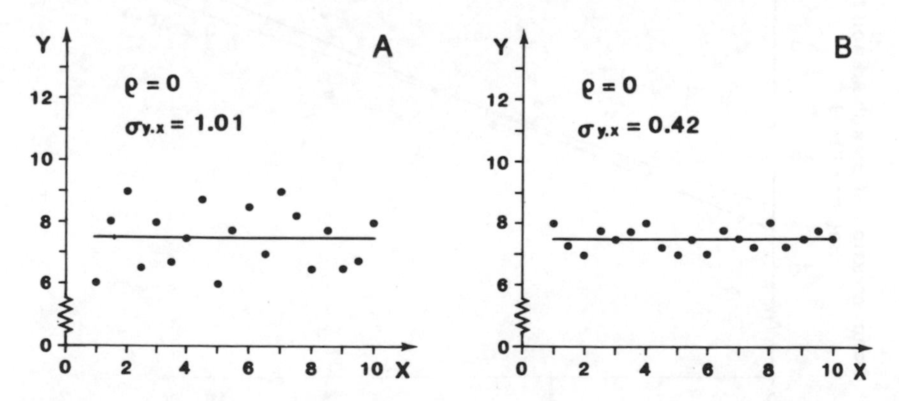

Fig. 19.10 - Two regression lines with zero slope but different degrees of scatter.

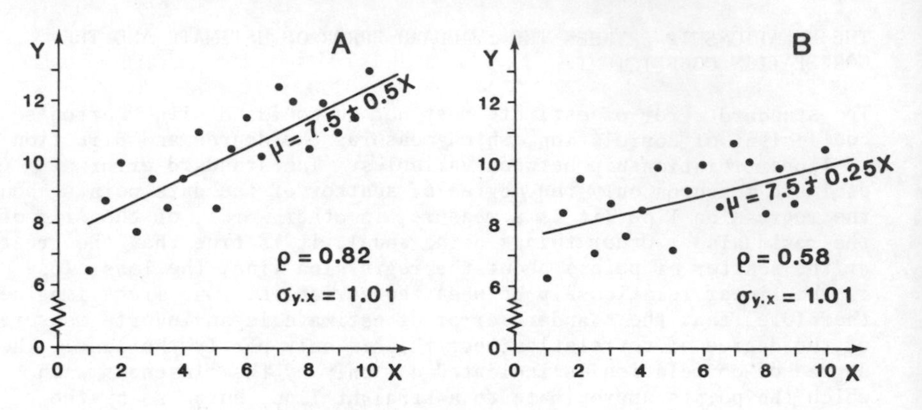

Fig. 19.11 - Two regression lines with the same intercept and degree of scatter but different slopes.

THE REGRESSION OF X ON Y

The regression line of Y on X minimises the sum of the squares of the vertical distances of the points from the regression line. The sum of the squares of the horizontal distances is not taken into account. As a matter of fact, it is impossible to fit a line so that both sums of squares are simultaneously a minimum, except in the special case where all the points are arranged exactly in line so that both sums of squares equal zero. If the points are not arranged exactly in a line, but are scattered, two regression lines can be fitted: the usual regression of Y on X, obtained by minimising the sum of the squares of the vertical distances, and the *regression of X on Y*, obtained by minimising the sum of the squares

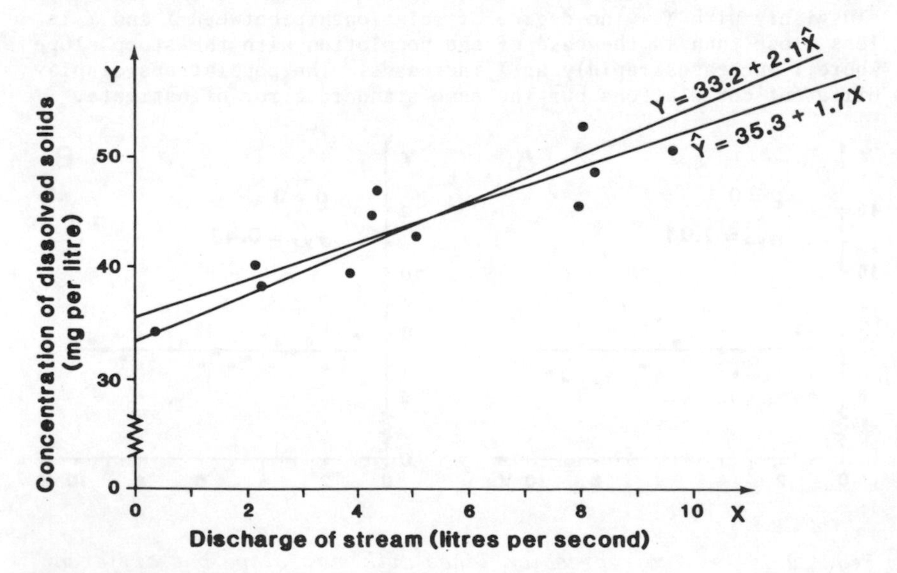

Fig. 19.12 - Linear regression of concentration of dissolved solids on discharge ($\hat{Y}$ = 35.3 + 1.7X) and linear regression of discharge on concentration of dissolved solids (Y = 33.2 + 2.1$\hat{X}$).

of the horizontal distances. The two regression lines for the stream data are shown in Fig. 19.12. Note that the gradients of the two lines are different, as are the intercepts.

The sample regression line of X on Y has an equation $Y = a' + b'\hat{X}$ or $\hat{X} = (Y - a')/b'$ where $\hat{X}$ is the predicted value of X corresponding to a given Y. The gradient of the line is

$$b' = \frac{n\Sigma Y^2 - (\Sigma Y)^2}{n\Sigma XY - \Sigma X\Sigma Y}$$

b' represents the increase in Y with unit change in X.

The line intersects the vertical or Y axis a' units from the origin where

$$a' = \frac{\Sigma Y}{n} - \frac{b'\Sigma X}{n} = \overline{Y} - b'\overline{X}$$

It crosses the regression line of Y on X at the point $(\overline{X},\overline{Y})$, the mean of the data.[1] The angle between the two regression lines decreases as the degree of correlation increases. Thus the more closely the data points approximate to a straight line the more slowly the regression lines diverge away from the mean. If the points are arranged exactly in line, the regression lines coincide. Conversely, if the points are distributed almost at random, the regression lines intersect more or less at right angles.[2]

The existence of two separate regression lines may appear confusing. Which regression line should be used to represent the relationship between two variables X and Y? If Y depends wholly on X, the correct regression to use to predict Y is Y on X since this regression measures the mean value of Y corresponding to the value of X. If there are other variables besides X causing the value of Y to vary, the mean value of Y is the value that best represents the dependency of Y on X. The mean value of Y is in this sense the most probable value. In order to predict the value of X from the value of Y, however, it is best to use the regression of X on Y, provided the pairs of values of X and Y are drawn at random.

Sometimes there are practical advantages to be gained from selecting the values of X and letting only the values of Y vary freely. In laboratory experiments for example it is often convenient to select a series of integer values of X such as 10, 20, 30 and 40. For instance, an experiment on frost weathering might be conducted by measuring the amount of disintegration undergone by rock specimens after 10 cycles of freezing and thawing, 20 cycles, 30 cycles, etc. Provided the Y values are not selected it is perfectly permissible to select the values of X. The linear regression of Y on X is not affected in any way. However, the regression of X on Y

---

[1]The case of the bivariate normal population is depicted in Fig. 17.10.

[2]Some books say that if the points are exactly random, the regression lines cross exactly at right angles to each other. The truth is that both lines vanish if there is complete randomness.

may be so greatly changed as to be meaningless. In these rather specialised circumstances it is best to use the regression of Y on X to predict X from Y, rather than the regression of X on Y.

THE REGRESSION LINE OF X ON Y : ALTERNATIVE FORMULAS

The correlation coefficient of a sample is

$$r = \frac{n\Sigma XY - \Sigma X\Sigma Y}{\sqrt{[n\Sigma X^2 - (\Sigma X)^2]}\sqrt{[n\Sigma Y^2 - (\Sigma Y)^2]}}$$

The regression coefficient of X on Y is

$$b' = \frac{n\Sigma Y^2 - (\Sigma Y)^2}{n\Sigma XY - \Sigma X\Sigma Y}$$

By combining the two equations one can readily show that

$$r = \frac{1}{b'}\left(\frac{s_y}{s_x}\right) \quad \text{and} \quad b' = \frac{1}{r}\left(\frac{s_y}{s_x}\right)$$

where s_y is the standard deviation of the Y values and s_x the standard deviation of the X values.

As explained earlier in the chapter

$$r = b\left(\frac{s_x}{s_y}\right) \quad \text{and} \quad b = r\left(\frac{s_y}{s_x}\right)$$

where b is the regression coefficient of Y on X. It follows that $r = \sqrt{(b/b')}$ and $b/b' = r^2$. The ratio of the slopes of the two regression lines is therefore the square of the correlation coefficient. The same simple relationship can be demonstrated for a population

$$\rho = \frac{1}{\beta'}\left(\frac{\sigma_y}{\sigma_x}\right) \quad \text{and} \quad \beta' = \frac{1}{\rho}\left(\frac{\sigma_y}{\sigma_x}\right)$$

Also $\rho = \beta\left(\frac{\sigma_x}{\sigma_y}\right)$ and $\beta = \rho\left(\frac{\sigma_y}{\sigma_x}\right)$

and therefore $\rho = \sqrt{(\beta/\beta')}$ and $\beta/\beta' = \rho^2$.

Because the ratio of the gradients of the two regression lines is equal to the square of the correlation coefficient, it follows that the angle between the two regression lines is inversely proportional to the correlation coefficient. If the correlation coefficient is +1 or -1, the ratio of the gradients is +1 and hence the two lines coincide.

In the case of a sample, if the regression coefficient for X on Y exceeds the coefficient for Y on X and both are positive quantities (as in Fig. 19.12), the acute angle T between the regression lines is given by the relation $\tan T = (b' - b)/(1 + bb')$. In the case of a population, the acute angle θ is given by $\tan \theta = (\beta' - \beta)/(1 + \beta\beta')$. Since $b' = b/r^2$ and $\beta' = \beta/\rho^2$ it follows that $\tan T = b(1 - r^2)/(b^2 + r^2)$ and $\tan \theta = \beta(1 - \rho^2)/(\beta^2 + \rho^2)$.

If the correlation is perfect the numerator on the right-hand side is zero, and therefore T (or θ) is also zero. The two regression lines coincide. The less perfect the correlation the more the lines diverge.

20 Confidence Limits and Significance Tests for Sample Regression Lines

THE LINEAR REGRESSION MODEL

A regression line calculated from sample data may well be affected by sampling error. Even if the sampling is random, the line is unlikely to be precisely the same as the regression line of the population from which the sample is taken. The equation $\hat{Y} = a + bX$ of the sample regression line is only an estimate of the equation $\mu = \alpha + \beta X$ of the regression line of the population.[1] It may be a bad estimate, especially if the sample is small or non-random. The regression coefficient b of the sample may differ greatly from the coefficient β of the population and a, the Y-intercept of the sample regression line, may be very different from α, the Y-intercept of the population regression line.

The regression equation of a sample that is greatly affected by sampling error may be of little or no practical value. On the other hand its practical value may be considerable if the sampling error is small. After calculating a sample regression equation one's next step should always be to try to determine how reliable the equation is likely to be. There are two alternatives: one can either calculate confidence limits for α and β or carry out significance tests on a and b.

It is convenient at this point to introduce the phrase *conditional distribution of Y values* to refer to the frequency distribution of all the Y values in a population that share a given X value. There are as many conditional distributions of Y values in a population as there are X values. The significance of the adjective "conditional" is simply that the Y values are partly conditional upon the value of X.

The methods commonly employed for calculating confidence limits and carrying out significance tests make four assumptions about the conditional distributions of the Y values in the parent population:

1. If the population is infinite in size, each conditional distribution of Y values has a mean μ lying on the straight line $\mu = \alpha + \beta x$, the population regression line. If the population is

[1] It is assumed in this chapter, as in the last, that one variable is cause while the other is its effect. Except where stated otherwise, the usual notation will be followed of denoting the causal variable by the letter X and the affected variable by the letter Y. No allowance will be made for possible systematic errors due to confounding.

finite, individual means may not lie exactly on the line, but any deviations are purely a matter of chance.

2. Each conditional distribution of Y values has the same standard deviation, $\sigma_{y.x}$, if the population is infinite. There may be chance variations in a finite population.

3. If the population is infinite, the deviation of any Y value from its mean μ is unrelated to the deviation of any other Y value, and unrelated to the value of any unexamined variable. Chance correlations may occur if the population is finite.

4. Each conditional distribution is exactly normal in shape if the population is infinite. Some non-normality will occur if the population is finite but this is due solely to chance.

Figure 20.1 illustrates three of these assumptions for four selected values of X in an infinite population. The mean, μ, of each conditional distribution lies on the regression line $\mu = \alpha + \beta x$. The standard deviation of each conditional distribution is the same, and each distribution is normal in shape.

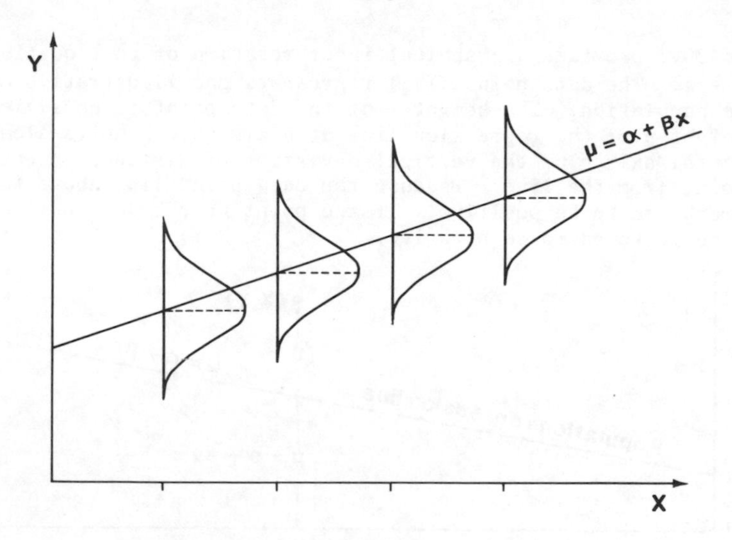

Fig. 20.1 - Conditional distributions of Y values for four selected values of X. The mean of each conditional distribution falls on the population regression line $\mu = \alpha + \beta X$. *The standard deviation of each distribution is the same, and each distribution is bounded by a normal curve.*

Each of the four assumptions has acquired a name. The first is known as the *linearity assumption*, because the underlying relationship between X and Y is presumed to be linear, not curvilinear. The second assumption is known as the *homoscedasticity assumption*. A population that fails to satisfy this requirement is said to exhibit *heteroscedasticity*. The third assumption is called the *independence assumption* because the deviations of the Y values are assumed to be independent of one another and independent of the values of variables not included in the regression analysis. The last assumption is called for obvious reasons the *normality assumption.*

Together the four assumptions constitute what is called the *linear regression model,* or *linear model* for short. The Y value of a member of the population is thought of as the sum of two parts: a part due to X (the *X component* or *linear term in X*) and a part due to the disturbing effect of variables other than X (the *error component* or *stochastic disturbance term*). The X component is represented by μ, the mean of the Y values of all members of the population having the same X value as the member in question. The error component or disturbance term is the deviation of the Y value of the member from the mean of the Y values, μ. Expressed as an equation

$$Y = \mu + \varepsilon$$

where $\mu = \alpha + \beta X$. The value of the error component or disturbance term ε is positive if Y is greater than μ, and negative if Y is less than μ. This value is imagined to be drawn randomly from a normal distribution with a mean of zero and a standard deviation $\sigma_{y.x}$. Because the value is drawn randomly it is independent of any other value.

Figure 20.2 provides a graphical interpretation of the equation $Y = \mu + \varepsilon$. The data point (X,Y) represents one illustrative member of the population. The height Y of the data point is equivalent to the height μ of the regression line at a distance X units along the horizontal axis plus the vertical deviation or distance ε separating the point from the line. Because the data point lies above the line ε is reckoned to be positive. If the point were below the line ε would be reckoned to be negative.

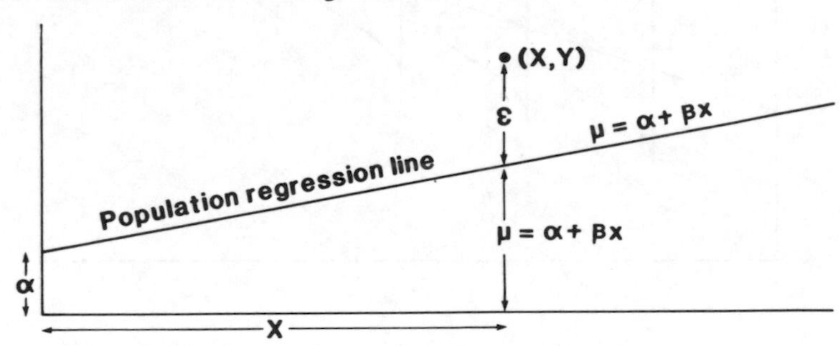

Fig. 20.2 - Diagram to demonstrate that $Y = \mu + \varepsilon = (\alpha + \beta X) + \varepsilon$.

It is important not to confuse ε with d. ε is the vertical deviation of a data point from the population regression line $\mu = \alpha + \beta X$, whereas d is the vertical deviation of the point from a sample regression line $\hat{Y} = a + bX$. By definition, $\varepsilon = Y - \mu$ whereas $d = Y - \hat{Y}$. $\hat{Y}$ is an estimator of μ and d is an estimator of ε. Both Y and d are subject to possible sampling error, but μ and ε are fixed.

The value of d can always be calculated for any sample point, but the value of ε cannot be calculated unless for some reason the equation of the population regression line is known.

The correlation model discussed in Chapter 17 can be thought of as a special case of the linear regression model. It assumes that both X and Y are distributed normally, whereas the linear model, while assuming that Y is normal, makes no assumption about X. The correlation model does not permit one to select particular X values when sampling. The linear regression model, on the other hand, allows one to adopt a fairly flexible sampling scheme: the values of X can be allowed to vary freely or they can be deliberately selected for their convenience. In a laboratory experiment, for example, one might deliberately select the round numbers 10, 20, 30 and 40 for X rather than allow X to vary randomly and assume inconvenient decimal numbers like 2.3679 and 15.7148. Neither the correlation nor the regression model permits particular Y values to be selected; these must always be allowed to vary freely.

The linear regression model, like the correlation model, is a mathematical fiction. The means of the Y values in real populations seldom lie precisely on a straight line, and the standard deviations of the Y values are not usually constant but tend to vary with X. What is more, the Y values are usually correlated amongst themselves and are not distributed randomly throughout the population as required by the model. Also no real population has Y values ranging from minus to plus infinity, and it is therefore impossible for the distribution of these values to be precisely normal.

The linear regression model is not so far-fetched as to lack practical applications. Despite the obvious inadequacies of the model, there can be no denying its usefulness. The number of real populations that have been found to approximate fairly closely to the model is surprisingly large. And when samples are drawn from such populations it is easy to calculate confidence limits and to carry out significance tests, as will shortly be demonstrated.

It is possible to prove that if a sampled population were to conform precisely with the first three assumptions of the linear regression model, a and b would be *unbiased* and *fully efficient* estimators of α and β whatever the size of the sample (Kendall and Stuart, 1973). An unbiased estimator is as likely to be too large as too small. A biased estimator, on the other hand, is, according to circumstances, more likely to be too large than too small, or more likely to be too small than too large. A fully efficient estimator has the smallest possible sampling variance or standard error. In other words, it is more likely to lie closer to the true value than any other estimator. Even if an estimator is unbiased, this does not necessarily mean that it is fully efficient.

Although it is extremely doubtful whether any real population meets the first three assumptions of the model precisely, this does not mean that a and b have no practical value as estimators of α and β. All the evidence suggests that a and b provide reasonably unbiased and efficient estimates if the first three assumptions are nearly true. However, if the assumptions are seriously violated, a and b may be very poor estimates of α and β. This problem is discussed at some length in the companion volume.

Non-mathematicians sometimes have difficulty in believing that the first three assumptions of the regression model are really needed.

So long as the Y values are free to vary, why should not a and b be unbiased and fully efficient estimators? After all, they are calculated in the same way as α and β. The formulas use different symbols but they are otherwise identical.

Though it may seem odd, the basic identity of the formulas does not ensure that a and b are the best possible estimators of α and β. What really matters is the distribution of the X and Y values in the population. The distribution may be such that the best estimators of α and β are not a and b but other measures, as will be explained in the companion volume.

All the methods of calculating confidence limits and testing significance that are in common use assume that a and b are unbiased and fully efficient estimators of α and β. This is the reason why the linearity, homoscedasticity and independence assumptions are included in the linear regression model. When a and b are appreciably biased or inefficient estimators of α and β there is no simple way of calculating confidence limits or testing significance.

Even if a and b are unbiased and fully efficient estimators of α and β, they are still likely to be affected by sampling error. The amount of sampling error can be estimated most simply if the values of Y corresponding to each value of X are normally distributed. If the values of Y are not normally distributed, significance testing and the calculation of confidence limits are more difficult, often even to the point of not being feasible. Like the first three assumptions, the normality assumption is included in the linear regression model to facilitate testing and to simplify the calculation of confidence limits. Unlike the first three assumptions, however, it does not ensure that a and b are unbiased and fully efficient estimators.[1]

The standard methods of calculating limits and of testing significance would yield exact results in the (impossible) event of the four assumptions of the linear model being true. Although the assumptions are not entirely realistic, the methods can be relied upon to yield reasonably accurate results provided that the discrepancies between the model and the sampled population are not too great.

Unfortunately, it may be difficult in a sampling survey to decide whether the discrepancies are large enough to matter. There must always be some uncertainty concerning the precise distributions of the X and Y values in the sampled population. If the distributions were precisely known, the sampling would be unnecessary. With only a sample to go on, the best one can do is to make assumptions about the distributions that seem reasonable, given the sample data. If the sample is very small, one may not be able to decide whether the linear model is appropriate or not.

The methods of calculating confidence limits and of testing significance described in this chapter assume (as does the linear model)

---

[1]Although lack of normality does not itself cause a and b to become biased estimators it may result from lack of independence which can cause bias. This problem is discussed in the companion volume.

that the sample values of X are accurate.[1] If the values of X are subject to measurement errors, a and b will be biased estimators of α and β, although the bias will only be large if the errors are large. It is not important that the Y values be accurate, so long as the errors are independent of X and have a normal distribution with zero mean so that they can be subsumed in the disturbance term ε. As explained earlier, ε is the part of the value of Y that is due to variables other than X. Any errors of measurement of Y (or errors of book-keeping etc. in respect of Y) contribute to the value of ε provided that they are independent of X.

The methods described in this chapter make use of formulas that are difficult to prove although they are easy enough to apply. No proofs will be given here, but they may be studied, if desired, in advanced text-books such as Kendall and Stuart (1973).

Alternative methods based on what is called analysis of variance are described in the companion volume.

THE ACCURACY WITH WHICH β IS ESTIMATED BY b

If an infinite number of samples of size n were to be drawn with individual replacements from the same population, the sampling distribution of the values of b would have a mean β and a standard deviation (standard error) σ_b given by the equation

$$\sigma_b = \frac{\sigma_{y.x}}{\sqrt{\{\Sigma X^2 - [(\Sigma X)^2/n]\}}}$$

where $\sigma_{y.x}$ is the standard deviation of the values of Y in the population about the population regression line $\mu = \alpha + \beta X$. If the Y values are distributed normally, the sampling distribution of the values of b is exactly normal. Even if the Y values are not normally distributed, the sampling distribution of b tends to become normal as n increases.

In most practical applications, only one sample is drawn from a population and $\sigma_{y.x}$ is not known precisely. Without $\sigma_{y.x}$, σ_b cannot be calculated. However, $s_{y.x}$ (the standard error of estimate or standard deviation of the Y values in the sample about the sample regression line $\hat{Y} = a + bX$) provides an unbiased estimate of $\sigma_{y.x}$. If the sample is of size n

$$s_b = \frac{s_{y.x}}{\sqrt{\{\Sigma X^2 - [(\Sigma X)^2/n]\}}}$$

where s_b is the estimated standard error.

[1] Opinion is divided as to whether this assumption should be included in the linear model or be thought of simply as a prerequisite. This book takes the second view arguing that almost every statistical test is founded on the assumption that the data are accurate and that it is unnecessary to make such a basic assumption explicit within the linear model.

Provided the requirements of the linear regression model are met, the quantity $|b - \beta|/s_b$ is distributed as t with n - 2 degrees of freedom. The 95% confidence limits associated with β are $b - t_{0.05}s_b$ and $b + t_{0.05}s_b$ or more concisely $b \pm t_{0.05}s_b$. β lies outside these limits only once in every 20 samples on average.

TO TEST WHETHER THE REGRESSION COEFFICIENT OF A POPULATION IS SOME SPECIFIED VALUE β

THE DATA

The population under study consists of paired values of two variables X and Y. The regression coefficient $b = (b\Sigma XY - \Sigma X\Sigma Y)/[n\Sigma X^2 - (\Sigma X)^2]$ is computed from a sample of n paired values drawn from the population. It is desired to test the hypothesis that the value of the population regression coefficient is β and not b.

PROCEDURE

1. Set up the null hypothesis: the sample regression coefficient b is misleading. The value of the regression coefficient of the population is actually β. The difference between b and β is due entirely to sampling error, in other words it is not statistically significant.

2. Calculate $s_{y.x}$, the standard deviation of the Y values of the sample points about the regression line $\hat{Y} = a + bX$

$$s_{y.x} = \sqrt{\frac{\Sigma(Y - \hat{Y})^2}{n - 2}} = \sqrt{\frac{n\Sigma Y^2 - (\Sigma Y)^2 - b(n\Sigma XY - \Sigma X\Sigma Y)}{n(n - 2)}}$$

3. Estimate s_b, the standard error of the regression coefficient

$$s_b = \frac{s_{y.x}}{\sqrt{\{\Sigma X^2 - [(\Sigma X)^2/n]\}}} = \frac{s_y}{s_x}\sqrt{\frac{(1 - r^2)}{(n - 2)}}$$

where s_y is the standard deviation of the Y values and s_x the standard deviation of the X values.

4. Calculate t using the equation

$$t = \frac{|b - \beta|}{s_b}$$

5. Calculate the number of degrees of freedom (df) associated with t

$$df = n - 2$$

6. Using Table E find P, the probability of obtaining this value of t, or one greater, through chance sampling (sampling error). P is the probability of obtaining the observed difference between b and β, or one greater, given an estimated standard error of s_b.

Note that the values of P listed in the table are for two-tailed tests. For a one-tailed test halve the value of P.

7. Decide whether to retain or reject the null hypothesis. If P is low, serious doubt is cast on the null hypothesis. Reject the null if P is equal to or less than whatever level of significance is judged appropriate, e.g. 0.05 or 0.01. Conclude that the difference between b and β is more than would ordinarily occur if the sample were drawn from a population with a regression coefficient of β. Assume that the value of the population regression coefficient is not β but some value nearer to b.

If P is greater than the level of significance reserve judgement, i.e. regard the null hypothesis as possibly, though not necessarily, correct.

EXAMPLE

To return to the stream data discussed in the last chapter, the regression coefficient b = 1.6859. The standard error of estimate (standard deviation of the Y values around the regression line) is

$$s_{y.x} = \sqrt{\frac{11(21640.99) - (484.3)^2 - 1.6859(1663.24)}{11(9)}}$$

$$= \sqrt{\frac{700.34368}{99}} = 2.65973$$

The estimated standard error of b is

$$s_b = \frac{2.65973}{\sqrt{\{381.95 - [(56.7)^2/11]\}}} = \frac{2.65973}{\sqrt{89.68727}} = 0.28085$$

Because the value of b is not very close to zero, and is so much larger than the value of s_b, the population coefficient β is hardly likely to be zero. A test of significance of the hypothesis that β = 0 is unnecessary. However, if a test were needed, it could be run as follows:

1. Null hypothesis: the sample is drawn from a population with a regression coefficient of zero.

2. Inspection of the data suggests that the linear regression model is fairly realistic. The deviations from the sample regression line are approximately normally distributed, and the deviations are seemingly independent of X. It will be assumed, therefore, that the requirements of the linear regression model are met, at least sufficiently to justify the calculations that follow. It will also be assumed that the values of X are free from measurement errors and they and the Y values are not distorted by confounding. How far these additional assumptions are valid is not known.

3. t = 1.6859/0.28085 = 6.0028

4. The degrees of freedom are 11 - 2 = 9.

5. Referring to Table E, the value of P is found to be approximately 0.0002. It is not realistic to assume that the regression coefficient of the population is zero. The evidence strongly suggests that the coefficient is greater than zero, i.e. that solute concentration is indeed related to discharge.

The 95% confidence limits for β are 1.6859 $\pm$ 2.262 (0.28085), that is to say 1.05 and 2.32. There is only a 1 in 20 chance that the value of β falls outside these limits.

ASSUMPTIONS AND LIMITATIONS

The test assumes that the requirements of the linear regression model are met. Thus the Y values corresponding to each value of X ought to be normally distributed. Each distribution of Y values ought to have the same standard deviation $\sigma_{y.x}$. The means of the different distributions of Y values ought all to fall on the population regression line $\mu = \alpha + \beta X$. Furthermore, the deviation of any Y value in the population from its mean μ ought to be independent of the deviation of any other Y value, and independent of the value of X. The deviations must not be correlated amongst themselves. Any violation of these assumptions may invalidate the test.

There are no special constraints on the values of X. Although the values must be accurate, they do not need to be selected randomly, but can be regularly spaced at convenient intervals, e.g. 5, 10, 15, etc. Only the Y values need to be free to vary randomly.

The test assumes that the sampling is with individual replacements or else the population is infinite in size. If the population is finite in size and the sampling is without replacements the value of P will be too big, thus favouring the null hypothesis. The amount of bias is unlikely to be large, however, if the sample represents less than about ten per cent of the population.

Although the test is commonly used to examine the hypothesis that β is zero, it is valid for any hypothesis concerning the value of β. When used to examine the hypothesis that β is zero, the test yields the same value of t as the test of the hypothesis that r is zero described in Chapter 17. It can be shown that the two tests are algebraically equivalent.

The test has often been claimed to be fairly robust in the face of non-normality. However, Kowalski's detailed investigation of the sampling distribution of r (see Chapter 17) demonstrates that even slight departures from normality can sometimes produce wildly inaccurate values of P. The test cannot therefore be relied upon when its basic assumptions are violated although it often gives reasonably accurate results in the face of violations and must not be dismissed out of hand.

THE ACCURACY WITH WHICH μ is ESTIMATED BY $\hat{Y}$

The linear regression model specifies that for a given value of X the mean of the Y values in the population lies on the population regression line. The mean is thus $\mu = \alpha + \beta X$, the height of the population regression line at the point X.

Sometimes it is necessary to estimate μ from sample data. An unbiased estimate is provided by $\hat{Y}$, the height of the sample regression line of the point X. The value of $\hat{Y}$ is liable to be affected by sampling error, however. The standard error associated with $\hat{Y}$ is estimated by the equation

$$s_{\hat{Y}} = s_{y.x}\sqrt{\left[\frac{1}{n} + \frac{n(X - \overline{X})^2}{n\Sigma(X - \overline{X})^2}\right]} = s_{y.x}\sqrt{\left[\frac{1}{n} + \frac{n(X - \overline{X})^2}{n\Sigma X^2 - (\Sigma X)^2}\right]}$$

where $s_{\hat{Y}}$ is the estimate of the true standard error σ_Y.

Provided the conditions of the linear regression model are met (and there is no confounding and no inaccuracy in the X values), the quantity $|\hat{Y} - \mu|/s_{\hat{Y}}$ follows the t-distribution with (n - 2) degrees of freedom, and hence the 95% confidence limits for μ are $\hat{Y} - t_{0.05}s_{\hat{Y}}$ and $Y + t_{0.05}s_{\hat{Y}}$.

Note that the size of the estimated standard error $s_{\hat{Y}}$ depends on the value of X. The more X deviates from the mean of the X values $\overline{X}$, the larger $s_{\hat{Y}}$ becomes. If the confidence limits are plotted for different values of X, it will be found that they are curved around the regression line, not straight as might be expected (Fig. 20.3). The curves emphasise the increasing difficulty of making accurate predictions of X with increasing distance from $\overline{X}$.

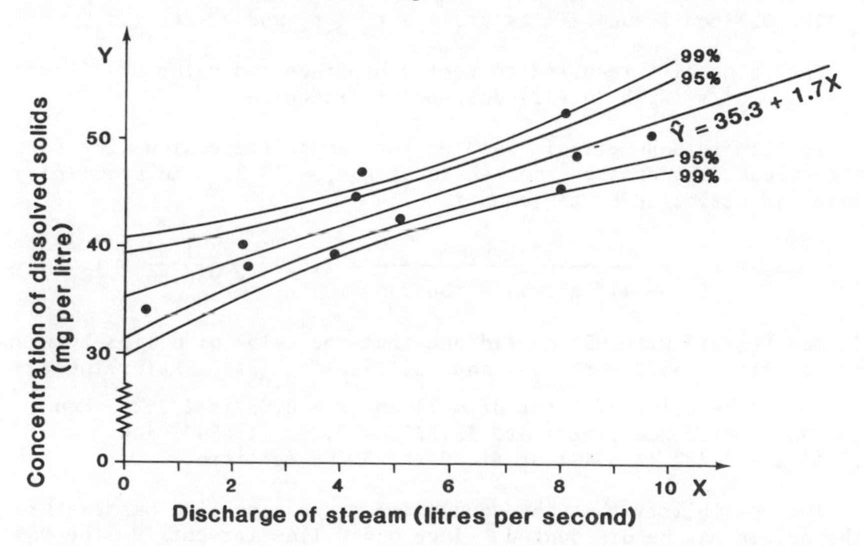

Fig. 20.3 - 95% and 99% confidence limits for μ for the stream data. The curves are portions of hyperbolas.

Suppose, for example, that an estimate is required of the mean concentration of dissolved solids when the discharge is 5.00 litres/sec and increasing. The mean concentration μ is estimated by $\hat{Y}$ = 35.3372 + 1.6859X = 35.3372 + 1.6859(5) = 43.7667 mg/litre. The standard error associated with this estimate is

$$s_{\hat{Y}} = s_{y.x}\sqrt{\left[\frac{1}{11} + \frac{11(5 - 5.1545)^2}{11(381.95) - (56.7)^2}\right]}$$

$$= 2.6597\sqrt{\left[\frac{1}{11} + \frac{11(0.02388)}{986.56}\right]}$$

$$= 2.6597\sqrt{0.09118} = 0.8031 \text{ mg/litre.}$$

The 95% confidence limits for μ are 43.7667 $\pm$ 2.262 (0.8031) or 41.95 and 45.58 mg/litre. One can be 95% certain that the true mean concentration lies between these limits.

THE ACCURACY WITH WHICH α IS ESTIMATED BY a

The population line $\mu = \alpha + \beta X$ crosses the Y-axis α units from the origin. In effect, α is the value of μ when X = 0. The sample estimate of α is $a = \overline{Y} - b\overline{X}$. The estimated standard error associated with a is

$$s_a = s_{y.x}\sqrt{\left[\frac{1}{n} + \frac{n(\overline{X})^2}{n\Sigma(X - \overline{X})^2}\right]} = s_{y.x}\sqrt{\frac{\Sigma X^2}{n\Sigma X^2 - (\Sigma X)^2}}$$

This is the equation for $s_{\hat{Y}}$ given in the preceding section but with X made equal to zero.

The 95% confidence limits are $a - t_{0.05}s_a$ and $a + t_{0.05}s_a$.

The value of t required to test a hypothesised value of α is $t = |a - \alpha|/s_a$ with (n - 2) degrees of freedom.

To turn to a numerical example: the sample regression line for the stream data crosses the Y-axis at a = 35.3372. The standard error is estimated by the equation

$$s_a = s_{y.x}\sqrt{\frac{381.95}{11(381.95) - (56.7)^2}} = 2.6597\sqrt{\frac{381.95}{986.56}} = 1.6549$$

It may be said with 95% confidence that the value of α lies between the limits $35.3372 - t_{0.05}s_a$ and $35.3372 + t_{0.05}s_a$. Referring to Table E, the value of t for df = 11 and P = 0.05 is 2.262. Hence the 95% confidence limits are 35.3372 - 2.262 (1.6549) and 35.3372 + 2.262 (1.6549) or 31.59 and 39.08 mg/litre.

The possibility that the population regression line passes through the origin may be discounted, since $\alpha = 0$ lies far outside the 95% confidence limits. No test of significance is necessary, but if it were the value of t would be

$$t = \frac{|35.3372 - 0|}{1.6549} = 21.353$$

with 11 degrees of freedom. The value of P would be vanishingly small.

*THE ACCURACY WITH WHICH Y IS ESTIMATED BY $\hat{Y}$

Sometimes it is necessary to estimate a value of Y from a value of X. There may exist, for instance, a member of the population whose X value alone has been measured yet whose Y value is also of interest. An unbiased estimate of an individual Y is provided by the sample regression equation $\hat{Y} = a + bX$. Beginners are sometimes perplexed that this equation is also used to estimate μ, the mean of the Y values corresponding to a given value of X. No contradiction exists, however. The linear regression model specifies that the Y values are symmetrically (normally) distributed around μ, and therefore μ provides an unbiased estimate of an individual Y. Also, because $\hat{Y}$ provides an unbiased estimate of μ, it provides an unbiased estimate of an individual Y. If $\hat{Y}$ is used to estimate an individual Y at a given X, an estimate of the standard error is

$$s_{\hat{Y}} = s_{y.x}\sqrt{\left[1 + \frac{1}{n} + \frac{n(X - \overline{X})^2}{n\Sigma X^2 - (\Sigma X)^2}\right]}$$

Compare this new equation for $s_{\hat{Y}}$ with that given earlier. Observe that $s_{\hat{Y}}$ is larger for an estimate of an individual Y than for an estimate of μ. The reason should be fairly obvious. An estimate of an individual Y is liable to be misleading because of two sources of sampling error:

1. The actual value of the individual Y may differ from the mean value μ.

2. In turn, the mean value μ may differ from the estimated value $\hat{Y}$.

An estimate of μ is subject only to the second of these sources of sampling error and consequently is associated with a smaller standard error than is an estimate of Y.

The quantity $|\hat{Y} - Y|/s_{\hat{Y}}$ is distributed as t with (n - 2) degrees of freedom. It follows therefore that the 95% confidence limits for Y are $\hat{Y} \pm t_{0.05}s_{\hat{Y}}$.

Imagine that on one occasion the discharge of the Welsh stream is measured, but not for some reason the solute concentration (perhaps the water sample is accidentally spilt during laboratory analysis). An estimate of the solute concentration is provided by the regression equation $\hat{Y} = 35.3372 + 1.6859X$. If the discharge is 5 litres per second and increasing, the estimated solute concentration (ignoring possible confounding, errors of measurement, etc.) is 35.3372 + 1.6859(5) = 43.7667 mg/litre. This estimate is precisely the same as that for the mean concentration on all the occasions when the discharge is 5 litres per second and increasing. The estimated standard error is larger, however, than that associated with the mean concentration

$$s_{\hat{Y}} = 2.6597\sqrt{\left[1 + \frac{1}{11} + \frac{11(5 - 5.1545)^2}{11(381.95) - (56.7)^2}\right]}$$

$$= 2.6597\sqrt{1.09118} = 2.7783.$$

The 95% confidence limits for the estimated solute concentration are $\hat{Y} - t_{0.05}s_{\hat{Y}}$ and $\hat{Y} + t_{0.05}s_{\hat{Y}}$, or $43.7667 \pm 2.262\ (2.7783)$. The limits are therefore 37.48 and 50.05 mg per litre.

*THE ACCURACY WITH WHICH X IS PREDICTED BY Y

If the pairs of values of X and Y have been selected randomly, the correct regression equation to use to predict X from Y is that of X on Y, not Y on X. The equation of X on Y takes the form $Y = a' + b'\hat{X}$ which when rearranged becomes $\hat{X} = (Y - a')/b'$. $\hat{X}$ is the predicted value of X corresponding to a given Y, while

$$b' = \frac{n\Sigma Y^2 - (\Sigma Y)^2}{n\Sigma XY - \Sigma X\Sigma Y} \quad \text{and} \quad a' = (\Sigma Y/n) - b'(\Sigma X/n) = \bar{Y} - b'\bar{X}$$

The sum of the squared horizontal deviations about the regression line is $\Sigma(X - \hat{X})^2$ where

$$\Sigma(X - \hat{X})^2 = \frac{n\Sigma X^2 - (\Sigma X)^2}{n} - \frac{(n\Sigma XY - \Sigma X\Sigma Y)^2}{n^2\Sigma Y^2 - n(\Sigma Y)^2}$$

The standard error of estimate $s_{x.y}$ is given by

$$s_{x.y} = \sqrt{\frac{\Sigma(X - \hat{X})^2}{n - 2}} = \sqrt{\left[\frac{n\Sigma X^2 - (\Sigma X)^2}{n} - \frac{(n\Sigma XY - \Sigma X\Sigma Y)^2}{n^2\Sigma Y^2 - n(\Sigma Y)^2}\right] \Big/ (n - 2)}$$

The 95% confidence limits for a value of $\hat{X}$ calculated from the regression of X on Y are

$$\hat{X} \pm t_{0.05}s_{x.y}\sqrt{\left[1 + \frac{1}{n} + \frac{n(Y - \bar{Y})^2}{n\Sigma Y^2 - (\Sigma Y)^2}\right]}$$

For the stream data, the discharge corresponding to a solute concentration of 50 mg/litre is estimated by the regression of X on Y to be $\hat{X} = (50 - 33.1668)/2.1070 = 7.9892$ litres per sec. Note that the estimate of solute concentration obtained by putting X equal to 7.9892 litres per sec in the equation for Y on X is $\hat{Y} = 35.3372 + 1.6859\ (7.9892) = 48.806$. The two equations are not identical.

The standard error of estimate $s_{x.y}$ is

$$s_{x.y} = \sqrt{\left\{\left[\frac{11(381.95) - (56.7)^2}{11} - \frac{[11(2647.55) - (56.7)(484.3)]^2}{11^2(21640.99) - 11(484.3)^2}\right] \Big/ (11 - 2)\right\}}$$

$$= \sqrt{\left\{\left[\frac{986.56}{11} - \frac{2766367.298}{38548.4}\right] \Big/ 9\right\}}$$

$$= \sqrt{(17.92379/9)} = 1.4112 \text{ litres per second.}$$

The 95% confidence limits for the value of $\hat{X}$ of 7.9892 are

$$7.9892 \pm 2.262\ (1.4112) \sqrt{\left[1 + \frac{1}{11} + \frac{11(50 - 44.02727)}{11(21640.99) - (484.3)^2}\right]}$$

$$= 7.9892 \pm 2.262\ (1.4112)(1.0534) = 7.9892 \pm 3.3626$$

$$= 4.63 \text{ to } 11.35 \text{ litres per second.}$$

If the values of X are selected non-randomly and only the values of Y are free to vary randomly, the regression of Y on X must be used to predict X from Y, instead of the regression of X on Y. The regression equation is $\hat{Y} = a + bX$. When used in reverse to predict X the equation becomes $\hat{X} = (Y - a)/b$ where $\hat{X}$ as before is the predicted value of X corresponding to a given Y. The 95% confidence limits are

$$\hat{X} + \frac{(\hat{X} - \bar{X})g \pm \frac{t_{0.05}s_{y.x}}{b} \sqrt{\frac{n(\hat{X} - \bar{X})}{n\Sigma X^2 - (\Sigma \bar{X})^2}} \sqrt{\left[\frac{n+1}{n}(1 - g)\right]}}{1 - g}$$

where g is given by

$$g = \frac{n(t_{0.05})^2 s_{y.x}^2}{b^2[n\Sigma X^2 - (\Sigma X)^2]} = \frac{n(t_{0.05})^2 s_{y.x}^2}{b(n\Sigma XY - \Sigma X\Sigma Y)}$$

The confidence limits may be unequal distances from X, and can under certain circumstances be non-existent (Williams, 1959).

*THE ACCURACY WITH WHICH $\sigma_{y.x}^2$ IS PREDICTED BY $s_{y.x}^2$

It is sometimes useful to calculate confidence limits for $\sigma_{y.x}^2$, the variance of the population Y values about the population regression line $\mu = \alpha + \beta X$. The 95% confidence limits for $\sigma_{y.x}^2$ are

$$\frac{(n-2)s_{y.x}^2}{\chi_{0.975}^2} \quad \text{and} \quad \frac{(n-2)s_{y.x}^2}{\chi_{0.25}^2}$$

(n - 2) degrees of freedom are associated with χ^2. The 99% confidence limits are

$$\frac{(n-2)s_{y.x}^2}{\chi_{0.995}^2} \quad \text{and} \quad \frac{(n-2)s_{y.x}^2}{\chi_{0.005}^2}$$

COMPARING THE REGRESSION LINES OF TWO SAMPLES

Sometimes two samples that provide data on the same pair of variables are available. Once the regression lines of the samples have been calculated they can be examined to see whether they differ in any way. Any differences that are found will need to be evaluated. Are they small and likely to be due to sampling error, or are they so large that it can be safely assumed that the regression lines of the sampled populations are different?

It is convenient to start by examining the scatter (or standard errors of estimate) of the data points about the regression lines. If the scatter varies from one line to another, is this likely to be due to sampling error or is it an indication that the samples are drawn from populations that have different degrees of scatter? The squared standard errors of estimate of the samples can be compared using an ordinary two-tailed F test as described in Chapter 14. The test procedure should by now be familiar, but a numerical example will be provided later in the present chapter to try to clear up any uncertainties that may persist.

Having examined the scatter, one can go on to examine the slopes of the regression lines or regression coefficients. If the regression lines have different slopes, is sampling error likely to be responsible or can one assume that the regression lines of the sampled populations have different slopes (regression coefficients)? This question can be answered quite easily using a test based on Student's t that is explained later in the chapter.

After comparing the slopes of the sample regression lines one may wish to go on to compare the intercepts. If the intercepts differ, is this likely to be because of sampling error or because the samples are drawn from populations whose regression lines differ in elevation? The answer to this question is fairly obvious if the sample regression lines have significantly different slopes. Indeed, the question hardly needs to be raised. The reason is simply that, if the population regression lines have different slopes, they must necessarily differ in elevation, except at the point where they cross. Unless the lines happen to cross at the Y-axis, there is no need to ask whether they have the same intercept. Only if the population regression lines have the same slope is it important to try and find out if the intercepts differ. If the intercepts prove to be the same, then the lines must be coincident.

A simple test is available for comparing the intercepts of sample regression lines. Like the test of slopes it is based on Student's t, and is described later in the chapter.

TO TEST WHETHER THE REGRESSION LINES OF TWO SAMPLES HAVE SIGNIFICANTLY DIFFERENT SLOPES OR INTERCEPTS

THE DATA

Two samples are drawn independently from two populations. One sample of size n has a regression equation $\hat{Y} = a_1 + b_1X$. The other sample of size n_2 has an equation $\hat{Y} = a_2 + b_2X$.

PROCEDURE (TEST OF SLOPES, i.e. REGRESSION COEFFICIENTS)

1. Set up the null hypothesis: the difference between the two regression coefficients b_1 and b_2 is due entirely to sampling error. The two samples are drawn from populations with the same regression coefficient β. Both b_1 and b_2 are estimates of β.

2. Calculate the square of the standard error of estimate of each

sample

$$s_{y.x}^{2} = \frac{\Sigma(Y - \hat{Y})^2}{n - 2}$$

Compare the squared standard errors using an F test.

3. If the squared standard errors are not significantly different, assume that the variances of the points about the population regression lines are equal. Estimate the common variance of the points using the formula

$$s^2 = \frac{(s_1^2)(n_1 - 2) + (s_2^2)(n_2 - 2)}{(n_1 - 2) + (n_2 - 2)}$$

where s_1^2 is the variance of the points in sample 1 about the sample regression line and s_2^2 is the variance of the points in sample 2. s^2 is an unbiased estimate provided that the variances of the points about the population regression lines really are equal.

4. Estimate the standard error of the difference between the two regression coefficients

$$s_{b_1-b_2} = s\sqrt{\left[\frac{1}{\Sigma(X_1 - \overline{X}_1)^2} + \frac{1}{\Sigma(X_2 - \overline{X}_2)^2}\right]}$$

$$= s\sqrt{\left[\frac{n_1}{n_1\Sigma X_1^2 - (\Sigma X_1)^2} + \frac{n_2}{n_2\Sigma X_2^2 - (\Sigma X_2)^2}\right]}$$

5. Calculate t using the equation $t = |b_1 - b_2|/s_{b_1-b_2}$.

6. Calculate the degrees of freedom, $df = n_1 + n_2 - 4$.

7. Using Table E find P, the probability of obtaining this value of t or one greater through chance sampling. P is the probability of obtaining the observed difference between the sample regression coefficients or one greater given the estimated standard error.

Note that the values of P listed in the table are for two-tailed tests. If a one-tailed test is needed, halve the value of P.

8. Decide whether to retain or reject the null hypothesis. If P is low, serious doubt is cast on the null hypothesis. Reject the null if P is equal to or less than whatever level of significance is judged appropriate, e.g. 0.05 or 0.01. Conclude that the difference between the sample regression coefficients is more than would ordinarily occur if the samples were drawn from populations with identical regression coefficients.

If P is greater than the level of significance, reserve judgement, i.e. regard the null hypothesis as possibly, though not necessarily, correct. Conclude that the two populations may have the same regression coefficient β. An unbiased estimate of β is given by

$$b = \frac{n_1(\Sigma X_1Y_1) + n_2(\Sigma X_2Y_2) - (\Sigma X_1\Sigma Y_1) - (\Sigma X_2\Sigma Y_2)}{n_1(\Sigma X_1{}^2) + n_2(\Sigma X_2{}^2) - (\Sigma X_1)^2 - (\Sigma X_2)^2}$$

Note that b is not the same as the regression coefficient of the combined samples.

PROCEDURE (TEST OF INTERCEPTS, i.e. ELEVATIONS)

9. Set up the null hypothesis: the difference between the two intercepts a_1 and a_2 is due entirely to sampling error. The population regression lines share the same intercept α. Both a_1 and a_2 are estimates of α.

10. Calculate Q

$$Q = \Sigma Y^2 - \frac{(\Sigma Y)^2}{n} - \frac{(n\Sigma XY - \Sigma X\Sigma Y)^2}{n[n\Sigma X^2 - (\Sigma X)^2]}$$

where $\Sigma Y^2 = \Sigma Y_1{}^2 + \Sigma Y_2{}^2$, $\Sigma Y = \Sigma Y_1 + \Sigma Y_2$, $\Sigma XY = \Sigma(X_1Y_1) + \Sigma(X_2Y_2)$, $\Sigma X^2 = \Sigma X_1{}^2 + \Sigma X_2{}^2$, $\Sigma X = \Sigma X_1 + \Sigma X_2$, and $n = n_1 + n_2$.

11. Calculate R

$$R = s^2(n - 4) + \frac{(b_1 - b_2)^2}{\left[\dfrac{n_1}{n_1\Sigma X_1{}^2 - (\Sigma X_1)^2} + \dfrac{n_2}{n_2\Sigma X_2{}^2 - (\Sigma X_2)^2}\right]}$$

Note that the square root of the quantity enclosed in square brackets has already been calculated in Step 4.

12. Calculate t

$$t = \sqrt{\frac{(Q - R)(n - 3)}{R}}$$

13. Calculate the degrees of freedom, df = n - 3.

14. Using Table E find P, the probability of obtaining the calculated value of t, or one greater, through chance sampling.

Note that the values of P listed in the table are for two-tailed tests. If a one-tailed test is needed, halve the value of P.

15. Decide whether to retain or reject the null hypothesis. If P is low, serious doubt is cast on the null hypothesis. Reject the null if P is equal to or less than whatever level of significance is judged appropriate, e.g. 0.05 or 0.01. Conclude that the difference between the sample intercepts is more than would ordinarily occur if the samples were drawn from populations where the regression lines have the same intercept.

If P is greater than the level of significance, reserve judgement, i.e. regard the null hypothesis as possibly, though not necessarily, correct. Conclude that the population regression lines may have the same intercept α. An unbiased estimate of α is given by $a = (\Sigma Y/n) - b(\Sigma X/n)$ where b is the unbiased estimate of β calculated in Step 8.

ASSUMPTIONS AND LIMITATIONS

Both t tests assume that the samples meet the requirements of the linear model. They also assume that the points are equally scattered about population regression lines. If the amount of scatter in the sampled populations appears to differ, the test of the slopes of the regression lines can be modified by calculating

$$t = \frac{|b_1 - b_2|}{\sqrt{\left[\dfrac{n_1(s_1^2)}{n_1\Sigma X_1^2 - (\Sigma X_1)^2} + \dfrac{n_2(s_2^2)}{n_2\Sigma X_2^2 - (\Sigma X_2)^2}\right]}}$$

If n_1 and/or n_2 is less than 20, Johnson and Leone (1964) recommend reducing the degrees of freedom to

$$df = \frac{1}{\left[\dfrac{c^2}{(n_1 - 2)} + \dfrac{(1 - c)^2}{(n_2 - 2)}\right]}$$

where

$$c = \frac{\left[\dfrac{n_1 s_1^2}{n_1\Sigma X_1^2 - (\Sigma X_1)^2}\right]}{\left[\dfrac{n_1(s_1^2)}{n_1\Sigma X_1^2 - (\Sigma X_1)^2} + \dfrac{n_2(s_2^2)}{n_2\Sigma X_2^2 - (\Sigma X_2)^2}\right]}$$

Many authors do not bother with this refinement, however.

The t test of the intercepts of the regression lines cannot be readily modified to allow for unequal scatter. Another limitation of the test is that it assumes that the slopes of the population regression lines are exactly equal. If the slopes differ, the test may give very inaccurate results. Unfortunately, even though the slopes of the sample regression lines may not be significantly different, one cannot be sure that the slopes of the population lines are the same. If the samples are only small, a real difference may be declared to be non-significant.

EXAMPLE

A number of streams drain into Llyn Ebyr in Central Wales besides the one discussed in previous chapters which Oxley (1974) has named "Ebyr North". Data for a second stream ("Ebyr South") are shown in Table 20.1.

TABLE 20.1 - DISCHARGE AND CONCENTRATIONS OF DISSOLVED SOLIDS DURING RISING STAGES OF A SMALL STREAM

	Discharge of stream (litres/sec) *X*	*Concentration of dissolved solids (mg/litre)* *Y*	X^2	Y^2	*XY*
	2.7	41.4	7.29	1713.96	111.78
	3.7	40.0	13.69	1600.00	148.00
	4.2	44.3	17.64	1962.49	186.06
	5.0	42.1	25.00	1772.41	210.50
	5.1	43.9	26.01	1927.21	223.89
	5.5	43.0	30.25	1849.00	236.50
	6.2	41.8	38.44	1747.24	259.16
	7.3	44.2	53.29	1953.64	322.66
	7.7	43.3	59.29	1874.89	333.41
	8.5	43.8	72.25	1918.44	372.30
	9.9	46.2	98.01	2134.44	457.38
	12.0	46.3	144.00	2143.69	555.60
Total	77.8	520.3	585.16	22597.41	3417.24

Source: Oxley (1974, p.146). The values of X and Y are taken from a graph and may not be entirely accurate.

The regression line for Ebyr South is $\hat{Y} = 39.82899 + 0.54437X$ whereas that for Ebyr North is $\hat{Y} = 35.33723 + 1.68590X$. Do the two lines have significantly different slopes and intercepts?

1. In the case of the first sample (Ebyr North), the standard error of estimate = 2.6597 (see Chapter 19). The square of the estimate is 7.0742.

In the case of the second sample (Ebyr South)

$$\Sigma(Y_2 - \hat{Y}_2)^2 = \left[\Sigma Y_2{}^2 - \frac{(\Sigma Y_2)^2}{n_1}\right] - \frac{\left[\Sigma X_2 Y_2 - \dfrac{\Sigma X_2 \Sigma Y_2}{n}\right]^2}{\left[\Sigma X_2{}^2 - \dfrac{(\Sigma X_2)^2}{n_2}\right]}$$

$$= \left[22597.41 - \frac{(520.3)^2}{12}\right] - \frac{\left[3417.24 - \dfrac{(77.8)(520.3)}{12}\right]^2}{\left[585.16 - \dfrac{(77.8)^2}{12}\right]}$$

$$= 14.13767$$

and the squared standard error of estimate is 14.13767/(12 - 2) = 1.41377.

2. The difference between the two squared standard errors is quite large. It can be tested for significance by calculating F

$$F = \frac{\text{larger squared standard error}}{\text{smaller squared standard error}} = \frac{7.07420}{1.41377} = 5.004$$

With (11 - 2) = 9 degrees of freedom in the numerator and (12 - 2) = 10 degrees in the denominator this is significant at the 0.05 level (two-tailed test). The scatter of points about the population regression line for Ebyr North is almost certainly greater than the scatter of points about the line for Ebyr South.

Because the scatter of the data points in the two populations must be presumed to be unequal the standard t test of the regression coefficients described in the last section is not strictly applicable, but nevertheless the computations will be carried out for illustrative purposes.

3. The pooled estimate of the common variance is

$$s^2 = \frac{63.66784 + 14.13767}{(11 - 2) + (12 - 2)} = \frac{77.80551}{19} = 4.09503$$

$$s = \sqrt{4.09503} = 2.02362$$

4. The estimated standard error of the difference between the regression coefficients is

$$s_{b_1 - b_2} = s\sqrt{\left[\frac{n_1}{n_1 \Sigma X_1^2 - (\Sigma X_1)^2} + \frac{n_2}{n_2 \Sigma X_2^2 - (\Sigma X_2)^2}\right]}$$

$$= 2.02362\sqrt{\left[\frac{11}{11(381.95) - (56.7)^2} + \frac{12}{12(585.16) - (77.8)^2}\right]}$$

$$= 2.02362\sqrt{\left[\frac{11}{986.56} + \frac{12}{969.08}\right]}$$

$$= 0.31043$$

5. $$t = \frac{1.6859 - 0.5444}{0.31043} = \frac{1.1415}{0.31043} = 3.677$$

6. The degrees of freedom are 11 + 12 - 4 = 19.

7. Referring to Table E, the value of P is found to lie between 0.002 and 0.001. The null hypothesis that the samples are drawn from populations with the same regression coefficient is scarcely tenable. The evidence suggests that the populations have different coefficients.

8. If the P value had been high, one might have been tempted to conclude that the two populations shared the same regression coefficient. An estimate of this common coefficient would have been given by

$$b = \frac{n_1(\Sigma X_1 Y_1) + n_2(\Sigma X_2 Y_2) - (\Sigma X_1 \Sigma Y_1) - (\Sigma X_2 \Sigma Y_2)}{n_1(\Sigma X_1^2) + n_2(\Sigma X_2^2) - (\Sigma X_1)^2 - (\Sigma X_2)^2}$$

$$= \frac{11(2647.55) + 12(3417.24) - (56.7)(484.3) - (77.8)(520.3)}{11(381.95) + 12(585.16) - (56.7)^2 - (77.8)^2}$$

$$= \frac{70129.93 - 67939.15}{11223.37 - 9267.73} = 1.1202.$$

If the two samples are combined into a single sample, the regression coefficient is 1.0525, which is slightly smaller.

9. As already mentioned, the standard t test of the regression coefficients is not really appropriate since the residual variances of the populations can be assumed to be unequal. The modified test now follows

$$t = \frac{1.6859 - 0.5444}{\sqrt{\left[\frac{11(7.07420)}{11(381.95) - (56.7)^2} + \frac{12(1.41377)}{12(585.16) - (77.8)^2}\right]}}$$

$$= \frac{1.1415}{\sqrt{0.07888 + 0.01751}} = 3.677$$

It is an accident that this is the same value of t (to 3 decimal places) as was calculated before.

The value of c is

$$c = \frac{0.07888}{0.07888 + 0.01751} = 0.81836$$

$$df = \frac{1}{\left[\frac{(0.81836)^2}{9} + \frac{(0.18164)^2}{10}\right]}$$

$$= 12.87 \text{ or } 13 \text{ approximately.}$$

According to Table E the value of P lies between 0.005 and 0.002. The null hypothesis can be rejected, though not as decisively as before.

10. Since the population regression lines can be assumed to have different slopes there is little point testing whether the intercepts of the sample regression lines are significantly different. The following calculations are intended purely for purposes of illustration.

When the samples are combined, $\Sigma X = \Sigma X_1 + \Sigma X_2 = 56.7 + 77.8 = 134.5$, $\Sigma Y = \Sigma Y_1 + \Sigma Y_2 = 484.3 + 520.3 = 1004.6$, $\Sigma X^2 = \Sigma X_1{}^2 + \Sigma X_2{}^2 = 381.95 + 585.16 = 967.11$, $\Sigma Y^2 = \Sigma Y_1{}^2 + \Sigma Y_2{}^2 = 21640.99 + 22597.41 = 44238.4$, and $\Sigma XY = \Sigma(X_1Y_1) + \Sigma(X_2Y_2) = 2647.55 + 3417.24 = 6064.79$. There are $n_1 + n_2 = 11 + 12 = 23$ pairs of measurements of X and Y.

11. $$Q = 44238.4 - \frac{(1004.6)^2}{23} - \frac{[23(6064.79) - 134.5(1004.6)]^2}{23[23(967.11) - (134.5)^2]}$$

$$= 159.17033$$

12. $$R = (4.09503)(19) + \frac{(1.68590 - 0.54437)^2}{\left[\frac{11}{(11(381.95) - (56.7)^2} + \frac{12}{12(585.16) - (77.8)^2}\right]}$$

$$= 77.80591 + \frac{(1.14153)^2}{\left[\frac{11}{986.56} + \frac{12}{969.08}\right]}$$

$$= 133.17870$$

13. $$t = \sqrt{\frac{(159.17033 - 133.17870)(20)}{133.17870}} = 1.976$$

14. With 20 degrees of freedom t falls just short of the 0.05 level of significance. The intercepts of the sample regression lines do not differ significantly.

Although the test has been carried out for illustrative purposes, it is not really needed, as has already been mentioned. Since the regression lines are most unlikely to be parallel, there is hardly any possibility of their intercepts being equal. The failure of the test to show that the intercepts are different could be the result of the small sizes of the samples. The test provides no positive evidence that the intercepts are the same. Note also that even if the test had shown a significant difference it would still be suspect since it assumes that the residual variances of the populations are equal, and also that the slopes (regression coefficients) are equal. Both assumptions are unrealistic.

Statistical Tables

TABLE A : RANDOM DIGITS

37694	42904	05468	16343	23804	04078	03815	62072	44907	72351
98375	04491	22041	04462	70774	09226	70114	42877	02355	23411
13927	26689	20931	14794	82022	26980	66106	07507	16607	11223
88828	46319	14784	39125	29973	58623	45174	80128	42085	24933
56719	84168	30141	23612	14440	12319	61475	80428	81622	06489
41812	07832	31220	22936	35222	49856	34092	22230	07244	43517
68088	56148	08510	07661	75972	77414	18920	63610	68886	23904
07791	14801	46983	66336	16107	63381	45677	66284	50309	97194
94659	76301	77639	84391	32280	14051	69736	76310	88197	49298
82833	40679	50344	18818	13855	04839	02213	60087	85590	58078
61957	36627	13061	55452	22265	71841	19102	28142	05815	35689
44093	52045	28472	83360	32417	12881	07368	23717	82920	88942
77913	53792	92250	82105	41782	51795	50825	21153	37632	39534
61459	63340	90388	44602	36945	38569	08446	85468	55193	00280
67690	43127	62906	91203	16284	55982	84552	77962	60549	80170
35513	72838	27518	31053	46514	33124	96101	30710	50530	24814
77675	72705	60813	29166	37423	08937	28034	17600	82816	60411
32946	57559	21369	83525	91255	52036	31466	73461	20668	53766
32240	86978	67793	24253	48419	30944	66217	61648	57008	45918
45672	61525	03817	49585	57441	36249	22770	28377	18176	35744
10084	54945	55643	35748	11243	67727	09679	62083	68323	59363
27130	29095	71173	02763	64276	73287	11376	32808	90940	85241
51246	02220	11879	03032	34710	92577	25530	36217	11243	58554
88915	98235	86051	64775	69839	46318	60568	06057	50163	38698
70045	00429	69728	73270	47280	17451	33812	65358	59590	82858
73672	23005	72383	21730	60517	37089	01091	81476	92329	01914
72432	30167	05616	18797	40404	74169	16207	95976	30969	09137
54099	37711	80311	66308	32373	38827	87826	53679	53849	61758
86607	98917	23044	24852	75949	26216	76853	50433	75500	65169
35853	18230	07223	70325	47606	36360	88988	04245	38991	64344
41489	79055	64821	36579	39395	81029	98870	94850	68502	28417
04486	30936	58884	25291	36888	61156	53724	30189	70473	87566
53893	64594	48171	72037	03767	89829	21007	25945	74363	79710
78085	31176	92110	83537	94396	76251	24945	06595	69702	09098
35095	10602	34852	53430	70250	18221	65259	16640	09874	41795
25216	94180	15495	82617	60927	83841	67645	12176	85935	84084
57777	16728	42433	38881	56355	45292	14433	68394	05084	00104
88802	31835	00282	64080	55865	62106	88319	55642	14210	21907
85109	51356	73978	26934	82951	92184	62627	64375	05824	37031
58930	46277	48139	94388	46477	93547	17469	17346	35608	65295
55992	95849	72230	89967	61066	39808	22111	64272	72512	78593
27011	20963	08419	48790	01876	97184	70154	32893	28168	76430
27975	07346	00747	90184	17526	30704	76075	24569	56315	92380
30074	73053	87403	38731	50397	69803	33137	55720	27257	95285
91457	31266	97838	08056	73668	32444	28184	58295	58174	60468
71330	89406	20856	83100	30252	83836	85652	53022	68947	57055
65119	66579	21529	03281	94866	34606	71225	54985	60049	15546
46510	72798	95997	27692	23111	61895	49339	65805	22560	09362
42516	98750	98239	27489	69209	28775	81973	19104	21423	99360
02704	68143	11565	99738	79911	70228	75679	49852	86051	57291

TABLE B : AREAS UNDER THE NORMAL CURVE IN THE TWO TAILS

Entries in the body of the table refer to the shaded areas in the diagram. These areas are listed as proportions of the total area under the normal curve.

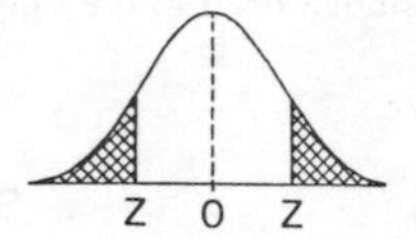

Second decimal place of Z

Z	.00	.01	.02	.03	.04	.05	.06	.07	.08	.09
0.0	1.0000	.9920	.9840	.9761	.9681	.9601	.9522	.9442	.9362	.9283
0.1	.9203	.9124	.9045	.8966	.8887	.8808	.8729	.8650	.8572	.8493
0.2	.8415	.8337	.8259	.8181	.8103	.8026	.7949	.7872	.7795	.7718
0.3	.7642	.7566	.7490	.7414	.7339	.7263	.7188	.7114	.7039	.6965
0.4	.6892	.6818	.6745	.6672	.6599	.6527	.6455	.6384	.6312	.6241
0.5	.6171	.6101	.6031	.5961	.5892	.5823	.5755	.5687	.5619	.5552
0.6	.5485	.5419	.5353	.5287	.5222	.5157	.5093	.5029	.4965	.4902
0.7	.4839	.4777	.4715	.4654	.4593	.4533	.4473	.4413	.4354	.4295
0.8	.4237	.4179	.4122	.4065	.4009	.3953	.3898	.3843	.3789	.3735
0.9	.3681	.3628	.3576	.3524	.3472	.3421	.3371	.3320	.3271	.3222
1.0	.3173	.3125	.3077	.3030	.2983	.2937	.2891	.2846	.2801	.2757
1.1	.2713	.2670	.2627	.2585	.2543	.2501	.2460	.2420	.2380	.2340
1.2	.2301	.2263	.2225	.2187	.2150	.2113	.2077	.2041	.2005	.1971
1.3	.1936	.1902	.1868	.1835	.1802	.1770	.1738	.1707	.1676	.1645
1.4	.1615	.1585	.1556	.1527	.1499	.1471	.1443	.1416	.1389	.1362
1.5	.1336	.1310	.1285	.1260	.1236	.1211	.1188	.1164	.1141	.1118
1.6	.1096	.1074	.1052	.1031	.1010	.0989	.0969	.0949	.0930	.0910
1.7	.0891	.0873	.0854	.0836	.0819	.0801	.0784	.0767	.0751	.0735
1.8	.0719	.0703	.0688	.0672	.0658	.0643	.0629	.0615	.0601	.0588
1.9	.0574	.0561	.0549	.0536	.0524	.0512	.0500	.0488	.0477	.0466
2.0	.0455	.0444	.0434	.0424	.0414	.0404	.0394	.0385	.0375	.0366
2.1	.0357	.0349	.0340	.0332	.0324	.0316	.0308	.0300	.0293	.0285
2.2	.0278	.0271	.0264	.0257	.0251	.0244	.0238	.0232	.0226	.0220
2.3	.0214	.0209	.0203	.0198	.0193	.0188	.0183	.0178	.0173	.0168
2.4	.0164	.0160	.0155	.0151	.0147	.0143	.0139	.0135	.0131	.0128
2.5	.0124	.0121	.0117	.0114	.0111	.0108	.0105	.0102	.00988	.00960
2.6	.00932	.00905	.00879	.00854	.00829	.00805	.00781	.00759	.00736	.00715
2.7	.00693	.00673	.00653	.00633	.00614	.00596	.00578	.00561	.00544	.00527
2.8	.00511	.00495	.00480	.00465	.00451	.00437	.00424	.00410	.00398	.00385
2.9	.00373	.00361	.00350	.00339	.00328	.00318	.00308	.00298	.00288	.00279

First decimal place of Z

Z	.0	.1	.2	.3	.4	.5	.6	.7	.8	.9
3	.00270	.00194	.00137	$.0^{3}967$	$.0^{3}674$	$.0^{3}465$	$.0^{3}318$	$.0^{3}216$	$.0^{3}145$	$.0^{4}962$
4	$.0^{4}633$	$.0^{4}413$	$.0^{4}267$	$.0^{4}171$	$.0^{4}108$	$.0^{5}680$	$.0^{5}422$	$.0^{5}260$	$.0^{5}159$	$.0^{6}958$
5	$.0^{6}573$	$.0^{6}340$	$.0^{6}199$	$.0^{6}116$	$.0^{7}666$	$.0^{7}380$	$.0^{7}214$	$.0^{7}120$	$.0^{8}663$	$.0^{8}364$
6	$.0^{8}197$	$.0^{8}106$	$.0^{9}565$	$.0^{9}298$	$.0^{9}155$	$.0^{10}803$	$.0^{10}411$	$.0^{10}208$	$.0^{10}105$	$.0^{11}520$

TABLE C : AREAS UNDER THE NORMAL CURVE FROM THE CENTRE TO THE RIGHT TAIL

Entries in the body of the table refer to the shaded area in the diagram.

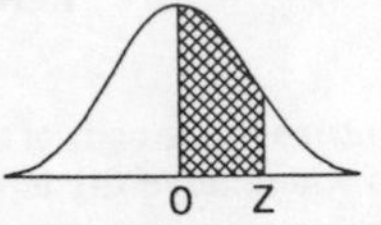

Second decimal place of Z

Z	.00	.01	.02	.03	.04	.05	.06	.07	.08	.09
0·0	0·0000	0·0040	0·0080	0·0120	0·0160	0·0199	0·0239	0·0279	0·0319	0·0359
0·1	0·0398	0·0438	0·0478	0·0517	0·0557	0·0596	0·0636	0·0675	0·0714	0·0754
0·2	0·0793	0·0832	0·0871	0·0910	0·0948	0·0987	0·1026	0·1064	0·1103	0·1141
0·3	0·1179	0·1217	0·1255	0·1293	0·1331	0·1368	0·1406	0·1443	0·1480	0·1517
0·4	0·1554	0·1591	0·1628	0·1664	0·1700	0·1736	0·1772	0·1808	0·1844	0·1879
0·5	0·1915	0·1950	0·1985	0·2019	0·2054	0·2088	0·2123	0·2157	0·2190	0·2224
0·6	0·2258	0·2291	0·2324	0·2357	0·2389	0·2422	0·2454	0·2486	0·2518	0·2549
0·7	0·2580	0·2612	0·2642	0·2673	0·2704	0·2734	0·2764	0·2794	0·2823	0·2852
0·8	0·2881	0·2910	0·2939	0·2967	0·2996	0·3023	0·3051	0·3078	0·3106	0·3133
0·9	0·3159	0·3186	0·3212	0·3238	0·3264	0·3289	0·3315	0·3340	0·3365	0·3389
1·0	0·3413	0·3438	0·3461	0·3485	0·3508	0·3531	0·3554	0·3577	0·3599	0·3621
1·1	0·3643	0·3665	0·3686	0·3708	0·3729	0·3749	0·3770	0·3790	0·3810	0·3830
1·2	0·3849	0·3869	0·3888	0·3907	0·3925	0·3944	0·3962	0·3980	0·3997	0·4015
1·3	0·4032	0·4049	0·4066	0·4082	0·4099	0·4115	0·4131	0·4147	0·4162	0·4177
1·4	0·4192	0·4207	0·4222	0·4236	0·4251	0·4265	0·4279	0·4292	0·4306	0·4319
1·5	0·4332	0·4345	0·4357	0·4370	0·4382	0·4394	0·4406	0·4418	0·4429	0·4441
1·6	0·4452	0·4463	0·4474	0·4484	0·4495	0·4505	0·4515	0·4525	0·4535	0·4545
1·7	0·4554	0·4564	0·4573	0·4582	0·4591	0·4599	0·4608	0·4616	0·4625	0·4633
1·8	0·4641	0·4649	0·4656	0·4664	0·4671	0·4678	0·4686	0·4693	0·4699	0·4706
1·9	0·4713	0·4719	0·4726	0·4732	0·4738	0·4744	0·4750	0·4756	0·4761	0·4767
2·0	0·4772	0·4778	0·4783	0·4788	0·4793	0·4798	0·4803	0·4808	0·4812	0·4817
2·1	0·4821	0·4826	0·4830	0·4834	0·4838	0·4842	0·4846	0·4850	0·4854	0·4857
2·2	0·4861	0·4864	0·4868	0·4871	0·4875	0·4878	0·4881	0·4884	0·4887	0·4890
2·3	0·4893	0·4896	0·4898	0·4901	0·4904	0·4906	0·4909	0·4911	0·4913	0·4916
2·4	0·4918	0·4920	0·4922	0·4925	0·4927	0·4929	0·4931	0·4932	0·4934	0·4936
2·5	0·4938	0·4940	0·4941	0·4943	0·4945	0·4946	0·4948	0·4949	0·4951	0·4952
2·6	0·4953	0·4955	0·4956	0·4957	0·4959	0·4960	0·4961	0·4962	0·4963	0·4964
[illegible]	[illegible]	[illegible]	·4967	0·4968	0·4969	0·4970	0·4971	0·4972	0·4973	0·4974
[illegible]	[illegible]	[illegible]	·4976	0·4977	0·4977	0·4978	0·4979	0·4979	0·4980	0·4981
[illegible]	[illegible]	[illegible]	·4982	0·4983	0·4984	0·4984	0·4985	0·4985	0·4986	0·4986
[illegible]	[illegible]	[illegible]	4987	0·4988	0·4988	0·4989	0·4989	0·4989	0·4990	0·4990
[illegible]	[illegible]	[illegible]	4991	0·4991	0·4992	0·4992	0·4992	0·4992	0·4993	0·4993
[illegible]	[illegible]	[illegible]	4994	0·4994	0·4994	0·4994	0·4994	0·4995	0·4995	0·4995
[illegible]	[illegible]	[illegible]	4995	0·4996	0·4996	0·4996	0·4996	0·4996	0·4996	0·4997
[illegible]	[illegible]	[illegible]	4997	0·4997	0·4997	0·4997	0·4997	0·4997	0·4997	0·4998
[illegible]	[illegible]	[illegible]	4998	0·4998	0·4998	0·4998	0·4998	0·4998	0·4998	0·4998
[illegible]	[illegible]	[illegible]	4999	0·4999	0·4999	0·4999	0·4999	0·4999	0·4999	0·4999
[illegible]	[illegible]	[illegible]	999	0·4999	0·4999	0·4999	0·4999	0·4999	0·4999	0·4999
[illegible]	[illegible]	[illegible]	999	0·4999	0·4999	0·4999	0·4999	0·4999	0·4999	0·4999
[illegible]	[illegible]	[illegible]	000	0·5000	0·5000	0·5000	0·5000	0·5000	0·5000	0·5000

[illegible]NATES OF THE NORMAL CURVE

First decimal place of Z

[illegible]	.0	.1	.2	.3	.4	.5	.6	.7	.8	.9
[illegible]	[illegible]	[illegible]	9802	.9560	.9231	.8825	.8353	.7827	.7262	.6670
[illegible]	[illegible]	[illegible]	4868	.4296	.3753	.3247	.2780	.2358	.1979	.1645
[illegible]	[illegible]	[illegible]	0889	.0710	.0561	.0439	.0341	.0261	.0198	.0149
[illegible]	[illegible]	[illegible]	0060	.0043	.0031	.0022	.0015	.0011	.0007	.0005
[illegible]	[illegible]	[illegible]	0002	.0001	.0001	.0000	.0000	.0000	.0000	.0000

TABLE E : VALUES OF t FOR SELECTED DEGREES OF FREEDOM (df) AND PROBABILITY P

df \ P	.20	.10	.05	.02	.01	.002	.001	.0002	.0001	.00002	.00001	.000002	.000001	.0000002
1	3.078	6.314	12.706	31.821	63.657	318.309	636.619	3,183.099	6,366.198	31,830.989	63,661.977	318,309.886	636,619.772	3,183,098.862
2	1.886	2.920	4.303	6.965	9.925	22.327	31.598	70.700	99.992	223.603	316.225	707.106	999.999	2,236.068
3	1.638	2.353	3.182	4.541	5.841	10.214	12.924	22.204	28.000	47.928	60.397	103.299	130.155	222.572
4	1.533	2.132	2.776	3.747	4.604	7.173	8.610	13.034	15.544	23.332	27.771	41.578	49.459	73.986
5	1.476	2.015	2.571	3.365	4.032	5.893	6.869	9.678	11.178	15.547	17.897	24.771	28.477	39.340
6	1.440	1.943	2.447	3.143	3.707	5.208	5.959	8.025	9.082	12.032	13.555	17.830	20.047	26.286
7	1.415	1.895	2.365	2.998	3.499	4.785	5.408	7.063	7.885	10.103	11.215	14.241	15.764	19.932
8	1.397	1.860	2.306	2.896	3.355	4.501	5.041	6.442	7.120	8.907	9.782	12.110	13.257	16.320
9	1.383	1.833	2.262	2.821	3.250	4.297	4.781	6.010	6.594	8.102	8.827	10.720	11.637	14.041
10	1.372	1.812	2.228	2.764	3.169	4.144	4.587	5.694	6.211	7.527	8.150	9.752	10.516	12.492
11	1.363	1.796	2.201	2.718	3.106	4.025	4.437	5.453	5.921	7.098	7.648	9.043	9.702	11.381
12	1.356	1.782	2.179	2.681	3.055	3.930	4.318	5.263	5.694	6.756	7.261	8.504	9.085	10.551
13	1.350	1.771	2.160	2.650	3.012	3.852	4.221	5.111	5.513	6.501	6.955	8.082	8.604	9.909
14	1.345	1.761	2.145	2.624	2.977	3.787	4.140	4.985	5.363	6.287	6.706	7.743	8.218	9.400
15	1.341	1.753	2.131	2.602	2.947	3.733	4.073	4.880	5.239	6.109	6.502	7.465	7.903	8.986
16	1.337	1.746	2.120	2.583	2.921	3.686	4.015	4.791	5.134	5.960	6.330	7.233	7.642	8.645
17	1.333	1.740	2.110	2.567	2.898	3.646	3.965	4.714	5.044	5.832	6.184	7.037	7.421	8.358
18	1.330	1.734	2.101	2.552	2.878	3.610	3.922	4.648	4.966	5.722	6.059	6.869	7.232	8.115
19	1.328	1.729	2.093	2.539	2.861	3.579	3.883	4.590	4.897	5.627	5.949	6.723	7.069	7.905
20	1.325	1.725	2.086	2.528	2.845	3.552	3.850	4.539	4.837	5.543	5.854	6.597	6.927	7.723
21	1.323	1.721	2.080	2.518	2.831	3.527	3.819	4.493	4.784	5.469	5.769	6.485	6.802	7.564
22	1.321	1.717	2.074	2.508	2.819	3.505	3.792	4.452	4.736	5.402	5.694	6.386	6.692	7.423
23	1.319	1.714	2.069	2.500	2.807	3.485	3.768	4.415	4.693	5.343	5.627	6.297	6.593	7.298
24	1.318	1.711	2.064	2.492	2.797	3.467	3.745	4.382	4.654	5.290	5.566	6.218	6.504	7.185
25	1.316	1.708	2.060	2.485	2.787	3.450	3.725	4.352	4.619	5.241	5.511	6.146	6.424	7.085
26	1.315	1.706	2.056	2.479	2.779	3.435	3.707	4.324	4.587	5.197	5.461	6.081	6.352	6.993
27	1.314	1.703	2.052	2.473	2.771	3.421	3.690	4.299	4.558	5.157	5.415	6.021	6.286	6.910
28	1.313	1.701	2.048	2.467	2.763	3.408	3.674	4.275	4.530	5.120	5.373	5.967	6.225	6.835
29	1.311	1.699	2.045	2.462	2.756	3.396	3.659	4.254	4.506	5.086	5.335	5.917	6.170	6.765
30	1.310	1.697	2.042	2.457	2.750	3.385	3.646	4.234	4.482	5.054	5.299	5.871	6.119	6.701
35	1.306	1.690	2.030	2.438	2.724	3.340	3.591	4.153	4.389	4.927	5.156	5.687	5.915	6.447
40	1.303	1.684	2.021	2.423	2.704	3.307	3.551	4.094	4.321	4.835	5.053	5.554	5.768	6.266
45	1.301	1.679	2.014	2.412	2.690	3.281	3.520	4.049	4.269	4.766	4.975	5.454	5.659	6.130
50	1.299	1.676	2.009	2.403	2.678	3.261	3.496	4.014	4.228	4.711	4.914	5.377	5.573	6.025
55	1.297	1.673	2.004	2.396	2.668	3.245	3.476	3.986	4.196	4.667	4.865	5.315	5.505	5.942
60	1.296	1.671	2.000	2.390	2.660	3.232	3.460	3.962	4.169	4.631	4.825	5.264	5.449	5.873
70	1.294	1.667	1.994	2.381	2.648	3.211	3.435	3.926	4.127	4.576	4.763	5.185	5.363	5.768
80	1.292	1.664	1.990	2.374	2.639	3.195	3.416	3.899	4.096	4.535	4.717	5.128	5.300	5.691
90	1.291	1.662	1.987	2.368	2.632	3.183	3.402	3.878	4.072	4.503	4.682	5.084	5.252	5.633
100	1.290	1.660	1.984	2.364	2.626	3.174	3.390	3.862	4.053	4.478	4.654	5.049	5.214	5.587
200	1.286	1.652	1.972	2.345	2.601	3.131	3.340	3.789	3.970	4.369	4.533	4.897	5.048	5.387
500	1.283	1.648	1.965	2.334	2.586	3.107	3.310	3.747	3.922	4.306	4.463	4.810	4.953	5.273
1,000	1.282	1.646	1.962	2.330	2.581	3.098	3.300	3.733	3.906	4.285	4.440	4.781	4.922	5.236
2,000	1.282	1.645	1.961	2.328	2.578	3.094	3.295	3.726	3.898	4.275	4.428	4.767	4.907	5.218
10,000	1.282	1.645	1.960	2.327	2.576	3.091	3.292	3.720	3.892	4.267	4.419	4.756	4.895	5.203
∞	1.282	1.645	1.960	2.326	2.576	3.090	3.291	3.719	3.891	4.265	4.417	4.753	4.892	5.199

Values of P are for a two-tailed test. For a one-tailed test they should be halved.

TABLE F : WILCOXON SIGNED–RANKS TEST

Entries in the body of the table are the probabilities for a **one**-tailed test. For a two-tailed test they should be doubled.

W \ n	1	2	3	4	5	6	7
0	.5000	.2500	.1250	.0625	.0313	.0156	.0078
1	1.0000	.5000	.2500	.1250	.0625	.0313	.0156
2		.7500	.3750	.1875	.0938	.0469	.0234
3		1.0000	.6250	.3125	.1563	.0781	.0391
4			.7500	.4375	.2188	.1094	.0547
5			.8750	.5625	.3125	.1563	.0781
6			1.0000	.6875	.4063	.2188	.1094
7				.8125	.5000	.2813	.1484
8				.8750	.5937	.3438	.1875
9				.9375	.6875	.4219	.2344
10				1.0000	.7812	.5000	.2891
11					.8437	.5781	.3438
12					.9062	.6562	.4063
13					.9375	.7187	.4688
14					.9687	.7812	.5312

W \ n	8	9	10	11	12	13	14
0	.0039	.0020	.0010	.0005	.0002	.0001	.0001
1	.0078	.0039	.0020	.0010	.0005	.0002	.0001
2	.0117	.0059	.0029	.0015	.0007	.0004	.0002
3	.0195	.0098	.0049	.0024	.0012	.0006	.0003
4	.0273	.0137	.0068	.0034	.0017	.0009	.0004
5	.0391	.0195	.0098	.0049	.0024	.0012	.0006
6	.0547	.0273	.0137	.0068	.0034	.0017	.0009
7	.0742	.0371	.0186	.0093	.0046	.0023	.0012
8	.0977	.0488	.0244	.0122	.0061	.0031	.0015
9	.1250	.0645	.0322	.0161	.0081	.0040	.0020
10	.1563	.0820	.0420	.0210	.0105	.0052	.0026
11	.1914	.1016	.0527	.0269	.0134	.0067	.0034
12	.2305	.1250	.0654	.0337	.0171	.0085	.0043
13	.2734	.1504	.0801	.0415	.0212	.0107	.0054
14	.3203	.1797	.0967	.0508	.0261	.0133	.0067
15	.3711	.2129	.1162	.0615	.0320	.0164	.0083
16	.4219	.2480	.1377	.0737	.0386	.0199	.0101
17	.4727	.2852	.1611	.0874	.0461	.0239	.0123
18	.5273	.3262	.1875	.1030	.0549	.0287	.0148
19	.5781	.3672	.2158	.1201	.0647	.0341	.0176
20	.6289	.4102	.2461	.1392	.0757	.0402	.0209
21	.6797	.4551	.2783	.1602	.0881	.0471	.0247
22	.7266	.5000	.3125	.1826	.1018	.0549	.0290
23	.7695	.5449	.3477	.2065	.1167	.0636	.0338
24	.8086	.5898	.3848	.2324	.1331	.0732	.0392
25	.8437	.6328	.4229	.2598	.1506	.0839	.0453
26	.8750	.6738	.4609	.2886	.1697	.0955	.0520
27	.9023	.7148	.5000	.3188	.1902	.1082	.0594
28	.9258	.7520	.5391	.3501	.2119	.1219	.0676
29	.9453	.7871	.5771	.3823	.2349	.1367	.0765
30	.9609	.8203	.6152	.4155	.2593	.1527	.0863
31	.9727	.8496	.6523	.4492	.2847	.1698	.0969
32	.9805	.8750	.6875	.4829	.3110	.1879	.1083
33	.9883	.8984	.7217	.5171	.3386	.2072	.1206
34	.9922	.9180	.7539	.5508	.3667	.2274	.1338
35	.9961	.9355	.7842	.5845	.3955	.2487	.1479
36	1.0000	.9512	.8125	.6177	.4250	.2709	.1629
37		.9629	.8389	.6499	.4548	.2939	.1788
38		.9727	.8623	.6812	.4849	.3177	.1955
39		.9805	.8838	.7114	.5151	.3424	.2131
40		.9863	.9033	.7402	.5452	.3677	.2316
41		.9902	.9199	.7676	.5750	.3934	.2508
42		.9941	.9346	.7935	.6045	.4197	.2708
43		.9961	.9473	.8174	.6333	.4463	.2915
44		.9980	.9580	.8398	.6614	.4730	.3129
45		1.0000	.9678	.8608	.6890	.5000	.3349
46			.9756	.8799	.7153	.5270	.3574
47			.9814	.8970	.7407	.5537	.3804
48			.9863	.9126	.7651	.5803	.4039
49			.9902	.9263	.7881	.6066	.4276
50			.9932	.9385	.8098	.6323	.4516
51			.9951	.9492	.8303	.6576	.4758
52			.9971	.9585	.8494	.6823	.5000

W \ n	15	16	17	18	19	20
0	.0000	.0000	.0000	.0000	.0000	.0000
1	.0001	.0000	.0000	.0000	.0000	.0000
2	.0001	.0000	.0000	.0000	.0000	.0000
3	.0002	.0001	.0000	.0000	.0000	.0000
4	.0002	.0001	.0001	.0000	.0000	.0000
5	.0003	.0002	.0001	.0000	.0000	.0000
6	.0004	.0002	.0001	.0001	.0000	.0000
7	.0006	.0003	.0001	.0001	.0000	.0000
8	.0008	.0004	.0002	.0001	.0000	.0000
9	.0010	.0005	.0003	.0001	.0001	.0000
10	.0013	.0007	.0003	.0002	.0001	.0000
11	.0017	.0008	.0004	.0002	.0001	.0001
12	.0021	.0011	.0005	.0003	.0001	.0001
13	.0027	.0013	.0007	.0003	.0002	.0001
14	.0034	.0017	.0008	.0004	.0002	.0001
15	.0042	.0021	.0010	.0005	.0003	.0001
16	.0051	.0026	.0013	.0006	.0003	.0002
17	.0062	.0031	.0016	.0008	.0004	.0002
18	.0075	.0038	.0019	.0010	.0005	.0002
19	.0090	.0046	.0023	.0012	.0006	.0003
20	.0108	.0055	.0028	.0014	.0007	.0004
21	.0128	.0065	.0033	.0017	.0008	.0004
22	.0151	.0078	.0040	.0020	.0010	.0005
23	.0177	.0091	.0047	.0024	.0012	.0006
24	.0206	.0107	.0055	.0028	.0014	.0007
25	.0240	.0125	.0064	.0033	.0017	.0008
26	.0277	.0145	.0075	.0038	.0020	.0010
27	.0319	.0168	.0087	.0045	.0023	.0012
28	.0365	.0193	.0101	.0052	.0027	.0014
29	.0416	.0222	.0116	.0060	.0031	.0016
30	.0473	.0253	.0133	.0069	.0036	.0018
31	.0535	.0288	.0153	.0080	.0041	.0021
32	.0603	.0327	.0174	.0091	.0047	.0024
33	.0677	.0370	.0198	.0104	.0054	.0028
34	.0757	.0416	.0224	.0118	.0062	.0032
35	.0844	.0467	.0253	.0134	.0070	.0036
36	.0938	.0523	.0284	.0152	.0080	.0042
37	.1039	.0583	.0319	.0171	.0090	.0047
38	.1147	.0649	.0357	.0192	.0102	.0053
39	.1262	.0719	.0398	.0216	.0115	.0060
40	.1384	.0795	.0443	.0241	.0129	.0068
41	.1514	.0877	.0492	.0269	.0145	.0077
42	.1651	.0964	.0544	.0300	.0162	.0086
43	.1796	.1057	.0601	.0333	.0180	.0096
44	.1947	.1156	.0662	.0368	.0201	.0107
45	.2106	.1261	.0727	.0407	.0223	.0120
46	.2271	.1372	.0797	.0449	.0247	.0133
47	.2444	.1489	.0871	.0494	.0273	.0148
48	.2622	.1613	.0950	.0542	.0301	.0164
49	.2807	.1742	.1034	.0594	.0331	.0181
50	.2997	.1877	.1123	.0649	.0364	.0200
51	.3193	.2019	.1218	.0708	.0399	.0220
52	.3394	.2166	.1317	.0770	.0437	.0242
53	.3599	.2319	.1421	.0837	.0478	.0266
54	.3808	.2477	.1530	.0907	.0521	.0291
55	.4020	.2641	.1645	.0982	.0567	.0319
56	.4235	.2809	.1764	.1061	.0616	.0348
57	.4452	.2983	.1889	.1144	.0668	.0379
58	.4670	.3161	.2019	.1231	.0723	.0413
59	.4890	.3343	.2153	.1323	.0782	.0448
60	.5110	.3529	.2293	.1419	.0844	.0487
61	.5330	.3718	.2437	.1519	.0909	.0527
62	.5548	.3910	.2585	.1624	.0978	.0570
63	.5765	.4104	.2738	.1733	.1051	.0615
64	.5980	.4301	.2895	.1846	.1127	.0664
65	.6192	.4500	.3056	.1964	.1206	.0715
66	.6401	.4699	.3221	.2086	.1290	.0768
67	.6606	.4900	.3389	.2211	.1377	.0825
68	.6807	.5100	.3559	.2341	.1467	.0884
69	.7003	.5301	.3733	.2475	.1562	.0947
70	.7193	.5500	.3910	.2613	.1660	.1012

TABLE G : UPPER TAIL OF THE F DISTRIBUTION

Probabilities (P) are for a two-tailed test, for a one-tailed test they should be halved. Entries in the body of the table are values of F.

DEGREES OF FREEDOM FOR NUMERATOR

DEGREES OF FREEDOM FOR DENOMINATOR	P	1	2	3	4	5	6	7	8	9	10	11	12	15	20	24	30	40	50	60	100	∞
4	0.50	1.81	2.00	2.05	2.06	2.07	2.08	2.08	2.08	2.08	2.08	2.08	2.08	2.08	2.08	2.08	2.08	2.08	2.08	2.08	2.08	2.08
	0.20	4.54	4.32	4.19	4.11	4.05	4.01	3.98	3.95	3.94	3.92	3.91	3.90	3.87	3.84	3.83	3.82	3.80	3.80	3.79	3.78	3.76
	0.10	7.71	6.94	6.59	6.39	6.26	6.16	6.09	6.04	6.00	5.96	5.94	5.91	5.86	5.80	5.77	5.75	5.72	5.70	5.69	5.66	5.63
	0.05	12.2	10.6	9.98	9.60	9.36	9.20	9.07	8.98	8.90	8.84	8.79	8.75	8.66	8.56	8.51	8.46	8.41	8.38	8.36	8.32	8.26
	0.02	21.2	18.0	16.7	16.0	15.5	15.2	15.0	14.8	14.7	14.5	14.4	14.4	14.2	14.0	13.9	13.8	13.7	13.7	13.7	13.6	13.5
	0.01	31.3	26.3	24.3	23.2	22.5	22.0	21.6	21.4	21.1	21.0	20.8	20.7	20.4	20.2	20.0	19.9	19.8	19.7	19.6	19.5	19.3
	0.002	74.1	61.2	56.2	53.4	51.7	50.5	49.7	49.0	48.5	48.0	47.7	47.4	46.8	46.1	45.8	45.4	45.1	44.9	44.7	44.5	44.0
	0.001	106	87.4	80.1	76.1	73.6	71.9	70.6	69.7	68.9	68.3	67.8	67.4	66.5	65.5	65.1	64.6	64.1	63.8	63.6	63.2	62.6
5	0.50	1.69	1.85	1.88	1.89	1.89	1.89	1.89	1.89	1.89	1.89	1.89	1.89	1.89	1.88	1.88	1.88	1.88	1.88	1.87	1.87	1.87
	0.20	4.06	3.78	3.62	3.52	3.45	3.40	3.37	3.34	3.32	3.30	3.28	3.27	3.24	3.21	3.19	3.17	3.16	3.15	3.14	3.13	3.10
	0.10	6.61	5.79	5.41	5.19	5.05	4.95	4.88	4.82	4.77	4.74	4.71	4.68	4.62	4.56	4.53	4.50	4.46	4.44	4.43	4.41	4.36
	0.05	10.0	8.43	7.76	7.39	7.15	6.98	6.85	6.76	6.68	6.62	6.57	6.52	6.43	6.33	6.28	6.23	6.18	6.14	6.12	6.08	6.02
	0.02	16.3	13.3	12.1	11.4	11.0	10.7	10.5	10.3	10.2	10.1	9.96	9.89	9.72	9.55	9.47	9.38	9.29	9.24	9.20	9.13	9.02
	0.01	22.8	18.3	16.5	15.6	14.9	14.5	14.2	14.0	13.8	13.6	13.5	13.4	13.1	12.9	12.8	12.7	12.5	12.5	12.4	12.3	12.1
	0.002	47.2	37.1	33.2	31.1	29.7	28.8	28.2	27.6	27.2	26.9	26.6	26.4	25.9	25.4	25.1	24.9	24.6	24.4	24.3	24.1	23.8
	0.001	63.6	49.8	44.4	41.5	39.7	38.5	37.6	36.9	36.4	35.9	35.6	35.2	34.6	33.9	33.5	33.1	32.7	32.5	32.3	32.1	31.6
6	0.50	1.62	1.76	1.78	1.79	1.79	1.78	1.78	1.78	1.77	1.77	1.77	1.77	1.76	1.76	1.75	1.75	1.75	1.75	1.74	1.74	1.74
	0.20	3.78	3.46	3.29	3.18	3.11	3.05	3.01	2.98	2.96	2.94	2.92	2.90	2.87	2.84	2.82	2.80	2.78	2.77	2.76	2.75	2.72
	0.10	5.99	5.14	4.76	4.53	4.39	4.28	4.21	4.15	4.10	4.06	4.03	4.00	3.94	3.87	3.84	3.81	3.77	3.75	3.74	3.71	3.67
	0.05	8.81	7.26	6.60	6.23	5.99	5.82	5.70	5.60	5.52	5.46	5.41	5.37	5.27	5.17	5.12	5.07	5.01	4.98	4.96	4.92	4.85
	0.02	13.7	10.9	9.78	9.15	8.75	8.47	8.26	8.10	7.98	7.87	7.79	7.72	7.56	7.40	7.31	7.23	7.14	7.09	7.06	6.99	6.88
	0.01	18.6	14.5	12.9	12.0	11.5	11.1	10.8	10.6	10.4	10.2	10.1	10.0	9.81	9.59	9.47	9.36	9.24	9.17	9.12	9.03	8.88
	0.002	35.5	27.0	23.7	21.9	20.8	20.0	19.5	19.0	18.7	18.4	18.2	18.0	17.6	17.1	16.9	16.7	16.4	16.3	16.2	16.0	15.7
	0.001	46.1	34.8	30.4	28.1	26.6	25.6	24.9	24.3	23.9	23.5	23.2	23.0	22.4	21.9	21.7	21.4	21.1	20.9	20.7	20.5	20.1
7	0.50	1.57	1.70	1.72	1.72	1.71	1.71	1.70	1.70	1.69	1.69	1.69	1.68	1.68	1.67	1.67	1.66	1.66	1.66	1.65	1.65	1.65
	0.20	3.59	3.26	3.07	2.96	2.88	2.83	2.78	2.75	2.72	2.70	2.68	2.67	2.63	2.59	2.58	2.56	2.54	2.52	2.51	2.50	2.47
	0.10	5.59	4.74	4.35	4.12	3.97	3.87	3.79	3.73	3.68	3.64	3.60	3.57	3.51	3.44	3.41	3.38	3.34	3.32	3.30	3.27	3.23
	0.05	8.07	6.54	5.89	5.52	5.29	5.12	4.99	4.90	4.82	4.76	4.71	4.67	4.57	4.47	4.42	4.36	4.31	4.28	4.25	4.21	4.14
	0.02	12.2	9.55	8.45	7.85	7.46	7.19	6.99	6.84	6.72	6.62	6.54	6.47	6.31	6.16	6.07	5.99	5.91	5.86	5.82	5.75	5.65
	0.01	16.2	12.4	10.9	10.0	9.52	9.16	8.89	8.68	8.51	8.38	8.27	8.18	7.97	7.75	7.65	7.53	7.42	7.35	7.31	7.22	7.08
	0.002	29.2	21.7	18.8	17.2	16.2	15.5	15.0	14.6	14.3	14.1	13.9	13.7	13.3	12.9	12.7	12.5	12.3	12.2	12.1	11.9	11.7
	0.001	37.0	27.2	23.5	21.4	20.2	19.3	18.7	18.2	17.8	17.5	17.2	17.0	16.5	16.0	15.7	15.5	15.2	15.1	15.0	14.7	14.4
8	0.50	1.54	1.66	1.67	1.66	1.66	1.65	1.64	1.64	1.64	1.63	1.63	1.62	1.62	1.61	1.60	1.60	1.59	1.59	1.59	1.58	1.58
	0.20	3.46	3.11	2.92	2.81	2.73	2.67	2.62	2.59	2.56	2.54	2.52	2.50	2.46	2.42	2.40	2.38	2.36	2.35	2.34	2.32	2.29
	0.10	5.32	4.46	4.07	3.84	3.69	3.58	3.50	3.44	3.39	3.35	3.31	3.28	3.22	3.15	3.12	3.08	3.04	3.02	3.01	2.97	2.93
	0.05	7.57	6.06	5.42	5.05	4.82	4.65	4.53	4.43	4.36	4.30	4.24	4.20	4.10	4.00	3.95	3.89	3.84	3.81	3.78	3.74	3.67
	0.02	11.3	8.65	7.59	7.01	6.63	6.37	6.18	6.03	5.91	5.81	5.73	5.67	5.52	5.36	5.28	5.20	5.12	5.07	5.03	4.96	4.86
	0.01	14.7	11.0	9.60	8.81	8.30	7.95	7.69	7.50	7.34	7.21	7.10	7.01	6.81	6.61	6.50	6.40	6.29	6.22	6.18	6.09	5.95
	0.002	25.4	18.5	15.8	14.4	13.5	12.9	12.4	12.0	11.8	11.5	11.4	11.2	10.8	10.5	10.3	10.1	9.92	9.80	9.73	9.57	9.34
	0.001	31.6	22.8	19.4	17.6	16.4	15.7	15.1	14.6	14.3	14.0	13.8	13.6	13.1	12.7	12.5	12.2	12.0	11.8	11.8	11.6	11.3
9	0.50	1.51	1.62	1.63	1.63	1.62	1.61	1.60	1.60	1.59	1.59	1.58	1.58	1.57	1.56	1.56	1.55	1.55	1.54	1.54	1.53	1.53
	0.20	3.36	3.01	2.81	2.69	2.61	2.55	2.51	2.47	2.44	2.42	2.40	2.38	2.34	2.30	2.28	2.25	2.23	2.22	2.21	2.19	2.16
	0.10	5.12	4.26	3.86	3.63	3.48	3.37	3.29	3.23	3.18	3.14	3.10	3.07	3.01	2.94	2.90	2.86	2.83	2.80	2.79	2.76	2.71
	0.05	7.21	5.71	5.08	4.72	4.48	4.32	4.20	4.10	4.03	3.96	3.91	3.87	3.77	3.67	3.61	3.56	3.51	3.47	3.45	3.40	3.33
	0.02	10.6	8.02	6.99	6.42	6.06	5.80	5.61	5.47	5.35	5.26	5.18	5.11	4.96	4.81	4.73	4.65	4.57	4.52	4.48	4.42	4.31
	0.01	13.6	10.1	8.72	7.96	7.47	7.13	6.88	6.69	6.54	6.42	6.31	6.23	6.03	5.83	5.73	5.62	5.52	5.45	5.41	5.32	5.19
	0.002	22.9	16.4	13.9	12.6	11.7	11.1	10.7	10.4	10.1	9.89	9.71	9.57	9.24	8.90	8.72	8.55	8.37	8.26	8.19	8.04	7.81
	0.001	28.0	19.9	16.8	15.1	14.1	13.3	12.8	12.4	12.1	11.8	11.6	11.4	11.0	10.6	10.4	10.2	9.94	9.80	9.71	9.53	9.26
10	0.50	1.49	1.60	1.60	1.59	1.59	1.58	1.57	1.56	1.56	1.55	1.55	1.54	1.53	1.52	1.52	1.51	1.51	1.50	1.50	1.49	1.48
	0.20	3.28	2.92	2.73	2.61	2.52	2.46	2.41	2.38	2.35	2.32	2.30	2.28	2.24	2.20	2.18	2.16	2.13	2.12	2.11	2.09	2.06
	0.10	4.96	4.10	3.71	3.48	3.33	3.22	3.14	3.07	3.02	2.98	2.94	2.91	2.85	2.77	2.74	2.70	2.66	2.64	2.62	2.59	2.54
	0.05	6.94	5.46	4.83	4.47	4.21	4.07	3.95	3.85	3.78	3.72	3.66	3.62	3.52	3.42	3.37	3.31	3.26	3.22	3.20	3.15	3.08
	0.02	10.0	7.56	6.55	5.99	5.64	5.39	5.20	5.06	4.94	4.85	4.77	4.71	4.56	4.41	4.33	4.25	4.17	4.12	4.08	4.01	3.91
	0.01	12.8	9.43	8.08	7.34	6.87	6.54	6.30	6.12	5.97	5.85	5.75	5.66	5.47	5.27	5.17	5.07	4.97	4.90	4.86	4.77	4.64
	0.002	21.0	14.9	12.6	11.3	10.5	9.92	9.52	9.20	8.96	8.75	8.58	8.44	8.13	7.80	7.64	7.47	7.30	7.19	7.12	6.98	6.76
	0.001	25.5	17.9	15.0	13.4	12.4	11.8	11.3	10.9	10.6	10.3	10.1	9.93	9.56	9.16	8.96	8.75	8.54	8.42	8.33	8.16	7.90
11	0.50	1.47	1.58	1.58	1.57	1.56	1.55	1.54	1.53	1.53	1.52	1.52	1.51	1.50	1.49	1.49	1.48	1.47	1.47	1.47	1.46	1.45
	0.20	3.23	2.86	2.66	2.54	2.45	2.39	2.34	2.30	2.27	2.25	2.23	2.21	2.17	2.12	2.10	2.08	2.05	2.04	2.03	2.00	1.97
	0.10	4.84	3.98	3.59	3.36	3.20	3.09	3.01	2.95	2.90	2.85	2.82	2.79	2.72	2.65	2.61	2.57	2.53	2.51	2.49	2.46	2.40
	0.05	6.72	5.26	4.63	4.28	4.04	3.88	3.76	3.66	3.59	3.53	3.47	3.43	3.33	3.23	3.17	3.12	3.06	3.03	3.00	2.96	2.88
	0.02	9.65	7.21	6.22	5.67	5.32	5.07	4.89	4.74	4.63	4.54	4.46	4.40	4.25	4.10	4.02	3.94	3.86	3.81	3.78	3.71	3.60
	0.01	12.2	8.91	7.60	6.88	6.42	6.10	5.86	5.68	5.54	5.42	5.32	5.24	5.05	4.86	4.76	4.65	4.55	4.49	4.45	4.36	4.23
	0.002	19.7	13.8	11.6	10.3	9.58	9.05	8.66	8.35	8.12	7.92	7.76	7.62	7.32	7.01	6.85	6.68	6.52	6.41	6.35	6.21	6.00
	0.001	23.6	16.4	13.6	12.2	11.2	10.6	10.1	9.76	9.48	9.24	9.04	8.88	8.52	8.14	7.94	7.75	7.55	7.43	7.35	7.18	6.93

TABLE G (continued)

DEGREES OF FREEDOM FOR NUMERATOR

DEGREES OF FREEDOM FOR DENOMINATOR	P	1	2	3	4	5	6	7	8	9	10	11	12	15	20	24	30	40	50	60	100	∞
12	0.50	1.46	1.56	1.56	1.55	1.54	1.53	1.52	1.51	1.51	1.50	1.50	1.49	1.48	1.47	1.46	1.45	1.45	1.44	1.44	1.43	1.42
	0.20	3.18	2.81	2.61	2.48	2.39	2.33	2.28	2.24	2.21	2.19	2.17	2.15	2.11	2.06	2.04	2.01	1.99	1.97	1.96	1.94	1.90
	0.10	4.75	3.89	3.49	3.26	3.11	3.00	2.91	2.85	2.80	2.75	2.72	2.69	2.62	2.54	2.51	2.47	2.43	2.40	2.38	2.35	2.30
	0.05	6.55	5.10	4.47	4.12	3.89	3.73	3.61	3.51	3.44	3.37	3.32	3.28	3.18	3.07	3.02	2.96	2.91	2.87	2.85	2.80	2.72
	0.02	9.33	6.93	5.95	5.41	5.06	4.82	4.64	4.50	4.39	4.30	4.22	4.16	4.01	3.86	3.78	3.70	3.62	3.57	3.54	3.47	3.36
	0.01	11.8	8.51	7.23	6.52	6.07	5.76	5.52	5.35	5.20	5.09	4.99	4.91	4.72	4.53	4.43	4.33	4.23	4.17	4.12	4.04	3.90
	0.002	18.6	13.0	10.8	9.63	8.89	8.38	8.00	7.71	7.48	7.29	7.14	7.01	6.71	6.40	6.25	6.09	5.93	5.83	5.76	5.63	5.42
	0.001	22.2	15.3	12.7	11.2	10.4	9.74	9.28	8.94	8.66	8.43	8.24	8.08	7.74	7.37	7.18	7.00	6.80	6.68	6.61	6.45	6.20
15	0.50	1.43	1.52	1.52	1.51	1.49	1.48	1.47	1.46	1.46	1.45	1.44	1.44	1.43	1.41	1.41	1.40	1.39	1.39	1.38	1.38	1.36
	0.20	3.07	2.70	2.49	2.36	2.27	2.21	2.16	2.12	2.09	2.06	2.04	2.02	1.97	1.92	1.90	1.87	1.85	1.83	1.82	1.79	1.76
	0.10	4.54	3.68	3.29	3.06	2.90	2.79	2.71	2.64	2.59	2.54	2.51	2.48	2.40	2.33	2.39	2.25	2.20	2.18	2.16	2.12	2.07
	0.05	6.20	4.76	4.15	3.80	3.58	3.41	3.29	3.20	3.12	3.06	3.01	2.96	2.86	2.76	2.70	2.64	2.59	2.55	2.52	2.47	2.40
	0.02	8.68	6.36	5.42	4.89	4.56	4.32	4.14	4.00	3.89	3.80	3.73	3.67	3.52	3.37	3.29	3.21	3.13	3.08	3.05	2.98	2.87
	0.01	10.8	7.70	6.48	5.80	5.37	5.07	4.85	4.67	4.54	4.42	4.33	4.25	4.07	3.88	3.79	3.69	3.59	3.52	3.48	3.39	3.26
	0.002	16.6	11.3	9.34	8.25	7.57	7.09	6.74	6.47	6.26	6.08	5.93	5.81	5.54	5.25	5.10	4.95	4.80	4.70	4.64	4.51	4.31
	0.001	19.5	13.2	10.8	9.48	8.66	8.10	7.68	7.36	7.11	6.91	6.75	6.60	6.27	5.93	5.75	5.58	5.40	5.29	5.21	5.06	4.83
20	0.50	1.40	1.49	1.48	1.47	1.45	1.44	1.43	1.42	1.41	1.40	1.39	1.39	1.37	1.36	1.35	1.34	1.33	1.33	1.32	1.31	1.29
	0.20	2.97	2.59	2.38	2.25	2.16	2.09	2.04	2.00	1.96	1.94	1.91	1.89	1.84	1.79	1.77	1.74	1.71	1.69	1.68	1.65	1.61
	0.10	4.35	3.49	3.10	2.87	2.71	2.60	2.51	2.45	2.39	2.35	2.31	2.28	2.20	2.12	2.08	2.04	1.99	1.97	1.95	1.91	1.84
	0.05	5.87	4.46	3.86	3.51	3.29	3.13	3.01	2.91	2.84	2.77	2.72	2.68	2.57	2.46	2.41	2.35	2.29	2.25	2.22	2.17	2.09
	0.02	8.10	5.85	4.94	4.43	4.10	3.87	3.70	3.56	3.46	3.37	3.29	3.23	3.09	2.94	2.86	2.78	2.69	2.64	2.61	2.54	2.42
	0.01	9.94	6.99	5.82	5.17	4.76	4.47	4.26	4.09	3.96	3.85	3.76	3.68	3.50	3.32	3.22	3.12	3.02	2.96	2.92	2.83	2.69
	0.002	14.8	9.95	8.10	7.10	6.46	6.02	5.69	5.44	5.24	5.08	4.94	4.82	4.56	4.29	4.15	4.01	3.86	3.77	3.70	3.58	3.38
	0.001	17.2	11.4	9.20	8.02	7.28	6.76	6.38	6.08	5.85	5.66	5.51	5.38	5.07	4.75	4.58	4.42	4.24	4.15	4.07	3.93	3.70
24	0.50	1.39	1.47	1.46	1.44	1.43	1.41	1.40	1.39	1.38	1.38	1.37	1.36	1.35	1.33	1.32	1.31	1.30	1.29	1.29	1.28	1.26
	0.20	2.93	2.54	2.33	2.19	2.10	2.04	1.98	1.94	1.91	1.88	1.85	1.83	1.78	1.73	1.70	1.67	1.64	1.62	1.61	1.58	1.53
	0.10	4.26	3.40	3.01	2.78	2.62	2.51	2.42	2.36	2.30	2.25	2.21	2.18	2.11	2.03	1.98	1.94	1.89	1.86	1.84	1.80	1.73
	0.05	5.72	4.32	3.72	3.38	3.15	2.99	2.87	2.78	2.70	2.64	2.59	2.54	2.44	2.33	2.27	2.21	2.15	2.11	2.08	2.02	1.94
	0.02	7.82	5.61	4.72	4.22	3.90	3.67	3.50	3.36	3.26	3.17	3.09	3.03	2.89	2.74	2.66	2.58	2.49	2.44	2.40	2.33	2.21
	0.01	9.55	6.66	5.52	4.89	4.49	4.20	3.99	3.83	3.69	3.59	3.50	3.42	3.25	3.06	2.97	2.87	2.77	2.70	2.66	2.57	2.43
	0.002	14.0	9.34	7.55	6.59	5.98	5.55	5.23	4.99	4.80	4.64	4.50	4.39	4.14	3.87	3.74	3.59	3.45	3.35	3.29	3.16	2.97
	0.001	16.2	10.6	8.52	7.39	6.68	6.18	5.82	5.54	5.31	5.13	4.98	4.85	4.55	4.25	4.09	3.93	3.76	3.66	3.59	3.44	3.22
30	0.50	1.38	1.45	1.44	1.42	1.41	1.39	1.38	1.37	1.36	1.35	1.35	1.34	1.32	1.30	1.29	1.28	1.27	1.26	1.26	1.25	1.23
	0.20	2.88	2.49	2.28	2.14	2.05	1.98	1.93	1.88	1.85	1.82	1.79	1.77	1.72	1.67	1.64	1.61	1.57	1.55	1.54	1.51	1.46
	0.10	4.17	3.32	2.92	2.69	2.53	2.42	2.33	2.27	2.21	2.16	2.13	2.09	2.01	1.93	1.89	1.84	1.79	1.76	1.74	1.70	1.62
	0.05	5.57	4.18	3.59	3.25	3.03	2.87	2.75	2.65	2.57	2.51	2.46	2.41	2.31	2.20	2.14	2.07	2.01	1.97	1.94	1.88	1.79
	0.02	7.56	5.39	4.51	4.02	3.70	3.47	3.30	3.17	3.07	2.98	2.91	2.84	2.70	2.55	2.47	2.39	2.30	2.25	2.21	2.13	2.01
	0.01	9.18	6.35	5.24	4.62	4.23	3.95	3.74	3.58	3.45	3.34	3.25	3.18	3.01	2.82	2.73	2.63	2.52	2.46	2.42	2.32	2.18
	0.002	13.3	8.77	7.05	6.12	5.53	5.12	4.82	4.58	4.39	4.24	4.11	4.00	3.75	3.49	3.36	3.22	3.07	2.98	2.92	2.79	2.59
	0.001	15.2	9.90	7.90	6.82	6.14	5.66	5.31	5.04	4.82	4.65	4.51	4.38	4.10	3.80	3.65	3.48	3.32	3.22	3.15	3.00	2.78
40	0.50	1.36	1.44	1.42	1.40	1.39	1.37	1.36	1.35	1.34	1.33	1.32	1.31	1.30	1.28	1.26	1.25	1.24	1.23	1.22	1.21	1.19
	0.20	2.84	2.44	2.23	2.09	2.00	1.93	1.87	1.83	1.79	1.76	1.73	1.71	1.66	1.61	1.57	1.54	1.51	1.48	1.47	1.43	1.38
	0.10	4.08	3.23	2.84	2.61	2.45	2.34	2.25	2.18	2.12	2.08	2.04	2.00	1.92	1.84	1.79	1.74	1.69	1.66	1.64	1.59	1.51
	0.05	5.42	4.05	3.46	3.13	2.90	2.74	2.62	2.53	2.45	2.39	2.33	2.29	2.18	2.07	2.01	1.94	1.88	1.83	1.80	1.74	1.64
	0.02	7.31	5.18	4.31	3.83	3.51	3.29	3.12	2.99	2.89	2.80	2.73	2.66	2.52	2.37	2.29	2.20	2.11	2.06	2.02	1.94	1.80
	0.01	8.83	6.07	4.98	4.37	3.99	3.71	3.51	3.35	3.22	3.12	3.03	2.95	2.78	2.60	2.50	2.40	2.30	2.23	2.18	2.09	1.93
	0.002	12.6	8.25	6.60	5.70	5.13	4.73	4.44	4.21	4.02	3.87	3.75	3.64	3.40	3.15	3.01	2.87	2.73	2.64	2.57	2.44	2.23
	0.001	14.4	9.25	7.33	6.30	5.64	5.19	4.85	4.59	4.38	4.21	4.07	3.95	3.68	3.39	3.24	3.08	2.92	2.82	2.74	2.60	2.37
60	0.50	1.35	1.42	1.41	1.38	1.37	1.35	1.33	1.32	1.31	1.30	1.29	1.29	1.27	1.25	1.24	1.22	1.21	1.20	1.19	1.17	1.15
	0.20	2.79	2.39	2.18	2.04	1.95	1.87	1.82	1.77	1.74	1.71	1.68	1.66	1.60	1.54	1.51	1.48	1.44	1.41	1.40	1.36	1.29
	0.10	4.00	3.15	2.76	2.53	2.37	2.25	2.17	2.10	2.04	1.99	1.95	1.92	1.84	1.75	1.70	1.65	1.59	1.56	1.53	1.48	1.39
	0.05	5.29	3.93	3.34	3.01	2.79	2.63	2.51	2.41	2.33	2.27	2.22	2.17	2.06	1.94	1.88	1.82	1.74	1.70	1.67	1.60	1.48
	0.02	7.08	4.98	4.13	3.65	3.34	3.12	2.95	2.82	2.72	2.63	2.56	2.50	2.35	2.20	2.12	2.03	1.94	1.88	1.84	1.75	1.60
	0.01	8.49	5.80	4.73	4.14	3.76	3.49	3.29	3.13	3.01	2.90	2.82	2.74	2.57	2.39	2.29	2.19	2.08	2.01	1.96	1.86	1.69
	0.002	12.0	7.76	6.17	5.31	4.76	4.37	4.09	3.87	3.69	3.54	3.43	3.31	3.08	2.83	2.69	2.56	2.41	2.31	2.25	2.11	1.89
	0.001	13.6	8.65	6.81	5.82	5.20	4.76	4.44	4.18	3.98	3.82	3.69	3.57	3.30	3.02	2.87	2.71	2.55	2.45	2.38	2.23	1.98
120	0.50	1.34	1.40	1.39	1.37	1.35	1.33	1.31	1.30	1.29	1.28	1.27	1.26	1.24	1.22	1.21	1.19	1.18	1.17	1.16	1.14	1.10
	0.20	2.75	2.35	2.13	1.99	1.90	1.82	1.77	1.72	1.68	1.65	1.62	1.60	1.55	1.48	1.45	1.41	1.37	1.34	1.32	1.27	1.19
	0.10	3.92	3.07	2.68	2.45	2.29	2.18	2.09	2.02	1.96	1.91	1.87	1.83	1.75	1.66	1.61	1.55	1.50	1.46	1.43	1.37	1.25
	0.05	5.15	3.80	3.23	2.89	2.67	2.52	2.39	2.30	2.22	2.16	2.10	2.05	1.95	1.82	1.76	1.69	1.61	1.56	1.53	1.45	1.31
	0.02	6.85	4.79	3.95	3.48	3.17	2.96	2.79	2.66	2.56	2.47	2.40	2.34	2.19	2.03	1.95	1.86	1.76	1.70	1.66	1.56	1.38
	0.01	8.18	5.54	4.50	3.92	3.55	3.28	3.09	2.93	2.81	2.71	2.62	2.54	2.37	2.19	2.09	1.98	1.87	1.80	1.75	1.64	1.43
	0.002	11.4	7.32	5.79	4.95	4.42	4.04	3.77	3.55	3.38	3.24	3.12	3.02	2.78	2.53	2.40	2.26	2.11	2.02	1.95	1.82	1.54
	0.001	12.8	8.10	6.34	5.39	4.79	4.37	4.07	3.82	3.63	3.47	3.34	3.22	2.96	2.67	2.53	2.38	2.21	2.11	2.01	1.88	1.60
∞	0.50	1.32	1.39	1.37	1.35	1.33	1.31	1.29	1.28	1.27	1.25	1.24	1.24	1.22	1.19	1.18	1.16	1.14	1.13	1.12	1.09	1.00
	0.20	2.71	2.30	2.08	1.94	1.85	1.77	1.72	1.67	1.63	1.60	1.57	1.55	1.49	1.42	1.38	1.34	1.30	1.26	1.24	1.18	1.00
	0.10	3.84	3.00	2.60	2.37	2.21	2.10	2.01	1.94	1.88	1.83	1.79	1.75	1.67	1.57	1.52	1.46	1.39	1.35	1.32	1.24	1.00
	0.05	5.02	3.69	3.12	2.79	2.57	2.41	2.29	2.19	2.11	2.05	1.99	1.94	1.83	1.71	1.64	1.57	1.48	1.43	1.39	1.30	1.00
	0.02	6.63	4.61	3.78	3.32	3.02	2.80	2.64	2.51	2.41	2.32	2.25	2.18	2.04	1.88	1.79	1.70	1.59	1.52	1.47	1.36	1.00
	0.01	7.88	5.30	4.28	3.72	3.35	3.09	2.90	2.74	2.62	2.52	2.43	2.36	2.19	2.00	1.90	1.79	1.67	1.59	1.53	1.40	1.00
	0.002	10.8	6.91	5.42	4.62	4.10	3.74	3.47	3.27	3.10	2.96	2.84	2.74	2.51	2.27	2.13	1.99	1.84	1.73	1.66	1.49	1.00
	0.001	12.1	7.60	5.91	5.00	4.42	4.02	3.72	3.48	3.30	3.14	3.02	2.90	2.65	2.37	2.22	2.07	1.91	1.79	1.71	1.53	1.00

TABLE H :

MANN–WHITNEY TEST, SMALL SAMPLES

n_a or $n_b = 3$

W	n_b or n_a = 3	4	5	6	7	8	9	10
0	.0500	.0286	.0179	.0119	.0083	.0061	.0045	.0035
1	.1000	.0571	.0357	.0238	.0167	.0121	.0091	.0070
2	.2000	.1143	.0714	.0476	.0333	.0242	.0182	.0140
3	.3500	.2000	.1250	.0833	.0583	.0424	.0318	.0245
4	.5000	.3143	.1964	.1310	.0917	.0667	.0500	.0385
5	.6500	.4286	.2857	.1905	.1333	.0970	.0727	.0559
6	.8000	.5714	.3929	.2738	.1917	.1394	.1045	.0804
7	.9000	.6857	.5000	.3571	.2583	.1879	.1409	.1084
8	.9500	.8000	.6071	.4524	.3333	.2485	.1864	.1434
9	1.0000	.8857	.7143	.5476	.4167	.3152	.2409	.1853
10		.9429	.8036	.6429	.5000	.3879	.3000	.2343
11		.9714	.8750	.7262	.5833	.4606	.3636	.2867
12		1.0000	.9286	.8095	.6667	.5394	.4318	.3462
13			.9643	.8690	.7417	.6121	.5000	.4056
14			.9821	.9167	.8083	.6848	.5682	.4685
15			1.0000	.9524	.8667	.7515	.6364	.5315
16				.9762	.9083	.8121	.7000	.5944
17				.9881	.9417	.8606	.7591	.6538
18				1.0000	.9667	.9030	.8136	.7133

n_a or $n_b = 4$

W	n_b or n_a = 4	5	6	7	8	9	10
0	.0143	.0079	.0048	.0030	.0020	.0014	.0010
1	.0286	.0159	.0095	.0061	.0040	.0028	.0020
2	.0571	.0317	.0190	.0121	.0081	.0056	.0040
3	.1000	.0556	.0333	.0212	.0141	.0098	.0070
4	.1714	.0952	.0571	.0364	.0242	.0168	.0120
5	.2429	.1429	.0857	.0545	.0364	.0252	.0180
6	.3429	.2063	.1286	.0818	.0545	.0378	.0270
7	.4429	.2778	.1762	.1152	.0768	.0531	.0380
8	.5571	.3651	.2381	.1576	.1071	.0741	.0529
9	.6571	.4524	.3048	.2061	.1414	.0993	.0709
10	.7571	.5476	.3810	.2636	.1838	.1301	.0939
11	.8286	.6349	.4571	.3242	.2303	.1650	.1199
12	.9000	.7222	.5429	.3939	.2848	.2070	.1518
13	.9429	.7937	.6190	.4636	.3414	.2517	.1868
14	.9714	.8571	.6952	.5364	.4040	.3021	.2268
15	.9857	.9048	.7619	.6061	.4667	.3552	.2697
16	1.0000	.9444	.8238	.6758	.5333	.4126	.3177
17		.9683	.8714	.7364	.5960	.4699	.3666
18		.9841	.9143	.7939	.6586	.5301	.4196
19		.9921	.9429	.8424	.7152	.5874	.4725
20		1.0000	.9667	.8848	.7697	.6448	.5275
21			.9810	.9182	.8162	.6979	.5804
22			.9905	.9455	.8586	.7483	.6334
23			.9952	.9636	.8929	.7930	.6823
24			1.0000	.9788	.9232	.8350	.7303

n_a or $n_b = 5$

W	n_b or n_a = 5	6	7	8	9	10
0	.0040	.0022	.0013	.0008	.0005	.0003
1	.0079	.0043	.0025	.0016	.0010	.0007
2	.0159	.0087	.0051	.0031	.0020	.0013
3	.0278	.0152	.0088	.0054	.0035	.0023
4	.0476	.0260	.0152	.0093	.0060	.0040
5	.0754	.0411	.0240	.0148	.0095	.0063
6	.1111	.0628	.0366	.0225	.0145	.0097
7	.1548	.0887	.0530	.0326	.0210	.0140
8	.2103	.1234	.0745	.0466	.0300	.0200
9	.2738	.1645	.1010	.0637	.0415	.0276
10	.3452	.2143	.1338	.0855	.0559	.0376
11	.4206	.2684	.1717	.1111	.0734	.0496
12	.5000	.3312	.2159	.1422	.0949	.0646
13	.5794	.3961	.2652	.1772	.1199	.0823
14	.6548	.4654	.3194	.2176	.1489	.1032
15	.7262	.5346	.3775	.2618	.1818	.1272
16	.7897	.6039	.4381	.3108	.2188	.1548
17	.8452	.6688	.5000	.3621	.2592	.1855
18	.8889	.7316	.5619	.4165	.3032	.2198
19	.9246	.7857	.6225	.4716	.3497	.2567
20	.9524	.8355	.6806	.5284	.3986	.2970
21	.9722	.8766	.7348	.5835	.4491	.3393
22	.9841	.9113	.7841	.6379	.5000	.3839
23	.9921	.9372	.8283	.6892	.5509	.4296
24	.9960	.9589	.8662	.7382	.6014	.4765

n_a or $n_b = 6$

W	n_b or n_a = 6	7	8	9	10
0	.0011	.0006	.0003	.0002	.0001
1	.0022	.0012	.0007	.0004	.0002
2	.0043	.0023	.0013	.0008	.0005
3	.0076	.0041	.0023	.0014	.0009
4	.0130	.0070	.0040	.0024	.0015
5	.0206	.0111	.0063	.0038	.0024
6	.0325	.0175	.0100	.0060	.0037
7	.0465	.0256	.0147	.0088	.0055
8	.0660	.0367	.0213	.0128	.0080
9	.0898	.0507	.0296	.0180	.0112
10	.1201	.0688	.0406	.0248	.0156
11	.1548	.0903	.0539	.0332	.0210
12	.1970	.1171	.0709	.0440	.0280
13	.2424	.1474	.0906	.0567	.0363
14	.2944	.1830	.1142	.0723	.0467
15	.3496	.2226	.1412	.0905	.0589
16	.4091	.2669	.1725	.1119	.0736
17	.4686	.3141	.2068	.1361	.0903
18	.5314	.3654	.2454	.1638	.1099
19	.5909	.4178	.2864	.1942	.1317
20	.6504	.4726	.3310	.2280	.1566
21	.7056	.5274	.3773	.2643	.1838
22	.7576	.5822	.4259	.3035	.2139
23	.8030	.6346	.4749	.3445	.2461
24	.8452	.6859	.5251	.3878	.2811

n_a or $n_b = 7$; n_a or $n_b = 8$; n_a or $n_b = 9$; $n_a = 10$, $n_b = 10$

W	n_a or n_b = 7: n_b or n_a = 7	8	9	10	n_a or n_b = 8: n_b or n_a = 8	9	10	n_a or n_b = 9: n_b or n_a = 9	10	n_a = 10: n_b = 10
0	.0003	.0002	.0001	.0001	.0001	.0000	.0000	.0000	.0000	.0000
1	.0006	.0003	.0002	.0001	.0002	.0001	.0000	.0000	.0000	.0000
2	.0012	.0006	.0003	.0002	.0003	.0002	.0001	.0001	.0000	.0000
3	.0020	.0011	.0006	.0004	.0005	.0003	.0002	.0001	.0001	.0000
4	.0035	.0019	.0010	.0006	.0009	.0005	.0003	.0002	.0001	.0001
5	.0055	.0030	.0017	.0010	.0015	.0008	.0004	.0004	.0002	.0001
6	.0087	.0047	.0026	.0015	.0023	.0012	.0007	.0006	.0003	.0002
7	.0131	.0070	.0039	.0023	.0035	.0019	.0010	.0009	.0005	.0002
8	.0189	.0103	.0058	.0034	.0052	.0028	.0015	.0014	.0007	.0004
9	.0265	.0145	.0082	.0048	.0074	.0039	.0022	.0020	.0011	.0005
10	.0364	.0200	.0115	.0068	.0103	.0056	.0031	.0028	.0015	.0008
11	.0487	.0270	.0156	.0093	.0141	.0076	.0043	.0039	.0021	.0010
12	.0641	.0361	.0209	.0125	.0190	.0103	.0058	.0053	.0028	.0014
13	.0825	.0469	.0274	.0165	.0249	.0137	.0078	.0071	.0038	.0019
14	.1043	.0603	.0356	.0215	.0325	.0180	.0103	.0094	.0051	.0026
15	.1297	.0760	.0454	.0277	.0415	.0232	.0133	.0122	.0066	.0034
16	.1588	.0946	.0571	.0351	.0524	.0296	.0171	.0157	.0086	.0045
17	.1914	.1159	.0708	.0439	.0652	.0372	.0217	.0200	.0110	.0057
18	.2279	.1405	.0869	.0544	.0803	.0464	.0273	.0252	.0140	.0073
19	.2675	.1678	.1052	.0665	.0974	.0570	.0338	.0313	.0175	.0093
20	.3100	.1984	.1261	.0806	.1172	.0694	.0416	.0385	.0217	.0116
21	.3552	.2317	.1496	.0966	.1393	.0836	.0506	.0470	.0267	.0144
22	.4024	.2679	.1755	.1148	.1641	.0998	.0610	.0567	.0326	.0177
23	.4508	.3063	.2039	.1349	.1911	.1179	.0729	.0680	.0394	.0216
24	.5000	.3472	.2349	.1574	.2209	.1383	.0864	.0807	.0474	.0262
25	.5492	.3894	.2680	.1819	.2527	.1606	.1015	.0951	.0564	.0315
26	.5976	.4333	.3032	.2087	.2869	.1852	.1185	.1112	.0667	.0376
27	.6448	.4775	.3403	.2374	.3227	.2117	.1371	.1290	.0782	.0446
28	.6900	.5225	.3788	.2681	.3605	.2404	.1577	.1487	.0912	.0526
29	.7325	.5667	.4185	.3004	.3992	.2707	.1800	.1701	.1055	.0615
30	.7721	.6106	.4591	.3345	.4392	.3029	.2041	.1933	.1214	.0716
31	.8086	.6528	.5000	.3698	.4796	.3365	.2299	.2181	.1388	.0827
32	.8412	.6937	.5409	.4063	.5204	.3715	.2574	.2447	.1577	.0952
33	.8703	.7321	.5815	.4434	.5608	.4074	.2863	.2729	.1781	.1088
34	.8957	.7683	.6212	.4811	.6008	.4442	.3167	.3024	.2001	.1237
35	.9175	.8016	.6597	.5189	.6395	.4813	.3482	.3332	.2235	.1399

TABLE I : MANN-WHITNEY TEST, MEDIUM SIZED SAMPLES.

One-tailed values of P

n_a	P	$n_b=2$	3	4	5	6	7	8	9	10	11	12	13	14	15	16	17	18	19	20
	.001	0	0	0	0	0	0	0	0	0	0	0	0	0	0	0	1	1	1	1
	.005	0	0	0	0	0	0	0	1	1	1	2	2	2	3	3	3	3	4	4
3	.01	0	0	0	0	0	1	1	2	2	2	3	3	3	4	4	5	5	5	6
	.025	0	0	0	1	2	2	3	3	4	4	5	5	6	6	7	7	8	8	9
	.05	0	1	1	2	3	3	4	5	5	6	6	7	8	8	9	10	10	11	12
	.10	1	2	2	3	4	5	6	6	7	8	9	10	11	11	12	13	14	15	16
	.001	0	0	0	0	0	0	0	0	1	1	1	2	2	2	3	3	4	4	4
	.005	0	0	0	0	1	1	2	2	3	3	4	4	5	6	6	7	7	8	9
4	.01	0	0	0	1	2	2	3	4	4	5	6	6	7	9	8	9	10	10	11
	.025	0	0	1	2	3	4	5	5	6	7	8	9	10	11	12	12	13	14	15
	.05	0	1	2	3	4	5	6	7	8	9	10	11	12	13	15	16	17	18	19
	.10	1	2	4	5	6	7	8	10	11	12	13	14	16	17	18	19	21	22	23
	.001	0	0	0	0	0	0	1	2	2	3	3	4	4	5	6	6	7	8	8
	.005	0	0	0	1	2	2	3	4	5	6	7	8	8	9	10	11	12	13	14
5	.01	0	0	1	2	3	4	5	6	7	8	9	10	11	12	13	14	15	16	17
	.025	0	1	2	3	4	6	7	8	9	10	12	13	14	15	16	18	19	20	21
	.05	1	2	3	5	6	7	9	10	12	13	14	16	17	19	20	21	23	24	26
	.10	2	3	5	6	8	9	11	13	14	16	18	19	21	23	24	26	28	29	31
	.001	0	0	0	0	0	0	2	3	4	5	5	6	7	8	9	10	11	12	13
	.005	0	0	1	2	3	4	5	6	7	8	10	11	12	13	14	16	17	18	19
6	.01	0	0	2	3	4	5	7	8	9	10	12	13	14	16	17	19	20	21	23
	.025	0	2	3	4	6	7	9	11	12	14	15	17	18	20	22	23	25	26	28
	.05	1	3	4	6	8	9	11	13	15	17	18	20	22	24	26	27	29	31	33
	.10	2	4	6	8	10	12	14	16	18	20	22	24	26	28	30	32	35	37	39
	.001	0	0	0	0	1	2	3	4	6	7	8	9	10	11	12	14	15	16	17
	.005	0	0	1	2	4	5	7	8	10	11	13	14	16	17	19	20	22	23	25
7	.01	0	1	2	4	5	7	8	10	12	13	15	17	18	20	22	24	25	27	29
	.025	0	2	4	6	7	9	11	13	15	17	19	21	23	25	27	29	31	33	35
	.05	1	3	5	7	9	12	14	16	18	20	22	25	27	29	31	34	36	38	40
	.10	2	5	7	9	12	14	17	19	22	24	27	29	32	34	37	39	42	44	47
	.001	0	0	0	1	2	3	5	6	7	9	10	12	13	15	16	18	19	21	22
	.005	0	0	2	3	5	7	8	10	12	14	16	18	19	21	23	25	27	29	31
8	.01	0	1	3	5	7	8	10	12	14	16	18	21	23	25	27	29	31	33	35
	.025	1	3	5	7	9	11	14	16	18	20	23	25	27	30	32	35	37	39	42
	.05	2	4	6	9	11	14	16	19	21	24	27	29	32	34	37	40	42	45	48
	.10	3	6	8	11	14	17	20	23	25	28	31	34	37	40	43	46	49	52	55
	.001	0	0	0	2	3	4	6	8	9	11	13	15	16	18	20	22	24	26	27
	.005	0	1	2	4	6	8	10	12	14	17	19	21	23	25	28	30	32	34	37
9	.01	0	2	4	6	8	10	12	15	17	19	22	24	27	29	32	34	37	39	41
	.025	1	3	5	8	11	13	16	18	21	24	27	29	32	35	38	40	43	46	49
	.05	2	5	7	10	13	16	19	22	25	28	31	34	37	40	43	46	49	52	55
	.10	3	6	10	13	16	19	23	26	29	32	36	39	42	46	49	53	56	59	63
	.001	0	0	1	2	4	6	7	9	11	13	15	18	20	22	24	26	28	30	33
	.005	0	1	3	5	7	10	12	14	17	19	22	25	27	30	32	35	38	40	43
10	.01	0	2	4	7	9	12	14	17	20	23	25	28	31	34	37	39	42	45	48
	.025	1	4	6	9	12	15	18	21	24	27	30	34	37	40	43	46	49	53	56
	.05	2	5	8	12	15	18	21	25	28	32	35	38	42	45	49	52	56	59	63
	.10	4	7	11	14	18	22	25	29	33	37	40	44	48	52	55	59	63	67	71
	.001	0	0	1	3	5	7	9	11	13	16	18	21	23	25	28	30	33	35	38
	.005	0	1	3	6	8	11	14	17	19	22	25	28	31	34	37	40	43	46	49
11	.01	0	2	5	8	10	13	16	19	23	26	29	32	35	38	42	45	48	51	54
	.025	1	4	7	10	14	17	20	24	27	31	34	38	41	45	48	52	56	59	63
	.05	2	6	9	13	17	20	24	28	32	35	39	43	47	51	55	58	62	66	70
	.10	4	8	12	16	20	24	28	32	37	41	45	49	53	58	62	66	70	74	79
	.001	0	0	1	3	5	8	10	13	15	18	21	24	26	29	32	35	38	41	43
	.005	0	2	4	7	10	13	16	19	22	25	28	32	35	38	42	45	48	52	55
12	.01	0	3	6	9	12	15	18	22	25	29	32	36	39	43	47	50	54	57	61
	.025	2	5	8	12	15	19	23	27	30	34	38	42	46	50	54	58	62	66	70
	.05	3	6	10	14	18	22	27	31	35	39	43	48	52	56	61	65	69	73	78
	.10	5	9	13	18	22	27	31	36	40	45	50	54	59	64	68	73	78	82	87
	.001	0	0	2	4	6	9	12	15	18	21	24	27	30	33	36	39	43	46	49
	.005	0	2	4	8	11	14	18	21	25	28	32	35	39	43	46	50	54	58	61
13	.01	1	3	6	10	13	17	21	24	28	32	36	40	44	48	52	56	60	64	68
	.025	2	5	9	13	17	21	25	29	34	38	42	46	51	55	60	64	68	73	77
	.05	3	7	11	16	20	25	29	34	38	43	48	52	57	62	66	71	76	81	85
	.10	5	10	14	19	24	29	34	39	44	49	54	59	64	69	75	80	85	90	95
	.001	0	0	2	4	7	10	13	16	20	23	26	30	33	37	40	44	47	51	55
	.005	0	2	5	8	12	16	19	23	27	31	35	39	43	47	51	55	59	64	68
14	.01	1	3	7	11	14	18	23	27	31	35	39	44	48	52	57	61	66	70	74
	.025	2	6	10	14	18	23	27	32	37	41	46	51	56	60	65	70	75	79	84
	.05	4	8	12	17	22	27	32	37	42	47	52	57	62	67	72	78	83	88	93
	.10	5	11	16	21	26	32	37	42	48	53	59	64	70	75	81	86	92	98	103
	.001	0	0	2	5	8	11	15	18	22	25	29	33	37	41	44	48	52	56	60
	.005	0	3	6	9	13	17	21	25	30	34	38	43	47	52	56	61	65	70	74
15	.01	1	4	8	12	16	20	25	29	34	38	43	48	52	57	62	67	71	76	81
	.025	2	6	11	15	20	25	30	35	40	45	50	55	60	65	71	76	81	86	91
	.05	4	8	13	19	24	29	34	40	45	51	56	62	67	73	78	84	89	95	101
	.10	6	11	17	23	28	34	40	46	52	58	64	69	75	81	87	93	99	105	111
	.001	0	0	3	6	9	12	16	20	24	28	32	36	40	44	49	53	57	61	66
	.005	0	3	6	10	14	19	23	28	32	37	42	46	51	56	61	66	71	75	80
16	.01	1	4	8	13	17	22	27	32	37	42	47	52	57	62	67	72	77	83	88
	.025	2	7	12	16	22	27	32	38	43	48	54	60	65	71	76	82	87	93	99
	.05	4	9	15	20	26	31	37	43	49	55	61	66	72	78	84	90	96	102	108
	.10	6	12	18	24	30	37	43	49	55	62	68	75	81	87	94	100	107	113	120
	.001	0	1	3	6	10	14	18	22	26	30	35	39	44	48	53	58	62	67	71
	.005	0	3	7	11	16	20	25	30	35	40	45	50	55	61	66	71	76	82	87
17	.01	1	5	9	14	19	24	29	34	39	45	50	56	61	67	72	78	83	89	94
	.025	3	7	12	18	23	29	35	40	46	52	58	64	70	76	82	88	94	100	106
	.05	4	10	16	21	27	34	40	46	52	58	65	71	78	84	90	97	103	110	116
	.10	7	13	19	26	32	39	46	53	59	66	73	80	86	93	100	107	114	121	128
	.001	0	1	4	7	11	15	19	24	28	33	38	43	47	52	57	62	67	72	77
	.005	0	3	7	12	17	22	27	32	38	43	48	54	59	65	71	76	82	88	93
18	.01	1	5	10	15	20	25	31	37	42	48	54	60	66	71	77	83	89	95	101
	.025	3	8	13	19	25	31	37	43	49	56	62	68	75	81	87	94	100	107	113
	.05	5	10	17	23	29	36	42	49	56	62	69	76	83	89	96	103	110	117	124
	.10	7	14	21	28	35	42	49	56	63	70	78	85	92	99	107	114	121	129	136
	.001	0	1	4	8	12	16	21	26	30	35	41	46	51	56	61	67	72	78	83
	.005	1	4	8	13	18	23	29	34	40	46	52	58	64	70	75	82	88	94	100
19	.01	2	5	10	16	21	27	33	39	45	51	57	64	70	76	83	89	95	102	108
	.025	3	8	14	20	26	33	39	46	53	59	66	73	79	86	93	100	107	114	120
	.05	5	11	18	24	31	38	45	52	59	66	73	81	88	95	102	110	117	124	131
	.10	8	15	22	29	37	44	52	59	67	74	82	90	98	105	113	121	129	136	144
	.001	0	1	4	8	13	17	22	27	33	38	43	49	55	60	66	71	77	83	89
	.005	1	4	9	14	19	25	31	37	43	49	55	61	68	74	80	87	93	100	106
20	.01	2	6	11	17	23	29	35	41	48	54	61	68	74	81	88	94	101	108	115
	.025	3	9	15	21	28	35	42	49	56	63	70	77	84	91	99	106	113	120	128
	.05	5	12	19	26	33	40	48	55	63	70	78	85	93	101	108	116	124	131	139
	.10	8	16	23	31	39	47	55	63	71	79	87	95	103	111	120	128	136	144	152

TABLE J : THE RIGHT TAIL OF THE CHI-SQUARED DISTRIBUTION

df \ P	.995	.99	.98	.975	.95	.90	.80	.75	.70	.50	.30	.25	.20	.10	.05	.025	.02	.01	.005	.001
1	$.0^{4}393$	$.0^{3}157$	$.0^{3}628$	$.0^{3}982$	.00393	.0158	.0642	.102	.148	.455	1.074	1.323	1.642	2.706	3.841	5.024	5.412	6.635	7.879	10.827
2	.0100	.0201	.0404	.0506	.103	.211	.446	.575	.713	1.386	2.408	2.773	3.219	4.605	5.991	7.378	7.824	9.210	10.597	13.815
3	.0717	.115	.185	.216	.352	.584	1.005	1.213	1.424	2.366	3.665	4.108	4.642	6.251	7.815	9.348	9.837	11.345	12.838	16.268
4	.207	.297	.429	.484	.711	1.064	1.649	1.923	2.195	3.357	4.878	5.385	5.989	7.779	9.488	11.143	11.668	13.277	14.860	18.465
5	.412	.554	.752	.831	1.145	1.610	2.343	2.675	3.000	4.351	6.064	6.626	7.289	9.236	11.070	12.832	13.388	15.086	16.750	20.517
6	.676	.872	1.134	1.237	1.635	2.204	3.070	3.455	3.828	5.348	7.231	7.841	8.558	10.645	12.592	14.449	15.033	16.812	18.548	22.457
7	.989	1.239	1.564	1.690	2.167	2.833	3.822	4.255	4.671	6.346	8.383	9.037	9.803	12.017	14.067	16.013	16.622	18.475	20.278	24.322
8	1.344	1.646	2.032	2.180	2.733	3.490	4.594	5.071	5.527	7.344	9.524	10.219	11.030	13.362	15.507	17.535	18.168	20.090	21.955	26.125
9	1.735	2.088	2.532	2.700	3.325	4.168	5.380	5.899	6.393	8.343	10.656	11.389	12.242	14.684	16.919	19.023	19.679	21.666	23.589	27.877
10	2.156	2.558	3.059	3.247	3.940	4.865	6.179	6.737	7.267	9.342	11.781	12.549	13.442	15.987	18.307	20.483	21.161	23.209	25.188	29.588
11	2.603	3.053	3.609	3.816	4.575	5.578	6.989	7.584	8.148	10.341	12.899	13.701	14.631	17.275	19.675	21.920	22.618	24.725	26.757	31.264
12	3.074	3.571	4.178	4.404	5.226	6.304	7.807	8.438	9.034	11.340	14.011	14.845	15.812	18.549	21.026	23.337	24.054	26.217	28.300	32.909
13	3.565	4.107	4.765	5.009	5.892	7.042	8.634	9.299	9.926	12.340	15.119	15.984	16.985	19.812	22.362	24.736	25.472	27.688	29.819	34.528
14	4.075	4.660	5.368	5.629	6.571	7.790	9.467	10.165	10.821	13.339	16.222	17.117	18.151	21.064	23.685	26.119	26.873	29.141	31.319	36.123
15	4.601	5.229	5.985	6.262	7.261	8.547	10.307	11.036	11.721	14.339	17.322	18.245	19.311	22.307	24.996	27.488	28.259	30.578	32.801	37.697
16	5.142	5.812	6.614	6.908	7.962	9.312	11.152	11.912	12.624	15,338	18.418	19.369	20.465	23.542	26.296	28.845	29.633	32.000	34.267	39.252
17	5.697	6.408	7.255	7.564	8.672	10.085	12.002	12.792	13.531	16.338	19.511	20.489	21.615	24.769	27.587	30.191	30.995	33.409	35.718	40.790
18	6.265	7.015	7.906	8.231	9.390	10.865	12.857	13.675	14.440	17.338	20.601	21.605	22.760	25.989	28.869	31.526	32.346	34.805	37.156	42.312
19	6.844	7.633	8.567	8.907	10.117	11.651	13.716	14.562	15.352	18.338	21.689	22.718	23.900	27.204	30.144	32.852	33.687	36.191	38.582	43.820
20	7.434	8.260	9.237	9.591	10.851	12.443	14.578	15.452	16.266	19.337	22.775	23.828	25.038	28.412	31.410	34.170	35.020	37.566	39.997	45.315
21	8.034	8.897	9.915	10.283	11.591	13.240	15.445	16.344	17.182	20.337	23.858	24.935	26.171	29.615	32.671	35.479	36.343	38.932	41.401	46.797
22	8.643	9.542	10.600	10.982	12.338	14.041	16.314	17.240	18.101	21.337	24.939	26.039	27.301	30.813	33.924	36.781	37.659	40.289	42.796	48.268
23	9.260	10.196	11.293	11.688	13.091	14.848	17.187	18.137	19.021	22.337	26.018	27.141	28.429	32.007	35.172	38.076	38.968	41.638	44.181	49.728
24	9.886	10.856	11.992	12.401	13.848	15.659	18.062	19.037	19.943	23.337	27.096	28.241	29.553	33.196	36.415	39.364	40.270	42.980	45.558	51.179
25	10.520	11.524	12.697	13.120	14.611	16.473	18.940	19.939	20.867	24.337	28.172	29.339	30.675	34.382	37.652	40.646	41.566	44.314	46.928	52.620
26	11.160	12.198	13.409	13.844	15.379	17.292	19.820	20.843	21.792	25.336	29.246	30.434	31.795	35.563	38.885	41.923	42.856	45.642	48.290	54.052
27	11.808	12.879	14.125	14.573	16.151	18.114	20.703	21.749	22.719	26.336	30.319	31.528	32.912	36.741	40.113	43.194	44.140	46.963	49.645	55.476
28	12.461	13.565	14.847	15.308	16.928	18.939	21.588	22.657	23.647	27.336	31.391	32.620	34.027	37.916	41.337	44.461	45.419	48.278	50.993	56.893
29	13.121	14.256	15.574	16.047	17.708	19.768	22.475	23.567	24.577	28.336	32.461	33.711	35.139	39.087	42.557	45.722	46.693	49.588	52.336	58.302
30	13.787	14.953	16.306	16.791	18.493	20.599	23.364	24.478	25.508	29.336	33.530	34.800	36.250	40.256	43.773	46.979	47.962	50.892	53.672	59.703
40	20.706	22.164	23.838	24.433	26.509	29.051	32.345	33.660	34.872	39.335	44.165	45.616	47.269	51.805	55.759	59.342	60.436	63.691	66.766	73.402
50	27.991	29.707	31.664	32.357	34.764	37.689	41.449	42.942	44.313	49.335	54.723	56.334	58.164	63.167	67.505	71.420	72.613	76.154	79.490	86.661
60	35.535	37.485	39.699	40.482	43.188	46.459	50.641	52.294	53.809	59.335	65.227	66.981	68.972	74.397	79.082	83.298	84.580	88.379	91.952	99.607
70	43.275	45.442	47.893	48.758	51.739	55.329	59.898	61.698	63.346	69.334	75.689	77.577	79.715	85.527	90.531	95.023	96.388	100.425	104.215	112.317
80	51.171	53.539	56.213	57.153	60.391	64.278	69.207	71.145	72.915	79.334	86.120	88.130	90.405	96.578	101.880	106.629	108.069	112.329	116.321	124.839
90	59.196	61.754	64.634	65.646	69.126	73.291	78.558	80.625	82.511	89.334	96.524	98.650	101.054	107.565	113.145	118.136	119.648	124.116	128.299	137.208
100	67.327	70.065	73.142	74.222	77.929	82.358	87.945	90.133	92.129	99.334	106.906	109.141	111.667	118.498	124.342	129.561	131.142	135.807	140.170	149.449

TABLE K : CHI-SQUARED DISTRIBUTION FOR 2 x 2 CONTINGENCY TABLES WITH SMALL n AND FREELY VARYING ROW AND COLUMN TOTALS

n = 3

χ^2	P
0	1.0000
0.75	0.5625
3.00	0.1875

n = 4

χ^2	P
0	1.0000
0.44	0.6719
1.33	0.4844
4.00	0.1094

n = 5

χ^2	P
0	1.0000
0.13	0.8789
0.31	0.6445
0.83	0.5664
1.87	0.3320
2.22	0.1758
5.00	0.0586

n = 6

χ^2	P
0	1.0000
0.23	0.7627
0.37	0.7334
0.60	0.6162
0.66	0.4990
1.20	0.4111
1.50	0.2939
2.40	0.2061
3.00	0.1475
6.00	0.0303

n = 7

χ^2	P
0	1.0000
0.05	0.9690
0.19	0.7639
0.46	0.5999
0.62	0.5486
0.87	0.4973
1.12	0.4290
1.21	0.3777
1.55	0.2751
2.09	0.2239
2.91	0.1213
3.73	0.1008
3.93	0.0496
7.00	0.0154

n = 8

χ^2	P
0	1.0000
0.03	0.8776
0.16	0.7751
0.17	0.7717
0.38	0.6691
0.53	0.6486
0.68	0.4435
0.88	0.4094
1.14	0.3632
1.60	0.3290
1.74	0.2607
1.90	0.2094
2.00	0.1889
2.66	0.1547
2.88	0.1035
3.42	0.0693
4.44	0.0624
4.80	0.0419
8.00	0.0075

n = 9

χ^2	P
0	1.0000
0.03	0.9345
0.08	0.8576
0.14	0.7423
0.22	0.7412
0.32	0.5874
0.56	0.5335
0.73	0.5181
0.89	0.5066
1.10	0.3720
1.14	0.2951
1.28	0.2874
1.40	0.2490
2.05	0.2336
2.25	0.1951
2.72	0.1387
3.21	0.1003
3.59	0.0772
3.93	0.0386
5.14	0.0366
5.62	0.0289
5.75	0.0135
9.00	0.0039

n = 10

χ^2	P
0	1.0000
0.02	0.8759
0.07	0.8471
0.10	0.7510
0.12	0.7125
0.27	0.7122
0.40	0.6373
0.47	0.5893
0.62	0.4675
0.74	0.3986
1.07	0.3890
1.11	0.3698
1.26	0.3601
1.40	0.3025
1.66	0.2997
1.83	0.1732
2.50	0.1571
2.59	0.1379
2.74	0.1352
2.85	0.1255
3.40	0.0935
3.59	0.0743
3.75	0.0623
4.28	0.0526
4.44	0.0334
5.83	0.0207
6.42	0.0180
6.66	0.0116
10.00	0.0019

TABLE L : LILLIEFORS' TEST STATISTIC

Selected probabilities for two-tailed tests

n \ P	.20	.15	.10	.05	.01
4	.300	.319	.352	.381	.417
5	.285	.299	.315	.337	.405
6	.265	.277	.294	.319	.364
7	.247	.258	.276	.300	.348
8	.233	.244	.261	.285	.331
9	.223	.233	.249	.271	.311
10	.215	.224	.239	.258	.294
11	.206	.217	.230	.249	.284
12	.199	.212	.223	.242	.275
13	.190	.202	.214	.234	.268
14	.183	.194	.207	.227	.261
15	.177	.187	.201	.220	.257
16	.173	.182	.195	.213	.250
17	.169	.177	.189	.206	.245
18	.166	.173	.184	.200	.239
19	.163	.169	.179	.195	.235
20	.160	.166	.174	.190	.231
25	.142	.147	.158	.173	.200
30	.131	.136	.144	.161	.187
Over 30	$\frac{.736}{\sqrt{n}}$	$\frac{.768}{\sqrt{n}}$	$\frac{.805}{\sqrt{n}}$	$\frac{.886}{\sqrt{n}}$	$\frac{1.031}{\sqrt{n}}$

TABLE M : PEARSON'S CORRELATION COEFFICIENT

Selected probabilities for two-tailed tests

df \ P	.1	.05	.02	.01	.001
1	.98769	.99692	.999507	.999877	.9999988
2	.90000	.95000	.98000	.990000	.99900
3	.8054	.8783	.93433	.95873	.99116
4	.7293	.8114	.8822	.91720	.97406
5	.6694	.7545	.8329	.8745	.95074
6	.6215	.7067	.7887	.8343	.92493
7	.5822	.6664	.7498	.7977	.8982
8	.5494	.6319	.7155	.7646	.8721
9	.5214	.6021	.6851	.7348	.8471
10	.4973	.5760	.6581	.7079	.8233
11	.4762	.5529	.6339	.6835	.8010
12	.4575	.5324	.6120	.6614	.7800
13	.4409	.5139	.5923	.6411	.7603
14	.4259	.4973	.5742	.6226	.7420
15	.4124	.4821	.5577	.6055	.7246
16	.4000	.4683	.5425	.5897	.7084
17	.3887	.4555	.5285	.5751	.6932
18	.3783	.4438	.5155	.5614	.6787
19	.3687	.4329	.5034	.5487	.6652
20	.3598	.4227	.4921	.5368	.6524
25	.3233	.3809	.4451	.4869	.5974
30	.2960	.3494	.4093	.4487	.5541
35	.2746	.3246	.3810	.4182	.5189
40	.2573	.3044	.3578	.3932	.4896
45	.2428	.2875	.3384	.3721	.4648
50	.2306	.2732	.3218	.3541	.4433
60	.2108	.2500	.2948	.3248	.4078
70	.1954	.2319	.2737	.3017	.3799
80	.1829	.2172	.2565	.2830	.3568
90	.1726	.2050	.2422	.2673	.3375
100	.1638	.1946	.2301	.2540	.3211

References

Aitchison, J. and Brown, J.A.C. (1957). *The lognormal distribution with special reference to its uses in economics*. Cambridge University Press, Cambridge.

Anderson, E.W. (1975). The measurement of linear rates of soil creep. B.G.R.G. Conference Paper, unpublished.

Anon. (1966). We wuz robbed. *Nature, 211*, No. 5050, 670.

Aspin, A.P. (1949). Tables for use in comparisons whose accuracy involves two variances, separately estimated. *Biometrika, 36*, 290-296.

Ayer, A.J. (1965). Chance. *Scientific American, 213*, No. 4, 44-56.

Barnett, V. (1974). *Elements of sampling theory*. Hodder and Stoughton, London.

Bartlett, M.S. (1936). The information available in small samples. *Proceedings of the Cambridge Philosophical Society, 32*, 560-566.

Bartlett, M.S. (1947). The use of transformations. *Biometrics, 3*, 39-52.

Bennett, B.M. and Horst, C. (1966). *Supplement to tables for testing significance in a 2 × 2 contingency table*. Cambridge University Press, Cambridge.

Benson, M.A. (1965). Spurious correlation in hydraulics and hydrology. *Journal of the Hydraulics Division, Proceedings of the American Society of Civil Engineers, 91*, 35-41.

Berry, B.J.L. (1962). Sampling, coding, and storing flood plain data. *United States, Department of Agriculture, Farm Economics Division, Agriculture Handbook*, no. 237.

Blalock, H.M. (1960). (2nd ed. 1972.) *Social statistics*. McGraw-Hill, New York.

Bliss, C.I. (1967). *Statistics in biology*. McGraw-Hill, New York.

Boneau, C.A. (1960). The effects of violations of assumptions underlying the t test. *Psychological Bulletin, 57*, 49-64.

Bortkiewicz, L. von (1898). *Das Gesetz der kleinen Zahlen*. Teubner, Leipzig.

Box, G.E.P., Hunter, W.G. and Hunter, J.S. (1978). *Statistics for experimenters: an introduction to design, data analysis and model building*. Wiley, New York.

Bradley, J.V. (1963). *Studies in research methodology IV. A sampling study of the Central Limit Theorem and the robustness of one-sample parametric tests*. A.M.R.L. Technical Documentary Report 63-29, Aerospace Medical Research Laboratories, Wright-Patterson Air Force Base, Ohio.

Bradley, J.V. (1964). *Studies in research methodology VI. The Central Limit Effect for a variety of populations and the robustness of Z, t, and F*. A.M.R.L. Technical Documentary Report 64-123, Aerospace Medical Research Laboratories, Wright-Patterson Air Force Base, Ohio.

Bradley, J.V. (1968). *Distribution-free statistical tests.* Prentice-Hall, New Jersey.

Bradley, J.V. (1971). A large-scale sampling study of the Central Limit Effect. *Journal of Quality Technology, 3,* 51-68.

Bradley, J.V. (1973). The central limit effect for a variety of populations and the influence of population moments. *Journal of Quality Technology, 5,* 171-177.

Bradley, J.V. (1976). *Probability; decision; statistics.* Prentice-Hall, New Jersey.

Brooks, C.E.P. and Carruthers, N. (1953). *Handbook of statistical methods in meteorology.* H.M.S.O., London.

Brown, G.S. (1957). *Probability and scientific inference.* Longmans, London.

Carnap, R. (1950). *Logical foundations of probability.* Routledge and Kegan Paul, London.

Centre for Urban Studies. (1969). *Housing in Camden.* Camden Borough Council, London.

Chaitin, G.J. (1975). Randomness and mathematical proof. *Scientific American, 232,* 47-53.

Chorley, R.J. (1957). Climate and morphometry. *Journal of Geology, 65,* 628-638.

Chorley, R.J. (1958). Aspects of the morphometry of a "poly-cyclic" drainage basin. *Geographical Journal, 124,* 370-374.

Chow, Ven Te (1954). The log-probability law and its engineering applications. *Proceedings of the American Society of Civil Engineers, 80,* Paper No. 536, 1-25.

Clark, W.A.V. and Avery, K.L. (1976). The effects of data aggregation in statistical analysis. *Geographical Analysis, 8,* 428-438.

Cliff, A.D. and Ord, J.K. (1975). The comparison of means when samples consist of spatially autocorrelated observations. *Environment and Planning, A, 7,* 725-734.

Cliff, A.D. and Ord, J.K. (1981). The effects of spatial autocorrelation on geographical modelling, pp. 108-137 in Craig, R.G. and Labovitz, M.L. (eds.), *Future trends in geomathematics.* Pion, London.

Cochran, W.G. (1952). The χ^2 test of goodness of fit. *Annals of Mathematical Statistics, 23,* 315-345.

Cochran, W.G. (1953). *Sampling techniques.* Wiley, New York.

Cochran, W.G. (1954). Some methods for strengthening the common χ^2 tests. *Biometrics, 10,* 417-451.

Conover, W.J. (1971). (2nd ed. 1980.) *Practical nonparametric statistics.* Wiley, New York.

Conover, W.J. (1973). On methods of handling ties in the Wilcoxon signed-rank test. *Journal of the American Statistical Association, 68,* 985-988.

Curry, L. (1964). The random spatial economy: an exploration in settlement theory. *Annals of the Association of American Geographers, 54,* 138-146.

Curry, L. (1966). A note on spatial association. *The Professional Geographer, 18,* 97-99.

Dacey, M.F. (1964). Modified Poisson probability law for point pattern more regular than random. *Annals of the Association of American Geographers, 54,* 559-565.

Dale, P.F. (1971). Children's reactions to maps and aerial photographs. *Area, 3,* 170-177.

Das, A.C. (1950). Two-dimensional systematic sampling and the associated stratified and random sampling. *Sankhya, 10,* 95-108.
Davis, J.C. (1973). *Statistics and data analysis in geology.* Wiley, New York.
Davis, K. (1969). *World urbanization 1950-1970.* Institute of International Studies, University of California, Berkeley.
Dixon, C. and Leach, B. (1977). *Sampling methods for geographical research.* Concepts and techniques in modern geography, 17. Geo. Abstracts, Norwich.
Dixon, C. and Leach, B. (1978). *Questionnaires and interviews in geographical research.* Concepts and techniques in modern geography, 18. Geo. Abstracts, Norwich.
Dixon, W.J. and Massey, F.J., Jr. (1957). *Introduction to statistical analysis.* McGraw-Hill, New York.
Draper, N.R. and Smith, H. (1966). *Applied regression analysis,* Wiley, New York.
Ebdon, D. (1977). *Statistics in geography.* Oxford University Press, Oxford.
Edwards, A.L. (1976). *An introduction to linear regression and correlation.* Freeman, San Francisco.
Ezekiel, M. and Fox, K.A. (1959). *Methods of correlation and regression analysis.* Wiley, New York.
Feller, W. (1957). (3rd ed. 1968.) *An introduction to probability theory and its applications.* Wiley, New York.
Ferguson, G.A. (1959). (5th ed. 1981.) *Statistical analysis in psychology and education.* McGraw-Hill, New York.
Ferguson, R. (1978). *Linear regression in geography.* Concepts and techniques in modern geography, 15. Geo. Abstracts, Norwich.
Finetti, B. de (1970). *Theory of probability.* Routledge and Kegan Paul, London.
Finney, D.J. (1941). On the distribution of a variate whose logarithm is normally distributed. *Journal of the Royal Statistical Society, Supplement B, 7,* 155-161.
Finney, D.J., Latscha, R., Bennett, B.M. and Hsu, P. (1963). *Tables for testing significance in a 2 × 2 contingency table.* Cambridge University Press, Cambridge.
Fisher, R.A. (1935). (7th ed. 1960.) *The design of experiments.* Oliver and Boyd, Edinburgh.
Fisher, R.A. and Yates, F. (1957). *Statistical tables for biological, agricultural, and medical research.* 5th ed. Oliver and Boyd, Edinburgh.
Flood Studies Report (1975). Natural Environment Research Council, London.
Galton, F. (1899). *Natural inheritance.* Macmillan, London.
Gardiner, V. (1973). Univariate distributional characteristics of some morphometric variables. *Geografiska Annaler* A, *54,* 147-153.
Gardiner, V. and Gardiner, G. (1978). *Analysis of frequency distributions.* Concepts and techniques in modern geography, 19. Geo. Abstracts, Norwich.
Gayen, A.K. (1951). The frequency distribution of the product-moment correlation coefficient in random samples of any size drawn from non-normal universes. *Biometrika, 38,* 219-247.
Geary, R.C. (1947). Testing for normality. *Biometrika, 34,* 209-242.
General Household Survey (1976). Office of Population Censuses and Surveys, Social Survey Division, H.M.S.O., London.

Gerrard, A.J.W. (1974). The geomorphological importance of jointing in the Dartmoor granite, pp. 39-51 in Brown, E.H. and Waters, R.S. (eds.), *Progress in Geomorphology*, Institute of British Geographers Special Publication 7.

Glass, R. (1970). Housing in Camden. *Town Planning Review*, *41*, 15-40.

Hacking, I. (1965). *Logic of statistical inference*. Cambridge University Press, Cambridge.

Haight, F.A. (1967). *Handbook of the Poisson distribution*. Wiley, New York.

Hansen, M.H., Hurwitz, W.N. and Madow, W.G. (1953). *Sample survey methods and theory*. Wiley, New York.

Hays, W.L. (1973). *Statistics for the social sciences*. 2nd ed. Holt, Rinehart and Winston, New York.

Hodges, J.L., Jr. and Lehmann, E. (1956). The efficiency of some nonparametric competitors of the t-test. *The Annals of Mathematical Statistics*, *27*, 324-335.

Holzinger, K.J. and Church, A.E.R. (1928). On the means of samples from a U-shaped population. *Biometrika*, *20A*, 361-388.

Hoyle, M.H. (1973). Transformations - an introduction and a bibliography. *International Statistical Review*, *41*, 203-223.

Hutchinson, P. (1974). *The climate of Zambia*. Occasional Study No. 7, Geographical Association.

Johnson, N.L. and Kotz, S. (1969). *Discrete distributions*. Houghton-Mifflin, New York.

Johnson, N.L. and Kotz, S. (1970). *Continuous univariate distributions*. Houghton-Mifflin, New York.

Johnson, N.L. and Kotz, S. (1972). *Continuous multivariate distributions*. Wiley, New York.

Johnson, N.L. and Leone, F.C. (1964). (2nd ed. 1977.) *Statistics and experimental design in engineering and the physical sciences*. Wiley, New York.

Jonkers, D.A. (1976). De stand van de Ooievaar in 1975. *Het Vogeljaar*, *24*, 261-266.

Keeble, D. (1978). Industrial decline in the inner city and conurbation. *Transactions of the Institute of British Geographers*, *3*, 101-114.

Kendall, M.G. and Stuart, A. (Vol. 1, 1st ed. 1958, 2nd ed. 1963, 4th ed. 1977; Vol. 2, 1st ed. 1961, 3rd ed. 1973; Vol. 3, 3rd ed. 1976.) *The advanced theory of statistics*. Griffin, London.

Koch, G.S., Jr. and Link, R.F. (1970). *Statistical analysis of geological data*. Wiley, New York.

Kowalski, C.J. (1972). On the effects of non-normality on the distribution of the sample product-moment correlation coefficient. *Applied Statistics*, *21*, 1-12.

Lack, D. (1966). *Population studies of birds* (Chapter 13, The White Stork in Germany). Oxford University Press, Oxford.

Lancaster, H.O. (1969). *The chi-squared distribution*. Wiley, New York.

Lehmann, E.L. (1975). *Nonparametrics; statistical methods based on ranks*. McGraw-Hill, New York.

Lewis, D. and Burke, C.J. (1949). The use and misuse of the chi-square test. *Psychological Bulletin*, *46*, 433-489.

Lloyd, E. (1980). *Probability*. Wiley, Chichester.

Mann, H.B. and Whitney, D.R. (1947). On a test of whether one of two random variables is stochastically larger than the other. *The Annals of Mathematical Statistics*, *18*, 50-60.

Matui, I. (1932). Statistical study of the distribution of scattered villages in two regions of the Tonami Plain, Toyama Prefecture. *Japanese Journal of Geology and Geography, 9*, 251-266.

Maxwell, J.C. (1967). Quantitative geomorphology of some mountain chaparral watersheds of Southern California, pp. 108-226 in Garrison, W.L. and Marble, D.F. (eds.), *Quantitative geography, Part II: Physical and cartographic topics*. Northwestern University, Studies in Geography, Number 14.

Meadows, D.H., Meadows, D.L., Randers, J. and Behrens III, W.W. (1972). *The limits to growth*. University Books, New York.

Mercer, W. and Hall, A.D. (1911). The experimental error of field trials. *Journal of Agricultural Science, 4*, 107-132.

Miller, V.C. (1953). *A quantitative geomorphic study of drainage basin characteristics in the Clinch Mountain area, Virginia and Tennessee*. Department of Geology, Columbia University, ONR Project NR 389-042, Technical Report 3.

Mises, R. von (1957). *Probability, statistics and truth*. Allen and Unwin, London.

Mogridge, M.J.H. (1969). Household income and household composition. *Centre for Environmental Studies Working Paper WP-29*.

Mood, A.M. (1954). On the asymptotic efficiency of certain nonparametric two-sample tests. *The Annals of Mathematical Statistics, 25*, 514-522.

Morisawa, M.E. (1957). Accuracy of determination of stream lengths from topographic maps. *Transactions of the American Geophysical Union, 33*, 86-88.

National Bureau of Standards (1949). *Tables of the binomial probability distribution*. U.S. Government Printing Office, Washington D.C.

Noether, G.E. (1967a). *Elements of nonparametric statistics*. Wiley, New York.

Noether, G.E. (1967b). Wilcoxon confidence intervals for location parameters in the discrete case. *Journal of the American Statistical Association, 62*, 184-188.

Norcliffe, G.B. (1977). *Inferential statistics for geographers*. Hutchinson, London.

Olkin, I. and Pratt, J.W. (1958). Unbiased estimation of certain correlation coefficients. *Annals of Mathematical Statistics, 29*, 201-211.

Olkin, I., Gleser, L.J. and Derman, C. (1980). *Probability models and applications*. Macmillan, New York.

Openshaw, S. and Taylor, P.J. (1979). A million or so correlation coefficients: three experiments on the modifiable areal unit problem, pp. 127-144 in Wrigley, N. (ed.), *Statistical applications in the spatial sciences*. Pion, London.

Openshaw, S. and Taylor, P.J. (1981). The modifiable areal unit problem, pp. 60-69 in Wrigley, N. and Bennett, R.J. (eds.), *Quantitative geography: a British view*. Routledge and Kegan Paul, London.

Ord, J.K. (1972). *Families of frequency distributions*. Griffin, London.

Ordnance Corps (1952). *Tables of the cumulative binomial probabilities*. Ordnance Corps Pamphlet PRDP 20-1, Washington, D.C.

Oxley, N.C. (1974). Suspended sediment delivery rates and the solute concentration of stream discharge in two Welsh catchments, pp. 141-

154 in Gregory, K.J. and Walling, D.E. (eds.), *Fluvial processes in instrumented watersheds*. Institute of British Geographers Special Publication 6.

Parr, J.B. and Suzuki, K. (1973). Settlement populations and the log-normal distribution. *Urban Studies, 10*, 335-352.

Parzen, E. (1960). *Modern probability theory and its applications*. Wiley, New York.

Payne, S.L. (1951). *The art of asking questions*. Princeton University Press, Princeton, New Jersey.

Pearson, E.S. (1929). Some notes on sampling tests with two variables. *Biometrika, 21*, 337-360.

Pearson, E.S. and Kendall, M.G. (1970). *Studies in the history of statistics and probability*. Hafner, New York.

Pearson, K. (1897). On a form of spurious correlation which may arise when indices are used in the measurement of organs. *Proceedings of the Royal Society of London, 60*, 489-502.

Pettijohn, F.J. (1949). (2nd ed. 1980.) *Sedimentary rocks*. Harper, New York.

Pratt, J.W. (1959). Remarks on zeros and ties in the Wilcoxon signed rank procedures. *Journal of the American Statistical Association, 54*, 655-667.

Pringle, D. (1976). Normality, transformations, and grid square data. *Area, 8*, 42-45.

Quenouille, M.H. (1949). Some problems of plane sampling. *Annals of Mathematical Statistics, 20*, 355-375.

Rand Corporation (1955). *A million random digits with 100,000 normal deviates*. Free Press, Glencoe, Illinois.

Rayner, J.N. (1960). *Temperature and wind frequency tables for North America and Greenland*. Arctic Meteorology Research Group Publication in Meteorology, no. 25, Montreal.

Reed, J.L. (1921). On the correlation between any two functions and its application to the general case of spurious correlation. *Journal of the Washington Academy of Science, 11*, 449-455.

Robinson, A.H. (1956). The necessity of weighting values in correlation analysis of areal data. *Annals of the Association of American Geographers, 46*, 233-236.

Robinson, W.S. (1950). Ecological correlations and the behaviour of individuals. *American Sociological Review, 15*, 351-357.

Rodda, J.C. (1970). Rainfall excesses in the United Kingdom. *Transactions of the Institute of British Geographers, 49*, 49-60.

Rowley, G. (1975). The redistribution of parliamentary seats in the United Kingdom: themes and opinions. *Area 7*, No. 1, 16-21.

Schumm, S.A. (1956). Evolution of drainage systems and slopes in badlands at Perth Amboy, New Jersey. *Bulletin of the Geological Society of America, 67*, 597-646.

Siegel, S. (1956). *Nonparametric statistics for the behavioral sciences*. McGraw-Hill, New York.

Siegel, S. and Tukey, J.W. (1960). A nonparametric sum of ranks procedure for relative spread in unpaired samples. *Journal of the American Statistical Association, 55*, 429-444 (*see also* Errata, 1961, 1005).

Silk, J. (1979). *Statistical concepts in geography*. Allen and Unwin, London.

Slakter, M.J. (1965). A comparison of the Pearson chi-square and Kol-

mogorov goodness-of-fit tests with respect to validity. *Journal of the American Statistical Association, 60*, 854-858.
Snedecor, G.W. and Cochran, W.G. (1967). *Statistical methods*. 6th ed., Iowa State University Press, Ames, Iowa.
Speight, J.G. (1971). Log-normality of slope distributions. *Zeitschrift für Geomorphologie, 15*, 290-311.
Staff of the Computation Laboratory (1955). *Tables of the cumulative binomial probability distribution*. Harvard University Press, Cambridge, Mass.
Strahler, A.N. (1950). Equilibrium theory of erosional slopes approached by frequency distribution analysis. *American Journal of Science, 248*, 673-696 and 800-814.
Strahler, A.N. (1954). Statistical analysis in geomorphic research. *Journal of Geology, 62*, 1-25.
Stuart, A. (1962). (2nd ed. 1976.) *Basic ideas of scientific sampling*. Griffin, London.
"Student" (1908). On the probable error of a mean. *Biometrika, 6*, 1-25.
Taylor, P.J. (1977). *Quantitative methods in geography: an introduction to spatial analysis*. Houghton-Mifflin, Boston.
Thomas, E.N. and Anderson, D.W. (1965). Additional comments on weighting values in correlation analysis of areal data. *Annals of the Association of American Geographers, 55*, 492-505.
Thompson, H.R. (1956). Distribution of distance to nth neighbour in a population of randomly distributed individuals. *Ecology, 37*, 391-394.
Till, R. (1974). *Statistical methods for the earth scientist*. Macmillan, London.
Tippett, L.H.C. (1952). *Technological applications of statistics*. Williams and Norgate, London.
Trenhaille, A.S. (1974). The geometry of shore platforms in England and Wales. *Transactions of the Institute of British Geographers, 62*, 129-142.
Trickett, W.H., Welch, B.L. and James, G.S. (1956). Further critical values for the two-means problem. *Biometrika, 43*, 203-205.
Wallis, W.A. and Roberts, H.V. (1956). *Statistics: a new approach*. Free Press, New York.
Weintraub, S. (1963). *Tables of the cumulative binomial probability distribution for small values of p*. Free Press, New York.
Welch, B.L. (1937). The significance of the difference between two means when the population variances are unequal. *Biometrika, 29*, 350-362.
Welch, B.L. (1947). The generalization of 'Student's' problem when several different population variances are involved. *Biometrika, 34*, 28-35.
Wilcoxon, F. (1945). Individual comparisons by ranking methods. *Biometrics, 1*, 80-83.
Wilkinson, R.K. and Sigsworth, E.M. (1972). Attitudes to the housing environment: an analysis of private and local authority households in Batley, Leeds and York. *Urban Studies, 9*, 193-209.
Williams, E.J. (1959). *Regression analysis*. Wiley, New York.
Wrigley, N. and Bennett, R.J. (1981). *Quantitative geography*. Routledge and Kegan Paul, London.
Yamane, T. (1964). *Statistics; an introductory analysis*. 2nd ed. Harper and Row, New York.

Yates, F. (1934). Contingency tables involving small numbers and the χ^2 test. *Journal of the Royal Statistical Society, 1*, 217-235.
Yates, F. (1937). Applications of sampling technique to crop estimation and forecasting. *Journal of the Manchester Statistical Society*, 1-25.
Yates, F. (1949). (4th ed. 1981). *Sampling methods for censuses and surveys*. Griffin, London.
Yule, G.U. (1926). Why do we sometimes get nonsense-correlations between time-series? - a study in sampling and the nature of time-series. *Journal of the Royal Statistical Society, 89*, 1-64.
Yule, G.U. and Kendall, M.G. (1950). *An introduction to the theory of statistics*. 14th ed. Griffin, London.

Further Reading

Useful introductory textbooks that deal with the statistical analysis of geographical data include Norcliffe (1977), Ebdon (1977), Taylor (1977), and Silk (1979). Equivalent basic reading for earth scientists is provided by Koch and Link (1970), Davis (1973) and Till (1974). Ferguson (1959), Blalock (1960) and Hays (1973) provide excellent elementary surveys for social scientists, while Snedecor and Cochran (1967) have produced a comprehensive work aimed at biologists.

Pearson and Kendall (1970) provide a detailed discussion of the history of statistical concepts and techniques, while Wrigley and Bennett (1981) closely examine the role of quantification in geography. Excellent texts on sampling theory include Yates (1949), Hansen *et al.* (1953), Stuart (1962), Barnett (1974), and Dixon and Leach (1977). Questionnaire surveys are well discussed by Payne (1951) and Dixon and Leach (1978).

Further reading on the meaning of probability and the construction of probability models is provided by Carnap (1950), Mises (1957), Feller (1957), Parzan (1960), Finetti (1970), Lloyd (1980), and Olkin *et al.* (1980), to cite only a sample of authors. The analysis of frequency distributions in geography is discussed by Gardiner and Gardiner (1978). For further material on probability distributions, such as the Poisson and the normal, see Haight (1967), Johnson and Kotz (1969, 1970, 1972) and Ord (1972).

The logic of significance testing is examined in detail in the companion volume to this book. Alternative significance tests, and ex information on the tests described here, can be found in Siegel (1956), Kendall and Stuart (1958, etc.), Snedecor and Cochran (1967), Bradley (1968), Conover (1971) and Lehmann (1975). Readers who wish to pursue the subject of transformations in depth will find Bartlett (1947) and Hoyle (1973) helpful.

Regression and correlation techniques are discussed by numerous authors. Ferguson (1978) gives a valuable introduction with geographical examples. Somewhat more advanced texts include Ezekiel and Fox (1959), Draper and Smith (1966), Snedecor and Cochran (1967), Bliss (1967), and Edwards (1976). The second volume of the present work covers many additional topics in regression and correlation.

Index